Energy, Environment, and Sustainability

AIMS AND SCOPE

This books series publishes cutting edge monographs and professional books focused on all aspects of energy and environmental sustainability, especially as it relates to energy concerns. The Series is published in partnership with the International Society for Energy, Environment, and Sustainability. The books in these series are edited or authored by top researchers and professional across the globe. The series aims at publishing state-of-the-art research and development in areas including, but not limited to:

- Renewable Energy
- Alternative Fuels
- Engines and Locomotives
- Combustion and Propulsion
- Fossil Fuels
- Carbon Capture
- Control and Automation for Energy
- Environmental Pollution
- Waste Management
- Transportation Sustainability

Review Process

The proposal for each volume is reviewed by the main editor and/or the advisory board. The chapters in each volume are individually reviewed single blind by expert reviewers (at least four reviews per chapter) and the main editor.

Ethics Statement for this series can be found in the Springer standard guidelines here https://www.springer.com/us/authors-editors/journal-author/journal-author-helpdesk/before-you-start/before-you-start/1330#c14214

More information about this series at https://link.springer.com/bookseries/15901

Gabriele Di Blasio · Avinash Kumar Agarwal ·
Giacomo Belgiorno · Pravesh Chandra Shukla
Editors

Clean Fuels for Mobility

Springer

Editors
Gabriele Di Blasio
Mechanical Engineering
National Research Council
Napoli, Italy

Giacomo Belgiorno
Advanced Engineering
PUNCH Torino
Turin, Italy

Avinash Kumar Agarwal
Mechanical Engineering
Indian Institute of Technology Kanpur
Kanpur, India

Pravesh Chandra Shukla
Mechanical Engineering
Indian Institute of Technology Bhilai
Raipur, India

ISSN 2522-8366 ISSN 2522-8374 (electronic)
Energy, Environment, and Sustainability
ISBN 978-981-16-8746-4 ISBN 978-981-16-8747-1 (eBook)
https://doi.org/10.1007/978-981-16-8747-1

This Springer imprint is published by the registered company Springer Nature Singapore Pte Ltd.
The registered company address is: 152 Beach Road, #21-01/04 Gateway East, Singapore 189721, Singapore

Preface

Fuels are essential in modern life for transportation and many other dependent purposes. Petroleum fuels cause pollution and are also available in a discrete manner on the crust of the earth. Limited reserves also restrict their use for the long term. Various clean fuel options such as alcohols, biodiesels, biogas, etc. have emerged as a way to sustainable transportation. This book is an attempt to gather information about various options of clean fuels and to assess their feasibility for sustainable transportation purposes.

The International Society for Energy, Environment and Sustainability (ISEES) was founded at the Indian Institute of Technology Kanpur (IIT Kanpur), India, in January 2014 to spread knowledge/awareness and catalyze research activities in the fields of Energy, Environment, Sustainability, and Combustion. Society's goal is to contribute to the development of clean, affordable, and secure energy resources and a sustainable environment for society and spread knowledge in the areas mentioned above, and create awareness about the environmental challenges the world is facing today. The unique way adopted by ISEES was to break the conventional silos of specializations (engineering, science, environment, agriculture, biotechnology, materials, fuels, etc.) to tackle the problems related to energy, environment, and sustainability in a holistic manner. This is quite evident by the participation of experts from all fields to resolve these issues. The ISEES is involved in various activities such as conducting workshops, seminars, conferences, etc. in the domains of its interests. Society also recognizes the outstanding works of young scientists, professionals, and engineers for their contributions in these fields by conferring them awards under various categories.

Fifth International Conference on "Sustainable Energy and Environmental Challenges" (V-SEEC) was organized under the auspices of ISEES from December 19 to 21, 2020, in virtual mode due to restrictions on travel because of the ongoing COVID-19 pandemic situation. This conference provided a platform for discussions between eminent scientists and engineers from various countries, including India, Spain, Austria, Bangladesh, Mexico, USA, Malaysia, China, UK, Netherlands, Germany, Israel, and Saudi Arabia. At this conference, eminent international speakers presented their views on energy, combustion, emissions, and alternative energy resources for sustainable development and a cleaner environment. The conference presented two high voltage plenary talks by Dr. VK Saraswat, Honorable Member, NITI Ayog, on "Technologies for Energy Security and Sustainability" and Prof. Sandeep Verma, Secretary, SERB, on "New and Equitable R&D Funding Opportunities at SERB."

The conference included nine technical sessions on topics related to energy and environmental sustainability. Each session had 6–7 eminent scientists from all over the world, who shared their opinion and discussed the trends for the future. The technical sessions in the conference included Emerging Contaminants: Monitoring and Degradation Challenges; Advanced Engine Technologies and Alternative Transportation Fuels; Future Fuels for Sustainable Transport; Sustainable Bioprocessing for Biofuel/Non-biofuel Production by Carbon Emission Reduction; Future of Solar Energy; Desalination and Wastewater Treatment by Membrane Technology; Biotechnology in Sustainable Development; Emerging Solutions for Environmental Applications; and Challenges and Opportunities for Electric Vehicle Adoption. 500+ participants and speakers from all over the world attended this 3-day conference.

The conference concluded with a high voltage panel discussion on "Challenges and Opportunities for Electric Vehicle Adoption," where the panelists were Prof. Gautam Kalghatgi (University of Oxford), Prof. Ashok Jhunjhunwala (IIT Madras), Dr. Kelly Senecal (Convergent Science), Dr. Amir Abdul Manan (Saudi Aramco), and Dr. Sayan Biswas (University of Minnesota, USA). Prof. Avinash K Agarwal, ISEES, moderated the panel discussion. This conference laid out the roadmap for technology development, opportunities, and challenges in Energy, Environment, and Sustainability domain. All these topics are very relevant for the country and the world in the present context. We acknowledge the support received from various agencies and organizations for the successful conduct of the Fifth ISEES conference V-SEEC, where these books germinated. We want to acknowledge SERB (special thanks to Dr. Sandeep Verma, Secretary) and our publishing partner Springer (special thanks to Ms. Swati Meherishi).

The editors would like to express their sincere gratitude to a large number of authors from all over the world for submitting their high-quality work on time and revising it appropriately at short notice. We would like to express our special gratitude to our prolific set of reviewers, Dr. Atul Dhar, Dr. Roberto Ianniello, Dr. Vikram Kumar, Dr. Chetan Patel, Dr. Nikhil Sharma, Dr. Giuseppe Di Luca, and Dr. Michele Pipicelli, who reviewed various chapters of this monograph and provided their valuable suggestions to improve the manuscripts.

This book provides an overview of clean fuels for sustainable mobility. The book initially deals with different types of alternative fuels, for example, ethanol, methanol, butanol, hydrogen, biogas, biodiesel, etc. Chapters include recent results and are focused on current trends in the automotive sector. In this book, readers will get information about various aspects of the clean fuels production process and formulation to improve the combustion characteristics and efficiency toward sustainability. Some of the important fuels like hydrogen, ammonia, natural gas, etc. are discussed in detail. Ammonia is discussed as an alternative option for marine transportation. Hydrogen is also discussed to assess its feasibility in combustion engines. The content of this book is oriented to the industrial and research community involved in fuels, combustion engines, and environmental research.

Naples, Italy — Gabriele Di Blasio
Kanpur, India — Avinash Kumar Agarwal
Turin, Italy — Giacomo Belgiorno
Raipur, India — Pravesh Chandra Shukla

Contents

Editors and Contributors

About the Editors

Dr. Gabriele Di Blasio is currently Research Scientist at the National Research Council of Italy. His main research interest is focused on advanced technologies and fuels for propulsion and energy conversion systems. He has lead various research projects in the field of internal combustion engine technology and fuel development. He has contributed to private and public projects in cooperation with universities, research centres and OEMs. He received his Ph.D. in Mechanical Engineering in 2012. Formerly Dr. Di Blasio worked as R&D responsible engineer in the industry sector leading projects on the development of dual fuel systems for heavy duty engines. He serves as an editor and reviewer for several indexed journals of national and international repute. He has authored over 70 publications in peer-reviewed journals, conference proceedings, books, book chapters and technical reports. He is a member of SAE International and SAE Engine committee.

Prof. Avinash Kumar Agarwal joined IIT Kanpur in 2001. He worked at the Engine Research Center, UW@Madison, the USA as a Post-Doctoral Fellow (1999–2001). His interests are IC engines, combustion, alternate and conventional fuels, lubricating oil tribology, optical diagnostics, laser ignition, HCCI, emissions, and particulate control, 1D and 3D Simulations of engine processes, and large-bore engines. Prof. Agarwal has published 435+ peer-reviewed international journal and conference papers, 70 edited books, 92 books chapters, and 12200+ Scopus and 19000+ Google Scholar citations. He is the associate principal editor of FUEL. He has edited "Handbook of Combustion" (5 Volumes; 3168 pages), published by Wiley VCH, Germany. Prof. Agarwal is a Fellow of SAE (2012), Fellow of ASME (2013), Fellow of ISEES (2015), Fellow of INAE (2015), Fellow of NASI (2018), Fellow of Royal Society of Chemistry (2018), and a Fellow of American Association of Advancement in Science (2020). He is the recipient of several prestigious awards such as Clarivate Analytics India Citation Award-2017 in Engineering and Technology, NASI-Reliance Industries Platinum Jubilee Award-2012; INAE Silver Jubilee Young Engineer Award-2012; Dr. C. V. Raman Young Teachers Award: 2011; SAE Ralph R. Teetor Educational Award-2008; INSA Young Scientist Award-2007; UICT Young Scientist Award-2007; INAE Young Engineer Award-2005. Prof. Agarwal received Prestigious CSIR Shanti Swarup Bhatnagar Award-2016 in Engineering Sciences. Prof. Agarwal is conferred upon Sir J C Bose National Fellowship (2019) by SERB for his outstanding contributions. Prof. Agarwal was a highly cited researcher (2018) and was in the top ten HCR from India among 4000 HCR researchers globally in 22 fields of inquiry.

Dr. Giacomo Belgiorno is Technology System Engineer in the Department of Advanced Engineering at PUNCH Torino S.p.A., formerly General Motors Global Propulsion Systems, Torino, Italy. He received his M.S. in Mechanical Engineering from University of Campania Luigi Vanvitelli in 2014, and Ph.D. in Energy Science and Engineering from University of Naples "Parthenope" in 2018. During the Ph.D. program, he worked at Istituto Motori as Research Associate and Guest Researcher in Lund University, Sweden. Subsequently, he worked in CNH Industrial. He has authored three book chapters and more than 20 conference and journal articles.

Dr. Pravesh Chandra Shukla is Assistant Professor in the Department of Mechanical Engineering at Indian Institute of Technology (IIT) Bhilai, India. Dr. Shukla received his Ph.D. from IIT Kanpur and has also worked as Senior Research Associate with the institute. He was a Postdoctoral Researcher in the Division of Combustion Engines, Department of Energy Sciences, Lund University, Sweden. He briefly worked in Ecole Centrale de Nantes, France, in the field of dual fuel combustion. He is a recipient of Young Scientist Award from the International Society for Energy, Environment and Sustainability. Dr. Shukla mainly works in the field of internal combustion engines and alternative fuels for transportation. He worked on the development of additives for high compression ratio heavy duty engines fueled with alcohol. He is involved in investigating the emission characteristics for alternative fuels like biodiesel, HVO and alcohols for conventional and advanced heavy duty compression ignition engines. He has published over 28 technical articles in journals of national and international repute and conference proceedings.

Contributors

Avinash Kumar Agarwal Department of Mechanical Engineering, Indian Institute of Technology Kanpur, Kanpur, India

Mariana S. T. Amândio CICECO—Aveiro Institute of Materials, Department of Chemistry, University of Aveiro, Campus Universitário de Santiago, Aveiro,

Portugal;
CIEPQPF, Department of Chemical Engineering, Faculty of Sciences and Technology, University of Coimbra, Pólo II, Coimbra, Portugal

Amaanuddin M. Azad CSIR-National Environmental Engineering Research Institute, CSIR-NEERI, Nagpur, Maharashtra, India;
Academy of Scientific and Innovative Research (AcSIR), Ghaziabad, Uttar Pradesh, India

Mattia Bartoli Center for Sustainable Future Technologies @POLITO, Fondazione Istituto Italiano di Tecnologia, Turin, Italy;
National Consortium for Materials Science and Technology (INSTM), Florence, Italy

Giacomo Belgiorno PUNCH Torino, Turin, Italy

Hemant Bherwani CSIR-National Environmental Engineering Research Institute, CSIR-NEERI, Nagpur, Maharashtra, India;
Academy of Scientific and Innovative Research (AcSIR), Ghaziabad, Uttar Pradesh, India

Gabriele Di Blasio Istituto di Scienze e Tecnologie per l'Energia e la Mobilità Sostenibili (STEMS), Consiglio Nazionale Delle Ricerche, Naples, Italy

Wojciech Czekała Poznań University of Life Sciences, Wojska Polskiego 28, Poznań, Poland

Amir Hossein Fakhari School of Mechanical Engineering, Iran University of Science and Technology, Tehran, Iran

Ayat Gharehghani School of Mechanical Engineering, Iran University of Science and Technology, Tehran, Iran

Mauro Giorcelli Center for Sustainable Future Technologies @POLITO, Fondazione Istituto Italiano di Tecnologia, Turin, Italy;
National Consortium for Materials Science and Technology (INSTM), Florence, Italy

Pravin Jagdale Center for Sustainable Future Technologies @POLITO, Fondazione Istituto Italiano di Tecnologia, Turin, Italy;
National Consortium for Materials Science and Technology (INSTM), Florence, Italy

Nikolaos C. Kokkinos Hephaestus Advanced Laboratory, Division of Petroleum Forensic Fingerprinting, Department of Chemistry, School of Science, International Hellenic University, Kavala, Greece

Nitin Labhasetwar CSIR-National Environmental Engineering Research Institute, CSIR-NEERI, Nagpur, Maharashtra, India

Rudrodip Majumdar National Institute of Advanced Studies (NIAS), Indian Institute of Science Campus, Bengaluru, Karnataka, India

Tamás Mizik Department of Agribusiness, Corvinus University of Budapest, Budapest, Hungary

Mohd Mubashshir Naved CSIR-National Environmental Engineering Research Institute, CSIR-NEERI, Nagpur, Maharashtra, India;
Academy of Scientific and Innovative Research (AcSIR), Ghaziabad, Uttar Pradesh, India

Jorge M. S. Rocha CIEPQPF, Department of Chemical Engineering, Faculty of Sciences and Technology, University of Coimbra, Pólo II, Coimbra, Portugal

Massimo Rovere National Consortium for Materials Science and Technology (INSTM), Florence, Italy;
Department of Applied Science and Technology, Politecnico di Torino, Turin, Italy

Kumar Saurabh National Institute of Advanced Studies (NIAS), Indian Institute of Science Campus, Bengaluru, Karnataka, India

Luísa S. Serafim CICECO—Aveiro Institute of Materials, Department of Chemistry, University of Aveiro, Campus Universitário de Santiago, Aveiro, Portugal

Pravesh Chandra Shukla Department of Mechanical Engineering, Indian Institute of Technology Raipur, Raipur, India

Alberto Tagliaferro National Consortium for Materials Science and Technology (INSTM), Florence, Italy;
Department of Applied Science and Technology, Politecnico di Torino, Turin, Italy

Ruggero Vigliaturo Department of Earth Sciences, University of Turin, Turin, Italy

Roshan Wathore CSIR-National Environmental Engineering Research Institute, CSIR-NEERI, Nagpur, Maharashtra, India;
Academy of Scientific and Innovative Research (AcSIR), Ghaziabad, Uttar Pradesh, India

Ana M. R. B. Xavier CICECO—Aveiro Institute of Materials, Department of Chemistry, University of Aveiro, Campus Universitário de Santiago, Aveiro, Portugal

Burak Zincir Maritime Faculty, Istanbul Technical University, İstanbul, Turkey

Part I
General

Chapter 1
Introduction to Clean Fuels for Mobility

Gabriele Di Blasio, Avinash Kumar Agarwal, Giacomo Belgiorno, and Pravesh Chandra Shukla

Mobility is one of the prime requirements in the modern world. Transportation scenario cannot be imagined without having automobiles and are an essential part of society. Exponential growth in the automobile and transportation sector, other problems have risen such as fuel scarcity, pollution etc. Petroleum fuel availability and supply is important for any country to sustain transportation. Heterogeneous distribution of fuel on earth's crust is a challenge to make it available for every part of the world. On the other hand, fossil fuels are infamous for environmental emissions. The emission of CO_2 is a huge problem in terms of global warming and climate change. Emission regulations are becoming stricter to control carbon monoxide, hydrocarbons, oxides of nitrogen and particulate matter. In this regard, there is a requirement of utilizing an alternative and clean fuel that can replace petroleum fuels partly or fully. This can have the potential to mitigate the problem related to fuel scarcity and environmental pollutions. This book is an attempt to summarize various sustainable and clean fuel options for transportation. The book is divided in three parts namely, (I) General, (II) Biofuels for Sustainable Mobility (III) Biogas for Sustainable Mobility. A total of 10 Chapters are included (excluding the Introduction Chapter) in this theme to explain various clean fuel strategies and utilization.

Under Part I, Chap. 2 are explained in a global manner stating the importance of clean fuel. It majorly explains about *Sustainable fuels in private transportation*

G. Di Blasio (✉)
Istituto di Scienze e Tecnologie per l'Energia e la Mobilità Sostenibili (STEMS), Consiglio Nazionale Delle Ricerche, Naples, Italy
e-mail: gabriele.diblasio@stems.cnr.it

A. K. Agarwal
Department of Mechanical Engineering, Indian Institute of Technology Kanpur, Kanpur, India

G. Belgiorno
PUNCH Torino, Turin, Italy

P. C. Shukla
Department of Mechanical Engineering, Indian Institute of Technology Raipur, Raipur, India

G. Di Blasio et al. (eds.), *Clean Fuels for Mobility*, Energy, Environment, and Sustainability, https://doi.org/10.1007/978-981-16-8747-1_1

and *Fuels for Sustainable Transport in India.* The 21st Century ushered in the era of the energy revolution in which alternative fuels are massively deployed in the energy sector as additives to fossil fuels or complementary alternatives to fossil fuels. Global warming caused by greenhouse gas emissions is one of the greatest challenges the world faces. Humanity is in the age of modern transportation, where a key issue is what can and should be used to fuel our vehicles. Moreover, the use of these gases requires special pumps at gas stations, as well as the modifications of the vehicles' fuel tanks. Ethanol offers a promising answer (mature technology, octane booster, less negative environmental effects, etc.) Biodiesel has similar characteristics to diesel engines, especially when used oils are recycled for production. Fast charging, a limited range, and the number of fuelling stations are the Achilles heels of electric cars. Besides, the plug-in vehicle fleet still has a marginal share within the global vehicle stock. The growth of biofuels will outpace electricity even in the next decade. Green hydrogen has high production costs, while biogas production is limited by the available raw materials. However, achieving higher sustainability in private transportation is essential. Chapter 2 provide an overview of the advancement of alternative fuels by showing their advantages and disadvantages and comparing them to fossil fuels. Chapter 3 mainly deals with *Fuels for Sustainable Transport in India.* Fossil fuels have a 98% share in energy consumption in the Indian transport sector. The Indian transport sector contributes to 13.2% of CO_2 emissions in India. In the quest for cleaner fuels for the surface transport sector, a number of transportation energy sources fuelling the vehicles (based on Internal Combustion Engines) have emerged as suitable alternatives in the 21st Century through decades of continued research and development efforts. Fuels such as natural gas (Compressed Natural Gas (CNG) or Liquefied Natural Gas (LNG)), Liquefied Petroleum Gas (LPG), hydrogen, and biofuels have been the front runners as potential alternatives to crude oil-based gasoline and diesel. These alternative fuels for cleaner transport in India are detailed and compared based on their physico-chemical properties, ignition characteristics, storage requirements, and safety parameters. CNG and LPG are identified to be useful in the fuel mix for the passenger segment in the short to medium-term future, while biodiesel is identified as a potential freight segment fuel in a similar timeframe. For the long-term future, ethanol is identified as a potential candidate for the passenger segment, while LNG and Hydrogen are found to be suitable for the freight segment.

Under Part II, Chaps. 4 and 5 discusses the *Alternative Refinery Process of Fuel Catalytic Upgrade in Aqueous Media* and *Ethanol Derived from Municipal Solid Waste for Sustainable Mobility.* There are severe disadvantages of using gasoline ether oxygenates (GEOs) in the refineries, for the final fuels to be in accordance with the current specifications. GEOs disperse swiftly contaminating the environment, as well as being vastly persistent, owning to its high water solubility and volatility, its low biodegradability, and its growing scale of use. An alternative fuel upgrade in aqueous media from refinery cuts could completely replace GEOs by producing in situ strong anti-knocking environmental friendly alcohol mixtures. The aforementioned process was successfully implemented by heterogenizing homogeneous catalysts, overcoming the separation difficulties of homogeneous catalysts and taking advantage of their very many benefits that are still kept inevitably apart from the petroleum industry. Chapter discusses the complication of the substrate composition and the ground-breaking characteristics of this alternative refinery process of fuel catalytic upgrade, the simulation model succeeded for the first time to develop a phase behaviour of the fuel and achieved an adequate average absolute relative deviation from the experimental data offering the opportunity for scaling up the chemical process. Chapter 5 deals with the discussion on rapid economic growth, especially in developing countries, directly impacts transport fuel demand, waste generation and greenhouse gas (GHG) emissions. there is an increasing need to supplement fossil fuel demand with sustainable alternative options while also addressing solid waste and GHG emissions. India generates the highest amount of annual municipal solid waste (MSW) (277 million tonnes out of the global 2.01 billion tonnes); this is estimated to double by 2050. This high MSW generation rate, inadequate management, unscientific landfilling and inefficient disposal practices is a serious concern to health and the environment. The organic fraction of MSW generated in India is estimated to be in the range of 40–60% and contains huge potential fuel value for waste to energy (WtE) options. Hence, there is an increasing need to supplement fossil fuel demand with sustainable alternative options while also addressing solid waste and GHG emissions. Few successful pilot projects as case studies will help getting several stakeholders together, which will be essential for taking this waste utilization option to a useful scale. *Bioethanol from wastes for mobility (from European perspective)* and *Bio-derived and waste fats use for the production of drop-in fuels* are the other topics dealt in this book under Chaps. 6 and 7. Currently, the world is facing an energy transition, with governments pushing for a switch from fossil to bio-based fuels. The transport sector is responsible for about 30% of the European energy consumption, which is still mainly obtained from burning fossil fuels, accounting for 25% of the greenhouse gas emissions. Consequently, the European Union established a Renewable Energy Directive II, setting a target of 3.6% blending for advanced biofuels in 2030. Biomass-derived fuels can play an important role, particularly for medium and long distances, being considered one of the main drivers to achieve decarbonisation targets in the transport and mobility sector, recognised as a roadmap for carbon neutrality. In countries such as Finland, Sweden, and Portugal, forestry-derived feedstocks dominate. Nevertheless, by-product and waste streams derived from other industrial sectors, namely dairy, brewery, pulp and paper, and food, have been evaluated. This chapter overviews the role of biofuels in the transport sector, with a particular focus on the European context. First, a brief introduction explains the main concepts and policies pushing the biofuels market. Then, the current scenario regarding bioethanol production from wastes in Europe is reviewed, listing the feedstocks and the biorefineries currently in operation. In summary, Chap. 6 discussed some of the main challenges regarding the production of cellulosic ethanol. Chapter 7 is mainly focused on drop-in fuels and discussed their production complexities and solutions. In this Chapter, a section is also dedicated to catalytic conversions and upgrading to provide a solid background to the reader. Chapter 8 deals with the explanation of *biodiesel as a clean fuel option for mobility*. In this Chapter, various clean fuels biodiesel options are being discussed. Various emission characteristics and low temperature combustion concepts are included in this Chapter. Conventional fossil fuel sources are limited and their combustion leads to emissions. For these reasons, more eco-friendly alternative fuels are needed. Biodiesel has been known as a suitable alternative fuel for the last few decades. This fuel is produced from various sources including vegetable oils, animal fats, and waste oils, which are all renewable. The use of biodiesel in conventional diesel engines leads to a considerable reduction in PM, HC, and CO emissions. the combination of biodiesel and RCCI combustion is useful both in terms of improving RCCI engine performance and in terms of solving the NO_x challenge in biodiesel combustion.

Part III, Chaps. 9 and 10 included the discussion on *ammonia for decarbonized maritime transportation* and *biogas as a sustainable and renewable energy source*. Carbon emissions are one of the important topics in maritime transportation recently since International Maritime Organization (IMO) implemented strict-er regulations. IMO announced Initial Greenhouse Gas (GHG) Strategy in 2018 for decarbonization of maritime transportation. International shipping consumed 300 million tons of high carbon content fossil fuels annually and it is increasing year by year. Maritime transportation constitutes 3.1% of the total global CO_2 emissions which depends on the usage of high carbon content fossil fuels. The Initial GHG Strategy aims to reach zero-carbon shipping in the future. One way to achieve this aim is the usage of alternative fuels with zero-carbon content. Ammonia is one of the alternative fuels that can be either used for fuel cells on ships and at marine diesel engines. Ammonia has a carbon-free and sulphur-free structure and it can be combusted by the dual-fuel combustion concept at marine diesel engines as same as other alternative marine fuels. The outcomes of Chap. 9 reveal that although there are some barriers to ammonia as a marine fuel, the existing experience of the maritime industry, supply chains and infrastructures of fertilizer industry, low modification requirement on engines, and achievement of significantly lower CO_2 and soot emissions, and almost zero SO_x emissions will facilitate the use of ammonia, and it can be one of the options for full decarbonized maritime transportation by 2050. Chapter 10 includes a discussion on the effect of rational biodegradable waste management will be energy production, directly related to reducing the amount of waste intended for landfills. Biogas is a renewable energy source with many advantages. Its production fits perfectly into the activities of sustainable development. The advantage of biogas is its many possible uses. The biogas can be used to produce electricity and heat in cogeneration and it can be used as fuel for vehicles or pump to the natural gas grid after purification to bio-methane. Biogas is a renewable energy source with many advantages. Towards the end, the current book deals with two important topics related to the utilization of *Natural Gas as a Clean fuel for mobility*. Natural gas is a considerably promising fuel for use as a transportation fuel due to its availability, extensive infrastructure for distribution, low cost, and cleaner combustion. Chapter 11 focuses on analysing the advantages, challenges, and different strategies of using NG in SI and CI engines. Using natural gas, SI engines can operate at a higher compression ratio than the conventional gasoline-fuelled engines, so they present higher thermal efficiency and while they increase NOx emissions, they produce lower levels of greenhouse gas emissions, carbon dioxide, unburned hydrocarbons, and carbon monoxide. CI engines can also use natural gas fuel in dual-fuel mode, where a high-Cetane fuel is injected along with natural gas to provide a source of ignition for the charge.

This monograph presents the different aspects of sustainable mobility. The topics are organized in three different parts: (I) General, (II) Biofuels for Sustainable Mobility, (III) Biogas for Sustainable Mobility. Specific topics covered in the monograph include:

1. Introduction to Clean Fuels for Mobility
2. Sustainable fuels in private transportation–present and future potential
3. Fuels for Sustainable Transport in India
4. Alternative Refinery Process of Fuel Catalytic Upgrade in Aqueous Media
5. Ethanol Derived from Municipal Solid Waste for Sustainable Mobility
6. Bioethanol from wastes for mobility: Europe on the road to sustainability
7. Bio-derived and waste fats use for the production of drop-in fuels
8. Biodiesel as a clean fuel for mobility
9. Ammonia for Decarbonized Maritime Transportation
10. Biogas as a sustainable and renewable energy source
11. Natural Gas as a Clean fuel for mobility.

Chapter 2
Sustainable Fuels in Private Transportation–Present and Future Potential

Tamás Mizik

2.1 Introduction

2.1.1 Greenhouse Gas Emissions

Global warming considerably changes the Earth's climate. Its primary source is the continuously increasing level of greenhouse gases (GHGs). GHGs include different gases and their main elements/categories are carbon dioxide (CO_2), methane (CH_4), sulfur dioxide (SO_2), nitrous oxide (N_2O), and fluorinated gases (chlorofluorocarbons (CFCs) and hydrochlorofluorocarbons (HCFs)). However, their global warming potential is different. This impact can be measured by the global warming potential where CO_2 is used as the unit value. Table 2.1 gives an overview of the global warming impact and the share in global emissions of these gases. Although CO_2 has the highest share of GHG emissions, fluorinated gases have the highest global warming potential.

As the currently available technologies are expensive, such as adsorption, absorption, and membrane separation in the case of CO_2 capture (Yoro and Daramola 2020), savings on GHG emissions are more important.

It should also be kept in mind that Earth has limited fossil energy sources. Moreover, their formation is a much slower process compared to their use. Therefore, humanity must switch to alternative, preferably renewable sources. Although the industry sector has become the largest greenhouse gas emitter during the previous decade, the share of transportation also shows an increasing trend (Fig. 2.1).

However, not only the composition of GHG emissions has changed over these years but also the size of global emissions has increased. The total GHG emissions increased by 23% (from 39 to 48 in gigatons of equivalent CO_2) between 2000

T. Mizik (✉)
Department of Agribusiness, Corvinus University of Budapest, Budapest, Hungary
e-mail: tamas.mizik@uni-corvinus.hu

G. Di Blasio et al. (eds.), *Clean Fuels for Mobility*, Energy, Environment, and Sustainability, https://doi.org/10.1007/978-981-16-8747-1_2

Table 2.1 Global warming impact and share of GHGs in global emissions

GHGs	Global warming potential (100 years)	Share in global emissions (2019)
Carbon dioxide (CO_2)	1	76
Methane (CH_4)	28	13
Sulfur dioxide (SO_2)	indirect GHG	7
Nitrous oxide (N_2O)	265	3
Fluorinated gases (CFCs, HCFs)	138 (CH_3CHF_2)–13,900 ($CClF_3$)	1

Source Author's composition based on Yoro and Daramola (2020) and Myhre et al. (2013)

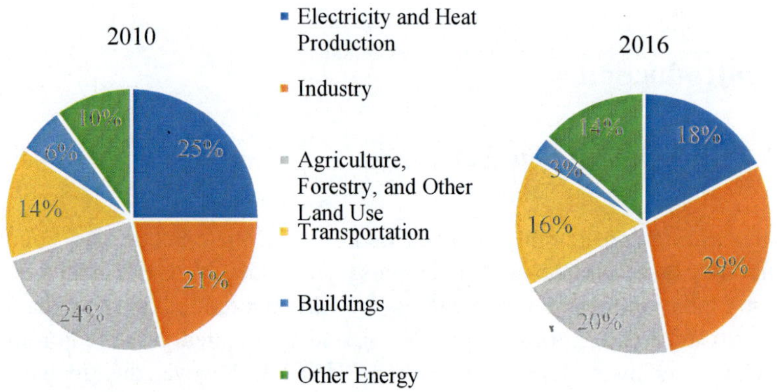

Fig. 2.1 Changes in the composition of GHG emissions. *Source* Author's composition based on IPCC (2014) and Ritchie (2020)

and 2010, and by an additional 6% (to 51 gigatons of equivalent CO_2) between 2010 and 2016 (CAT 2021). Although air pollutants are also important elements of sustainability (Van Fan et al. 2018), they are excluded from this analysis.

2.1.2 Global Vehicle Fleet and Its Characteristics

The size of the global vehicle fleet was increasing continuously in the last two decades and is expected to increase in the next three decades too (Fig. 2.2). However, its growth rate seems to be decreasing. This rate was between 2.14 and 5.14% before 2020, while 2.37–0.11% annual growth is expected after 2020 (Fig. 2.2). Another important fact that can be identified on Fig. 2.2 is the dominancy of gasoline-fueled cars even in 2050. The "Hybrid and fully electric cars" refers to three categories: battery electric vehicle (BEV), hybrid electric vehicle (HEV), and plug-in hybrid electric vehicle (PHEV). The first category contains fully electric cars, while HEVs and PHEVs are partly electric cars as they have an internal combustion engine.

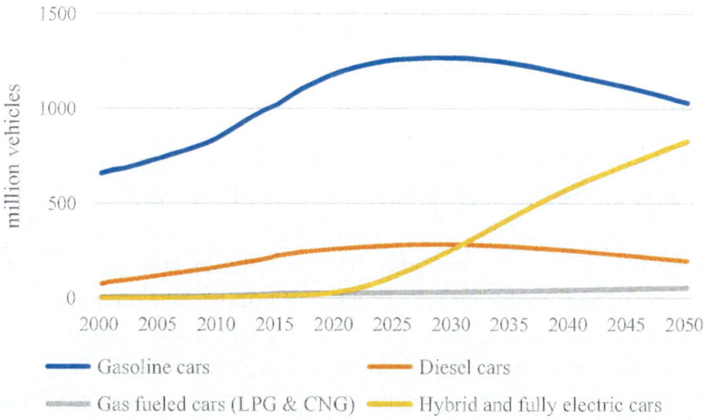

Fig. 2.2 Global vehicle fleet. *Source* Author's composition based on GECF (2020)

As seen on Fig. 2.2, cars with electric engines will rapidly increase in the next three decades and their share is expected to reach almost 40% by 2050. This means 827 million cars out of the predicted 2108 million vehicle fleet.

Besides the number of vehicles, their average age is also important. Newer cars do not only provide higher safety but also better fuel economy; therefore, they have lower GHG emissions. Table 2.2 provides data on these characteristics of the global vehicle fleet. China has the fastest growing vehicle market with the lowest average age, while the Russian cars are the oldest on average among the seven largest passenger car owner countries.

Table 2.2 Size and age of the global vehicle fleet, 2019

Country	Passenger cars (million)	Average fleet age (years)
EU28	261.553	10.8
China	161.025	4.9
USA	118.520	11.8
Japan	61.943	8.7
Russia	48.711	13.1*
Brazil	39.507	9.8
Mexico	29.547	12.0**
Rest of the world	289.214	–
Total	943.112	9.6[e]

*2018 value. It was presumably higher in 2019, but no comparable data was available
**2020 value. It was presumably lower in 2019, but no comparable data was available
[e]Estimated value calculated by the seven largest passenger car owner countries' data
Source Author's composition based on ACEA (2019), Autostat (2018), GlobalFleet (2021), IHS Markit (2019), Nationmaster (2019), statista (2021a, b, c)

A decrease in average vehicle age is important as fuel efficiency shows an inverse relationship with age. However, this development seems to be slowing down. According to the Energy Information Administration's data, the average fuel efficiency of the on-road new cars is expected to increase from 6.69 l/100 km (2020) to 5.52 l/100 km by 2050 (EIA 2021a). This forecasts less than 1% annual decrease; therefore, more attention should be paid to those fuels that are less polluting and more sustainable. However, even attractive future technologies may have adverse effects, e.g. gasoline-fueled cars may become cheaper and inexpensive; therefore, they will be back in use again (Keith et al. 2019). It should also be noted that vehicles using alternative fuels are generally more expensive; however, they provide less utility and convenience compared to their conventional competitors (NRC 2013).

Autonomous vehicles provide a promising option; however, only in the long run (Liu et al. 2019). Besides cleaner vehicles that are analyzed in the following subchapters, other opportunities, e.g. different forms of transport sharing and the development of public transport systems, may significantly reduce GHG emissions (Fan et al. 2017). In general, public transport is more energy-efficient compared to private vehicles (Wang et al. 2018). However, achieving a notable reduction in GHG emissions requires other components, such as the promotion of clean technologies by subsidies and research and development activities (Santos 2017).

The topic itself is extremely diverse; therefore, the following subsections concentrate only on passenger vehicles and road transportation. The contribution of this analysis to the existing literature is twofold:

- it provides a detailed overview of the five analyzed alternative fuels, namely ethanol, biodiesel, electricity, green hydrogen, and biogas, from sustainability aspect,
- it compares them by their characteristics, advantages, and disadvantages.

2.2 Direct Fuel Substitutes

The two major direct fuel substitutes are ethanol and biodiesel. They can be either blended with gasoline and diesel, respectively or used purely. Blending rates may vary between 1 and 100%; however, high-level blends require certain modifications of the engines and their connecting parts due to the different characteristics of biofuels compared to either gasoline or diesel.

The use of biofuels is mainly driven by the blending mandates. These rates are different from country to country, and they are somehow related to the countries' biofuels production potential.

2.2.1 Ethanol

Ethanol is a renewable biofuel, and its use is almost GHG emissions neutral. During its burning, only the previously absorbed CO_2 is released; however, CO_2 emissions during the whole process (e.g. sowing, harvesting, processing, transportation) should be taken into account. Another advantage of ethanol is its mature technology, both dry and wet milling processes are safe and efficient. Additionally, the octane number of ethanol is higher than that of traditional gasoline (IEA-AMF 2021); therefore, it can be used as an octane booster.

The most important question of ethanol production is the raw material. The current production is dominated by first generation ethanol, which raises the debate on "food vs. fuel". Its major feedstocks are maize and sugarcane which can be used for human consumption as well. Table 2.3 gives an overview of ethanol generations, their potential feedstocks, production methods, and CO_2 balances.

Non-food raw materials are used for advanced ethanol (2nd–4th generations) production alongside an even better CO_2 balance (neutral or even negative). However, advanced ethanol production is currently too expensive due to its high input intensity, including energy and water (Darda et al. 2019).

Blending ethanol with fuel up to 10% requires no modifications on the new vehicles (Singh and Walia 2016). However, high-level ethanol blends (above 30%) require special vehicles such as flex-fuel vehicles (FFVs). These vehicles are popular in Brazil, e.g., 84% of the new passenger cars were FFVs in 2019 (ANFAVEA 2020), where the blending mandate is the highest among the most significant ethanol producers. Table 2.4 provides an overview of the major ethanol producer countries and their blending mandates.

In the case of first generation or conventional ethanol, price is determined by the cost of raw material. This was 58% on average for a representative Iowa (corn-based, dry-milling) plant based on data from 2007 to 2021 (CARD 2021). High corn or sugarcane yields can make it significantly lower; therefore, achieving the

Table 2.3 Ethanol generations and their characteristics

Generations	Potential feedstocks	Production method	CO_2 balance
1st	Wheat, maize, sugar beet, sugarcane (high starch or sugar content)	Fermentation	Positive
2nd	Agricultural wastes, various grasses and trees (lignocellulosic biomass)	Hydrolysis and fermentation	Basically neutral
3rd	Microalgae, macroalgae (microorganism)	Hydrolysis and fermentation	Negative
4th	Microalgae, macroalgae (genetically modified microorganisms)	Hydrolysis and fermentation	Negative

Source Author's composition based on Alalwan et al. (2019)

Table 2.4 The five largest ethanol producers and their blending mandates by volume, 2019

Countries	Ethanol production (million liters)	Share in global production (%)	Blending mandate
USA	59,719	54.35	10–15
Brazil	32,441	29.53	27
EU28	5451	4.96	10*
China	3407	3.10	10
Canada	1893	1.72	5–8.5

*According to the Renewable Energy Directive II, the maximum cap on first generation biofuels is 7%, while the actual country-level blending mandate values are generally lower than 10% (Lieberz 2020)

Source Author's composition based on OECD-FAO (2021) and Lane (2021)

highest possible yields is the simplest way of becoming more cost-efficient. Price competitiveness of ethanol production is closely linked to the oil price, which is the major substitute for ethanol. It should be noted that ethanol has an energy content of about 2/3 that of gasoline (Alalwan et al. 2019). Regarding gasoline price, crude oil was its major cost element and accounted for 56% of its total production cost between 2011 and 2020 on average (EIA 2021b). These ratios (2/3–100% in the case of ethanol and 56–100% in the case of crude oil) are used for calculating adjusted prices in Table 2.5 to have comparable results, i.e., gasoline to ethanol with the same energy content.

Based on the adjusted prices, all the significant producers, except for China, can produce ethanol for a lower price than that of gasoline. Canada was the most efficient producer in 2019 followed by the USA and Brazil.

However, these calculations are sensitive to changes in ethanol price caused by exchange rates (national currencies to USD) and crude oil price. For instance, even without any changes in the yield of the raw material, its price, or production technology, ethanol price would be lower if the given country's currency depreciated, while the opposite happens in the case of appreciation against the USD. In the case of gasoline, lower crude oil prices result immediately in lower gasoline prices. For

Table 2.5 Producer prices of the largest ethanol producers and oil price, 2019

Countries	Ethanol price (USD/L)	Adjusted price (USD/L)
USA	0.41	0.62
Brazil	0.44	0.67
EU28	0.47	0.71
China	0.49	0.74
Canada	0.33	0.50
Oil (world)	0.40 (crude oil)	0.72 (gasoline)

Source Author's composition based on OECD-FAO (2021) and IRS (2021)

example crude oil price was 64.07 USD/barrel in 2019 (OECD-FAO 2021). However, Brent crude oil price went down to 41.96 USD/barrel in 2020 (statista 2021d), while the share of crude oil cost in gasoline production went down to 43% (EIA 2021b), resulting in a 0.61 USD/L gasoline price.

2.2.2 *Biodiesel*

Biodiesel has similar characteristics to ethanol. This also provides a renewable solution by using mature technology. Another similarity is the prevalence of first generation technology, so the "food vs fuel" debate also applies for biodiesel. It can be differentiated between three generations and only algal synthesis provides the opportunity of negative CO_2 balance (Table 2.6).

There are, basically, two biodiesel production methods. With cold press extraction, roughly 80% of the oil can be extracted, while this goes up to 99% in the case of hot press extraction (Jaeger and Siegel 2008). This is the reason why large biodiesel refineries use the latter.

As biodiesel is closer to petroleum diesel in nature than ethanol to gasoline, higher blends can be used without any need for modifications. According to many studies, blends up to 20% biodiesel can be used without any problems and even higher-level blends are safe to use; however, they may cause warranty problems (Ogunkunle and Ahmed 2019). With the highest share of diesel cars globally, the EU is the largest biodiesel producer. Despite the diesel emission scandal of Volkswagen, still 42.3% of the passenger cars in use were diesel in 2019 (ACEA 2021). Due to its huge palm resources, Indonesian production is growing rapidly, which is further accelerated by the highest blending mandate among the largest biodiesel producers (Table 2.7).

Compared to ethanol, feedstock cost has an even higher share in biodiesel production cost. Its share is 75% on average but can be up to 90% (Gebremariam and Marchetti 2018). This also points out how the process can be made more cost-efficient, more specifically, by using cheaper raw materials, e.g. used vegetable oil. Biodiesel has only a slightly, approximately 10%, lower energy content in comparison with petroleum diesel (EC 2015). This results in a narrower gap between biodiesel price and its adjusted form. Regarding petroleum diesel price, the share of crude oil is larger compared to gasoline price (EIA 2021c) which results in the higher price

Table 2.6 Biodiesel generations and their characteristics

Generations	Potential feedstocks	Production method	CO_2 balance
1st	Rapeseed, sunflower, palm, soybean, animal fat	Transesterification	Positive
2nd	Jatropha and nonedible oils	Transesterification	Positive
3rd	Algae and seaweeds	Algal synthesis	Negative

Source Mizik and Gyarmati (2021)

Table 2.7 The five largest biodiesel producers and their blending mandates by volume, 2019

Countries	Biodiesel production (million liters)	Share in global production (%)	Blending mandate (%)
EU28	14,600	31.19	10*
Indonesia	9031	19.29	30
USA	7380	15.76	5
Brazil	5800	12.39	12
Argentina	2500	5.34	10

*According to the Renewable Energy Directive II, the maximum cap on first generation biofuels is 7%, while the actual country-level blending mandate values are generally lower than 10% (Lieberz 2020)

Source Author's composition based on OECD-FAO (2021) and Lane (2021)

Table 2.8 Producer prices of the largest biodiesel producers and oil price, 2019

Countries	Biodiesel price (USD/L)	Adjusted price (USD/L)
EU28	0.75	0.83
Indonesia	0.68	0.76
USA	0.42	0.47
Brazil	0.64	0.71
Argentina	0.60	0.66
Oil (world)	0.40 (crude oil)	0.84 (petroleum diesel)

Source Author's composition based on OECD-FAO (2021), IRS (2021), statista (2021e), and EIA (2021c)

of the final product. Table 2.8 summarizes the producer prices of the five largest producers in comparison with the world oil price.

The different cost structure of petroleum diesel results in a higher final price that made all the largest biodiesel producers cost competitive in 2019. The soybean-based US production provides the cheapest biodiesel, followed by Argentina and Brazil. Although soybean has lower oil content compared to palm (Indonesia) or rapeseed (European Union), Argentina and Brazil also use this raw material. The European Union is the world leading biodiesel producer, but its producer price is the highest among the other competitors. These calculations are also sensitive to exchange rate changes, and due to the lower share of crude oil price in the cost of the final product, even more sensitive to crude oil price changes.

2.3 Other Alternative Fuels

Electricity, green hydrogen, and biogas are indirect fuel substitutes as their use require significant modifications on the vehicles (plug-in system, batteries, and electric motors for electricity; plug-in system and special fuel tank for gases). However, there is a significant difference between them. While the share of electric vehicles is expected to grow rapidly in the forthcoming decades, gas-fueled cars will remain marginal within the global vehicle fleet (Fig. 2.2).

2.3.1 Electricity

Electric vehicles are definitely the future, but their share is expected to be marginal even for a couple of more years. However, as seen on Fig. 2.3, the growth of electric vehicles (EV) will be continuous and will end up achieving almost 40% of the total vehicle fleet by 2050. According to the Gas Exporting Countries Forum Global Gas Model (GECF 2020), the major driver of this growth will be the sale of hybrid electric vehicles (HEVs). They will give almost 50% of the EVs in 2050.

Electric engines have two advantages over internal combustion engines (petroleum diesel and gasoline), namely the higher engine efficiency (the so-called brake thermal efficiency) and the cheaper raw material (electricity). Mass produced gasoline engines have 41% peak thermal efficiency, while diesel engines can reach 50% (Zhao et al. 2020). These values are much lower in low gears and at low speed that is very common in the urban areas. Contrary to this, electric engines have about 95% efficiency (Kalghatgi 2018). This huge difference explains the expected market success of hybrid electric vehicles where the internal combustion engine can work

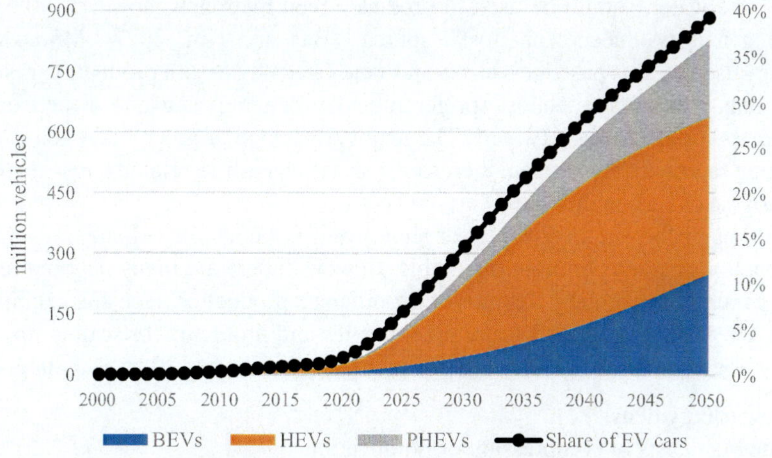

Fig. 2.3 Global EV fleet. *Source* Author's composition based on GECF (2020)

close to its peak brake thermal efficiency during the charging of batteries. In addition to high efficiency, electricity is cheaper than any other alternatives. For example, this costs 13.5 US cents per kilowatt-hour (kWh) in the USA in 2021 (EIA 2021d), while an average electric car consumes roughly 15 kWh/100 km (Noura et al. 2018). This adds up to only 2 USD/100 km with no travel emission. However, it does matter how the electricity is produced, because the sustainability of electric cars is determined by the source of the electricity. In other words, electric cars are as green as the electricity they use.

Although electric cars represent the future but their forecasted widespread pose many questions, potential problems, and barriers (CTT 2018; Kalghatgi 2018):

- incentives for EV purchase will no longer be sustainable which decreases their value for money.
- governments should find alternatives to replace fuel taxes on gasoline and petroleum diesel.
- there is still a high need for smart charging, as well as fast charging, solutions.
- grid stability should be guaranteed, especially before the share of EVs starts to rapidly increase. Widespread of EVs increases overall energy consumption and leads to even higher peak demands.
- the current battery cell production technology should be improved (next-generation technology).
- battery recycling will be an even bigger challenge.

2.3.2 Green Hydrogen

Hydrogen can be used in two ways: burning in an internal combustion engine and transforming into electricity with hydrogen fuel cells. Hydrogen can be extracted from many different raw materials. Renewable sources, such as biomass or (waste)water, should be used to produce green hydrogen; however, 90% of the hydrogen is produced from fossil sources (Bae and Kim 2017). Moreover, the energy used for this process is also important. Green hydrogen production is sustainable only if renewable energy (preferably solar or wind) is used for the extraction (biomass) or electrolysis (water). The major advantage of green hydrogen is that its burning requires only oxygen; therefore, there is no carbon-related emission (carbon monoxide or carbon dioxide).

Although the use of green hydrogen provides better fuel efficiency, its higher production cost may entirely offset this. However, there are many ongoing research and experiments aiming for more cost-efficient production (see for example Liu et al. 2021). There are still many prerequisites and problems for scaling up the use of hydrogen as an alternative fuel (Steenberghen and Lopez 2008; Kalghatgi 2018):

- low energy density
- the high costs of compression or liquification
- the high relative cost of hydrogen-fueled vehicles compared to its alternatives
- the lack of transport infrastructure

- low-density fueling network
- environmental benefits occur only if renewable energies are used for the transformation process of the green hydrogen.

As seen on Fig. 2.2, the widespread of gas-fueled private vehicles are not anticipated. Hydrogen is more promising for those types of transportation where electrification is less feasible, such as buses, taxis, trains, and heavy-duty road vehicles (EC 2020). Interest in this technology at the country level depends on the share of (imported) fossil energy sources, e.g., South Asian countries are especially vulnerable to climate security (Liu et al. 2021).

2.3.3 Biogas

The term "bio" in biogas refers to its non-fossil nature and sustainability. Biogas can be produced from various sources such as solid waste landfills, sewage sludge (wastewater), livestock manure, and organic waste (EPA 2020). However, it should be highlighted that biomass feedstocks are often decentralized which limits production capacities and makes logistics more difficult and expensive (Mustafi and Agarwal 2020).

Using biogas as a fuel requires the removal of different contaminations, such as water, hydrogen sulphide, activated carbon, nitrogen, ammonia, siloxanes, and different particulates (Papacz 2011). In addition to cleaning, upgrading of biogas is also necessary. Upgrading means the removal of CO_2, which is up to 40% of the biogas, to increase calorific value and therefore, driving distance with the same amount of biogas (Persson 2007). Due to its low energy density, which is similar to hydrogen, biogas should be either compressed or liquified. Both processes are costly.

However, if renewable energy is used for the biogas upgrade and compression, CO_2 emissions will be reduced by as high as 99% compared to fossil fuels (Papacz 2011). Higher fuel efficiency is another advantage of biogas over its fossil rivals. According to Tabar et al. (2017), the use of CNG (compressed natural gas) results in approximately 11–39% lower fuel consumption compared to gasoline. This can be applied to biogas too, as there is no significant difference between CNG and biogas when both are transformed into fuel. Moreover, higher biogas production can help countries, especially developing countries, to achieve higher energy self-dependency (Kumar Ghosh and Mandal 2018).

Based on a literature review, Nevzorova and Kutcherov (2019) provided a detailed list of the different barriers of wider biogas implementation. Table 2.9 summarizes their results.

In addition to the barriers above, pipeline injection problems may also arise, and not to forget that fossil natural gas is much cheaper (EPA 2020). Moreover, at the vehicle level, compressed gas requires reinforced; therefore, heavier tank compared to fossil-fueled vehicles, while liquified gas should be stored at -162 °C which also results in a more expensive tank than that of conventional vehicles (Kalghatgi 2018). Not only the biogas plants are expensive but (bio)gas-fueled vehicles are also more

Table 2.9 Categorization of selected barriers of wider biogas implementation

Technical – lack of fueling stations – availability of feedstock – lack of skills and trainings at plant level	Economic – high investment and maintenance costs of the biogas plants – lack of research and development	Market – cheaper fossil fuels – competing alternatives, e.g., ethanol or electricity
Institutional – lack of political support – lack of private sector participation	Socio-cultural – lack of public participation and consumer interest – stigmatization, lack of knowledge and information	Environmental – noise and odor pollution – need for water resources – potential gas leakage

Source Author's composition based on Nevzorova and Kutcherov (2019)

expensive compared to the competing technologies. Due to the marginal share of biogas in transportation, further enhancement of this requires strong political will and public support, such as in Sweden where developing biogas is supported at municipal and state levels (Persson 2007).

2.4 Comparison of the Analyzed Alternative Fuels

The analyzed alternative fuels, together with their fossil rivals, will be compared by two aspects: travel economy and advantages-disadvantages. The first includes monetary terms (Per Gasoline Gallon Equivalent (GGE) and Per Million British Thermal Units (MBtu)), while the second refers to efficiency-related issues (fuel and cost efficiency).

2.4.1 Travel Economy

Environmental sustainability is important and becoming even more important in the future. From this aspect, although at a different extent, every analyzed alternative fuel performs better compared to fossil fuels. However, it is also important to know about their costs. Table 2.10 provides a comparison of fossil fuels together with their analyzed alternatives.

The second column of Table 2.10 shows the major argument for fossil fuels, both gasoline and diesel are cheaper when fuels are measured in GGE. This ranking is almost the same for the third column when the measurement is MBtu. However, all the alternatives provide higher fuel efficiency, except for ethanol. This difference can be up to 150%. Cost efficiency is the sum of the CGE multiplied by fuel efficiency. This value tells us how much money the different fuels cost on the same distance. It can be stated that electricity provides the cheapest way of transportation. BEVs can be operated with a minimum of 65% less cost compared to gasoline-fueled vehicles.

Table 2.10 Comparison of the different fuels (US prices)

Fuels	GGE	MBtu	Fuel efficiency	Cost efficiency
Gasoline	$2.32	$20.30	100%	100%
Diesel	$2.35	$20.51	+20%	−5%
Ethanol (E85)	$2.65	$30.25	−28%	+18%
Biodiesel (B100)	$3.11	$26.5	+10%	+20%
Electricity	$4.50	$39.56	min. +150%	min. −65%
Biogas*	$5.96	$52.12	+11–39%	+57 → 129%
Green hydrogen	$16.33	$24.43–53.74	+60%	−28% → +59%

*This is calculated from CNG data by using that renewable natural gas is at least 2.72 times more expensive compared to fossil natural gas (EPA 2020)
Source Author's composition based on AFDC, EC (2020), EIA (2021d), USDE (2021a; b, c), and Verhelst et al. (2009)

The other competitor can be green hydrogen; however, its cost efficiency largely depends on the cost of raw material and electricity used for hydrogen production.

There are two more aspects that should be considered. On the one hand, technology provides many opportunities for internal combustion engines, such as upgrade and/or diversification of fuels, intelligent networking technologies (e.g. the use of traffic information), higher efficiency, and more simplified and cheaper usage (Zhao et al. 2020). This means that fossil fuels may become more cost efficient; therefore, they could provide more value for money. On the other hand, most of the alternative fuels represent newer technologies compared to fossil fuels; therefore, further technological development and cheaper products can be expected. The overall balance between the different fuels will be determined by the extent of these changes.

2.4.2 Advantages and Disadvantages of Alternative Fuels

This subchapter provides a summary by collecting advantages and disadvantages of the analyzed alternative fuels (Table 2.11). Their common characteristics, such as all of them are renewable, have lower GHG emissions compared to fossil fuels, provide higher self-dependency, are not indicated.

2.5 Summary and Conclusions

Global warming is one of the greatest challenges of humanity caused mainly by the increasing GHG emissions. The fossil resources of the Earth are limited. Transportation encompasses both, as it contributes to GHG emissions and uses mostly fossil fuels. The global vehicle fleet is increasing and most of them are expected to run on gasoline even in 2050. Although newer cars generally have better fuel efficiency, the average age of the entire vehicle fleet is also increasing.

Table 2.11 Advantages and disadvantages of the analyzed alternative fuels

Alternative fuels	Advantages	Disadvantages
Ethanol	Octane booster Mature technology High-density fueling network Easy fuel transportation Possible negative CO_2 balance	First generation—food versus fuel Advanced ethanol—expensive 1/3 lower energy content High-level blends require modifications Low brake thermal efficiency
Biodiesel	Octane booster Mature technology High-density fueling network Easy fuel transportation Possible negative CO_2 balance	First generation—food versus fuel Advanced ethanol—expensive 10% lower energy content High-level blends require modifications Low brake thermal efficiency
Electricity	High engine efficiency Cheap electricity Cost efficiency The cheapest travel option No travel emission	Low-density fueling network Grid stability Lack of smart/fast charging options Need for next-generation batteries
Biogas	Use of biomass or (waste)water High fuel efficiency Potential cost efficiency Further technological development	Expensive (production process, fuel, and vehicle) Low-density fueling network Lack of transport infrastructure
Green hydrogen	Cheap and abundant feedstocks Better fuel efficiency Further technological development	Need for decentralized production Expensive (plant, production process, fuel, tank) Low-density fueling network Pipeline injection problems

Source Author's composition

Biofuels are direct fuel substitutes and can be mixed with fuels or used purely. Currently, first generation technologies are used both for ethanol and biodiesel production. This means that the feedstock used for production of biofuel can be used for food production as well. For example, wheat, maize, sugar beet, and sugarcane can be used for ethanol production, while rapeseed, sunflower, palm, and soybean for biodiesel production. Despite their clear environmental advantages (lower GHG emissions), their use is driven by country-level blending mandates and production costs. The latter depends on the (crude) oil prices and its processing costs. The most significant advantages of biofuels are their mature production technology, easy use as either as a fuel substitute or pure biofuel, and more environmentally friendly nature compared to fossil fuels.

The other alternative fuels analyzed in this subchapter are electricity, green hydrogen, and biogas. Electric vehicles represent the future, but the share of hybrid and fully electric cars is expected to reach 40% only by 2050. They have higher engine efficiency and use cheaper raw materials. However, the sustainability of electric vehicles depends on how the electricity was produced, i.e. was it green or not? Limited travel range and charging infrastructure, grid stability problems, and the battery cell production technology are the bottlenecks of this technology. Moreover, recycling used batteries will be a huge future challenge. Hydrogen can only be green if renewable energy is used for its production. Currently, the vast majority of hydrogen production is based on fossil resources. The major advantage of green hydrogen is its clean burning with no carbon-related emission. However, hydrogen has many disadvantages, such as low energy density, low-density fueling networks, and among other things, hydrogen-fueled vehicles are expensive. This is a more promising fuel for public transport, e.g. buses, trains, and heavy-duty road vehicles. To use biogas, it should be cleaned, upgraded, and, either compressed or liquified. These are costly processes. In general, biogas is similar to (green) hydrogen in nature providing similar benefits and suffering from the same drawbacks.

Every analyzed alternative fuel provides environmental benefits compared to fossil fuels. From a purely sustainability point of view, they are much better. Except for ethanol, they provide higher fuel efficiency. In addition to these advantages, using electricity is the cheapest way of transportation. There are at least two aspects that should be considered. First, fossil fueling is a mature technology; therefore, it is easier and more convenient to use with high fueling infrastructure and cheaper vehicles. Second, alternative fuels provide the opportunity for huge technological advancements, which may make these technologies much cheaper. However, it should be kept in mind that fossil fuels are relatively cheap with high energy density. This makes it hard to replace them for a while, but humanity needs to switch to alternative fuels as soon as possible to be able to deal with global warming.

The major conclusions of this research are the followings:

- alternative fuels provide many advantages, especially from the environmental point of view (lower or no GHG emissions)
- currently, electricity is the cheapest alternative in private transportation
- the most common barriers to using alternative technologies are the low density fueling infrastructure and more expensive vehicles
- continuous technological advancements may make alternative fuels more competitive.

For the future, there are many potential research paths. This study concentrated on environmental aspects and partially on economic ones. Social and political dimensions could be analyzed too. Moreover, technological advancements of alternative fuels, as well as other alternatives, could also be researched.

References

ACEA (2019) ACEA report: vehicles in use—Europe 2019. European Automobile Manufacturers' Association, Brussels, Belgium

ACEA (2021) ACEA report: vehicles in use—Europe, Jan 2021. European Automobile Manufacturers' Association, Brussels, Belgium

Alalwan HA, Alminshid AH, Aljaafari HA (2019) Promising evolution of biofuel generations: subject review. Renew Energy Focus 28:127–139. https://doi.org/10.1016/j.ref.2018.12.006

ANFAVEA (2020) Brazilian automotive industry yearbook 2020. Brazilian Automotive Industry Association, São Paulo, Brazil

Autostat (2018) The average age of cars in Russia has exceeded 13 years. https://eng.autostat.ru/news/15457/

Bae C, Kim J (2017) Alternative fuels for internal combustion engines. Proc Combust Inst 36(3):3389–3413. https://doi.org/10.1016/j.proci.2016.09.009

CARD (2021) Historical ethanol operating margins. Center for Agricultural and Rural Development, Iowa State University. https://www.card.iastate.edu/research/biorenewables/tools/hist_eth_gm.aspx

CAT (2021) Climate action tracker emissions gap. https://climateactiontracker.org/global/cat-emissions-gaps/

CTT (2018) Report 1.10.2018 on deployment of infrastructure for alternative fuels in the European Union: time to act! (2018/2023(INI)). Committee on Transport and Tourism https://www.europarl.europa.eu/doceo/document/A-8-2018-0297_EN.html

Darda S, Papalas T, Zabaniotou A (2019) Biofuels journey in Europe: currently the way to low carbon economy sustainability is still a challenge. J Clean Prod 208:575–588. https://doi.org/10.1016/j.jclepro.2018.10.147

EC (2015) Directive (EU) 2015/1513 of the European Parliament and of the council of 9 September 2015 amending Directive 98/70/EC relating to the quality of petrol and diesel fuels and amending Directive 2009/28/EC on the promotion of the use of energy from renewable sources. Official Journal of the European Union, 239, 1–29

EC (2020) A hydrogen strategy for a climate-neutral Europe. COM(2020) 301 final, Brussels, Belgium

EIA (2021a) Annual energy review 2021. Table 7. Transportation sector key indicators and delivered energy consumption. Energy Information Administration, Washington, DC, USA https://www.eia.gov/outlooks/aeo/data/browser/#/?id=7-AEO2021&region=0-0&cases=ref2021&start=2019&end=2050&f=A&linechart=ref2021-d113020a.5-7-AEO2021&sourcekey=0

EIA (2021b) Gasoline explained. Factors affecting gasoline prices. U.S. Energy Information Administration. https://www.eia.gov/energyexplained/gasoline/factors-affecting-gasoline-prices.php

EIA (2021c) Gasoline and diesel fuel update. U.S. Energy Information Administration. https://www.eia.gov/petroleum/gasdiesel/

EIA (2021d) Short-term energy outlook data browser. U.S. Energy Information Administration. https://www.eia.gov/outlooks/steo/data/browser/#/?v=19&f=A&s=&id=&maptype=0&ctype=linechart

EPA (2020) An overview of renewable natural gas from biogas. United States Environmental Protection Agency, Washington, DC, USA

Fan JL, Wang JX, Li F, Yu H, Zhang X (2017) Energy demand and greenhouse gas emissions of urban passenger transport in the Internet era: a case study of Beijing. J Clean Prod 165:177–189. https://doi.org/10.1016/j.jclepro.2017.07.106

Gebremariam SN, Marchetti JM (2018) Economics of biodiesel production. Energy Convers Manage 168:74–84. https://doi.org/10.1016/j.enconman.2018.05.002

GECF (2020) GECF global gas outlook 2050 synopsis. Gas exporting countries forum global gas model, Feb 2020

GlobalFleet (2021) Mexico. https://www.globalfleet.com/en/wikifleet/mexico

IEA-AMF (2021) Ethanol properties. In: International energy agency technology collaboration programme on advanced motor fuels. https://www.iea-amf.org/content/fuel_information/ethanol/e10/ethanol_properties#octane_numbers

IHS Markit (2019) Average age of cars and light trucks in U.S. rises again in 2019 to 11.8 years. IHS Markit. https://news.ihsmarkit.com/prviewer/release_only/slug/automotive-average-age-cars-and-light-trucks-us-rises-again-2019-118-years-ihs-markit-

IPCC (2014) Climate change 2014: mitigation of climate change. In: Edenhofer O, Pichs-Madruga R, Sokona Y, Farahani E, Kadner S, Seyboth K, Adler A, Baum I, Brunner S, Eickemeier P, Kriemann B, Savolainen J, Schlömer S, von Stechow C, Zwickel T, Minx JC (eds) Contribution of working group III to the fifth assessment report of the intergovernmental panel on climate change. Cambridge University Press, Cambridge, United Kingdom and New York, NY, USA

IRS (2021) Yearly average currency exchange rates. https://www.irs.gov/individuals/international-taxpayers/yearly-average-currency-exchange-rates

Jaeger WK, Siegel R (2008) Economics of oilseed crops and their biodiesel potential in Oregon's Willamette Valley. Special report 1081. Oregon State University Extension Service, Corvallis, OR, USA

Kalghatgi G (2018) Is it really the end of internal combustion engines and petroleum in transport? Appl Energy 225:965–974

Keith DR, Houston S, Naumov S (2019) Vehicle fleet turnover and the future of fuel economy. Environ Res Lett 14(2):021001. https://doi.org/10.1088/1748-9326/aaf4d2

Kumar Ghosh S, Mandal S (2018) Evaluation of biogas as an alternative driving force of electrically operated vehicles: a case study. Int J Eng 31(5):834–840

Lane J (2021) The digest's biofuels mandates around the world 2021. https://www.biofuelsdigest.com/bdigest/2021/01/06/the-digests-biofuels-mandates-around-the-world-2021/

Lieberz S (2020) Biofuel mandates in the EU by member state in 2020. Report number: E42020-0025, United States Department of Agriculture, Foreign Agricultural Service, Washington, DC, USA

Liu F, Zhao F, Liu Z, Hao H (2019) Can autonomous vehicle reduce greenhouse gas emissions? A country-level evaluation. Energy Policy 132:462–473. https://doi.org/10.1016/j.enpol.2019.06.013

Liu J, Abbas Q, Alharthi M, Mohsin M, Rasul F, Iqbal N (2021) Managerial policy and economic analysis of wind-generated renewable hydrogen for light-duty vehicles: green solution of energy crises. Environ Sci Pollut Res 28(9):10642–10653. https://doi.org/10.1007/s11356-020-11018-1

Mizik T, Gyarmati G (2021) Economic and sustainability of biodiesel production—a systematic literature review. Clean Technol 3(1):19–36. https://doi.org/10.3390/cleantechnol3010002

Myhre G, Shindell D, Bréon F-M, Collins W, Fuglestvedt J, Huang J, Koch D, Lamarque J-F, Lee D, Mendoza B, Nakajima T, Robock A, Stephens G, Takemura T, Zhang H (2013) Anthropogenic and natural radiative forcing. In: Stocker TF, Qin D, Plattner G-K, Tignor M, Allen SK, Boschung J, Nauels A, Xia Y, Bex V, Midgley PM (eds) Climate change: the physical science basis. Contribution of working group I to the fifth assessment report of the intergovernmental panel on climate change. Cambridge University Press, Cambridge, United Kingdom and New York, NY, USA

Mustafi NN, Agarwal AK (2020) Biogas for transport sector: current status, barriers, and path forward for large-scale adaptation. In: Singh A, Sharma Y, Mustafi N, Agarwal A (eds) Alternative fuels and their utilization strategies in internal combustion engines. energy, environment, and sustainability. Springer, Singapore, pp 229–271. https://doi.org/10.1007/978-981-15-0418-1_13

Nationmaster (2019) Top countries in global passenger cars in use. https://www.nationmaster.com/nmx/ranking/global-passenger-cars-in-use

Nevzorova T, Kutcherov V (2019) Barriers to the wider implementation of biogas as a source of energy: a state-of-the-art review. Energ Strat Rev 26:100414. https://doi.org/10.1016/j.esr.2019.100414

NRC (2013) Transitions to alternative vehicles and fuels. In: National research council. The National Academies Press, Washington, DC, USA.https://doi.org/10.17226/18264

Noura N, Erradi I, Desreveaux A, Bouscayrol A (2018) Comparison of the energy consumption of a diesel car and an electric car. In: 2018 IEEE vehicle power and propulsion conference, institute of electrical and electronics engineers, Piscataway, NJ, USA, pp 1–6.https://doi.org/10.1109/VPPC.2018.8605020

OECD-FAO (2021) OECD-FAO agricultural outlook 2020–2029. https://stats.oecd.org/Index.aspx?DataSetCode=HIGH_AGLINK_2019

Ogunkunle O, Ahmed NA (2019) A review of global current scenario of biodiesel adoption and combustion in vehicular diesel engines. Energy Rep 5:1560–1579. https://doi.org/10.1016/j.egyr.2019.10.028

Papacz W (2011) Biogas as vehicle fuel. J KONES Powertrain Transport 18(1):403–410

Persson M (2007) Biogas upgrading and utilization as vehicle fuel. In: Proceedings future of biogas in Europe—III, University of Southern Denmark, Bioenergy Department, Esbjerg, Denmark, pp 59–64

Ritchie H (2020) Sector by sector: where do global greenhouse gas emissions come from. Our World in Data. https://ourworldindata.org/ghg-emissions-by-sector

Santos G (2017) Road transport and CO_2 emissions: what are the challenges? Transp Policy 59:71–74. https://doi.org/10.1016/j.tranpol.2017.06.007

Singh RS, Walia AK (2016) Biofuels: historical perspectives and public opinions. In: Singh RS, Pandey A, Gnansounou E (eds) Biofuels: production and future perspectives. CRC Press, Boca Raton, FL, USA, pp 21–42

statista (2021a): Average age of the passenger vehicles in China from 2015 to 2022. https://www.statista.com/statistics/865046/china-passenger-vehicles-average-age/

statista (2021b) Average age of passenger cars in use in Japan from fiscal year 2010 to 2019. https://www.statista.com/statistics/680051/japan-passenger-car-average-age/

statista (2021c) Average age of passenger cars in circulation in Brazil from 2014 to 2019. https://www.statista.com/statistics/831150/passenger-car-fleet-average-age-brazil/

statista (2021d) Average annual Brent crude oil price from 1976 to 2021. https://www.statista.com/statistics/262860/uk-brent-crude-oil-price-changes-since-1976/

statista (2021e) Average exchange rate of Indonesian rupiahs (IDR) to U.S. dollars (USD) from 2009 to 2020. https://www.statista.com/statistics/995840/indonesia-exchange-rate-between-rupiahs-and-us-dollar/

Steenberghen T, Lopez E (2008) Overcoming barriers to the implementation of alternative fuels for road transport in Europe. J Clean Prod 16(5):577–590. https://doi.org/10.1016/j.jclepro.2006.12.001

Tabar AR, Hamidi AA, Ghadamian H (2017) Experimental investigation of CNG and gasoline fuels combination on a 1.7 L bi-fuel turbocharged engine. Int J Energy Environ Eng 8(1):37–45. https://doi.org/10.1007/s40095-016-0223-3

USDE (2021a) Clean cities alternative fuel price report. U.S. Department of Energy. https://www.fueleconomy.gov/feg/evtech.shtml

USDE (2021b) All-electric vehicles. U.S. Department of Energy, Washington, DC, USA

USDE (2021c) Fuel properties comparison. U.S. Department of Energy. https://afdc.energy.gov/fuels/properties?properties=energy_ratio,energy_comparison,energy_content_per_gallon,energy_content_higher_value,physical_state

Van Fan Y, Perry S, Klemeš JJ, Lee CT (2018) A review on air emissions assessment: transportation. J Clean Prod 194:673–684. https://doi.org/10.1016/j.jclepro.2018.05.151

Verhelst S, Maesschalck P, Rombaut N, Sierens R (2009) Efficiency comparison between hydrogen and gasoline, on a bi-fuel hydrogen/gasoline engine. Int J Hydrogen Energy 34(5):2504–2510. https://doi.org/10.1016/j.ijhydene.2009.01.009

Wang A, Stogios C, Gai Y, Vaughan J, Ozonder G, Lee S, Posen ID, Miller EJ, Hatzopoulou M (2018) Automated, electric, or both? Investigating the effects of transportation and technology scenarios on metropolitan greenhouse gas emissions. Sustain Cities Soc 40:524–533. https://doi.org/10.1016/j.scs.2018.05.004

Yoro KO, Daramola MO (2020) CO_2 emission sources, greenhouse gases, and the global warming effect. In: Rahimpour MR, Farsi M, Makarem MA (eds) Advances in carbon capture. methods, technologies and applications. Woodhead Publishing, Sawston, United Kingdom, pp 3–28. https://doi.org/10.1016/B978-0-12-819657-1.00001-3

Zhao F, Chen K, Hao H, Liu Z (2020) Challenges, potential and opportunities for internal combustion engines in China. Sustainability 12(12):4955. https://doi.org/10.3390/su12124955

Chapter 3
Fuels for Sustainable Transport in India

Kumar Saurabh and Rudrodip Majumdar

Abbreviations

CNG	Compressed natural gas
CO_2	Carbon dioxide
EBP	Ethanol blended petrol
EV	Electric vehicle
FC	Fuel cell
FCEV	Fuel cell electric vehicle
FCV	Fuel cell vehicle
FFE	Flex-fuel engine
FY	Fiscal year
GDP	Gross domestic product
GGE	Gasoline gallon equivalency
GHG	Green house gas
H-CNG	Hydrogen–compressed natural gas
HHV	Higher heating value
HP	Horsepower
HTE	High temperature electrolysis
ICE	Internal combustion engine
LHV	Lower heating value
LNG	Liquefied natural gas
LPG	Liquefied petroleum gas
MIE	Minimum ignition energy

K. Saurabh · R. Majumdar (✉)
National Institute of Advanced Studies (NIAS), Indian Institute of Science Campus, Bengaluru, Karnataka 560012, India
e-mail: rudrodip@nias.res.in

K. Saurabh
Manipal Academy of Higher Education (MAHE), Manipal, Karnataka 576104, India

G. Di Blasio et al. (eds.), *Clean Fuels for Mobility*, Energy, Environment, and Sustainability, https://doi.org/10.1007/978-981-16-8747-1_3

NTP	Normal temperature and pressure
R&D	Research and development
RON	Research octane number
SDG	Sustainable development goal
SI	Spark ignition
TTW	Tank to wheel
WHTM	Wheel to miles
WTM	Well to miles
WTT	Well to tank
WTW	Well to wheel

Symbols

Atm	Atmosphere
cm	Centimetre
cm/s	Centimetre/second
cm^2/s	Centimetre squared/second
CO_{2eq}/MJ	Carbon dioxide equivalent/Mega Joule
g/mol	Gram/mole
K	Kelvin
kg/m^3	Kilogram/metre cubed
kJ/kg	Kilo Joule/kilogram
km	Kilo metre
kPa	Kilo Pascal
mJ	Milli-Joule
mm^2/s	Millimetre squared/second
MJ/kg	Mega Joule/kilogram
MJ/l	Mega Joule/litre
MJ/m^3	Mega Joule/metre cubed
Wh/kg	Watt-hour/kilogram
Wh/l	Watt-hour/litre
vol%	Volume percentage

3.1 Introduction

The transport sector is directly linked to the second target of Sustainable Development Goal (SDG) 7, which aims to increase the share of renewable and clean energy sources in the global energy mix. In addition, the transport sector is expected to contribute heavily towards the third target of SDG 7, i.e., to double the global rate of improvement in energy efficiency (UNSDG Homepage 2021). India's transport sector has a share of 7% in India's Gross Domestic Product (GDP) (Ministry of Statistics and Program Implementation, Government of India (GoI) 2020), and 98% of the Indian transport sector runs on fossil fuels (International Energy Agency (IEA) 2021b). In 2017, 13.5% of the CO_2 emissions in India were attributed to the transport sector (International Energy Agency (IEA) NITI-Aayog, Government of India (GoI) 2020). The Government of India seeks to reduce the carbon intensity of fuels, measured as CO_{2eq}/MJ, by substituting oil-based products with low Green House Gas (GHG) emitting sustainable fuels (Sims et al. 2014). The road transport sector in India had the highest energy consumption among all the modes of transport, with an 84.5% share in Fiscal Year (FY) 2017–18 (Ministry of Statistics and Program Implementation, Government of India (GoI) 2020; International Energy Agency (IEA) 2021a). The road transport was responsible for nearly three-fourths (74%) of the CO_2 emissions (coming from all modes) in FY 2018–19, making it the most polluted mode of transport in India (International Energy Agency 2021b). India's roads carry 71% of India's freight transport against only 18% by the Indian Railways (NITI-Aayog, Government of India (GoI) 2021a). While light, medium and heavy freight carrying commercial vehicles constitute only 5% of the on-road vehicles, these freight carrying commercial vehicles account for more than 50% of the energy consumption as well as CO_2 emissions attributed to the road transport sector (International Energy Agency 2021b; Ministry of Road Transport Highways, Government of India (GoI) 2019).

An energy source can be considered as an alternative to the incumbent fuels (gasoline and diesel) in the transport sector if it is sustainable (leads to reduced level of emissions) and reliable (safe and portable and offers an adequate riding range). The scope of this chapter is limited to the in-depth analysis of the physico-chemical properties of the existing and futuristic alternate fuels pertaining to only surface transport for both passenger and freight segments in India. This book chapter seeks to provide a broad understanding about the technical feasibility of the futuristic fuel options for Indian road transport sector for the researchers belonging to diverse backgrounds and the practitioners alike. This chapter would provide a baseline for the policymakers in particular. The alternate fuels discussed in this chapter are ICE based fuel options, viz. CNG, LNG, LPG, Hydrogen, ethanol, and blended biofuels that are likely to be part of the diversified fuel mix in the envisaged sustainable road transport in India.

3.1.1 Sustainability and Reliability of a Fuel

The life cycle analysis of a fuel is referred to as Well to Wheel (WTW), which refers to the emissions emanating from the journey of a fuel from its exploration phase to the consumption phase where it is consumed in powering a vehicle (Gupta et al. 2016). The WTW journey is further divided into two parts: the first part is the Well to Tank (WTT) or the 'fuel part', and the second part is called the Tank to Wheel

(TTW) part or the 'vehicle part'. In the WTT part, the raw form of the fuel is extracted from the natural reserves, and it undergoes several stages of processing to become a refined fuel ready for consumption. After that, the refined fuel is transported and distributed through retail outlets to be finally loaded into a vehicle tank. In the TTW phase, a fuel's electrical or chemical energy changes its form and gets converted into mechanical energy during the motor or engine operation by the vehicle propulsion system. In a review article published (Hänggi et al. 2019) in 2019, an extension to WTW analysis of a fuel has been discussed with an additional Wheel to Miles (WHTM) part after the TTW segment. In this part, the energy source is manifested in terms of the vehicle's speed once the wheels start rolling upon being powered by the vehicle's propulsion system. Once the propulsion system has powered the wheels, the speed and distance covered by the vehicle are independent of the vehicle and the corresponding fuel characteristics. All the three parts in the Well to Miles (WTM) journey, represented in Fig. 3.1, impact the sustainability of a fuel as a viable potential alternative.

The reliability of a fuel is dependent upon two major factors i.e., technical feasibility of an energy source as a fuel and the supply-chain management of the fuel. The technical viability of a fuel is established by its physico-chemical properties, ignition parameters, and safety attributes. The commercial viability of a fuel is established based on the *4As* (Availability, Accessibility, Acceptability and Affordability) of energy security (Winzer 2012). These *4As* are applicable with varying degrees to all the three stages of the fuel life cycle. However, as seen in Fig. 3.2, in the WTT phase of the life cycle, all the *4As* play an important role in establishing the reliability of a fuel. 'Availability' is the most important dimension of energy security among all

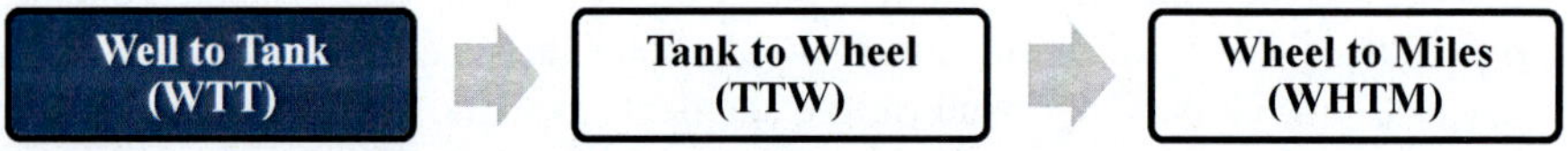

Fig. 3.1 Journey of a fuel from Well to Miles

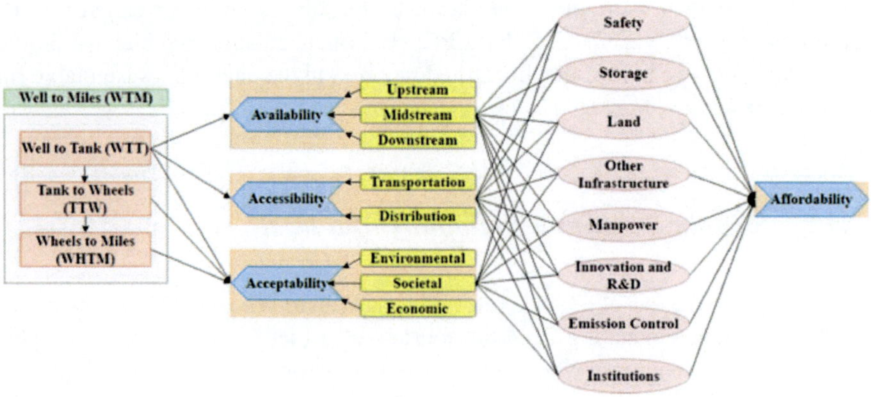

Fig. 3.2 Factors establishing the reliability of a fuel

the 4A's as nearly all studies around defining energy security identify it to be central to energy security (Ang et al. 2015). As per a study, the technical maturity of a fuel is also considered as an important dimension of 'Availability' (Ren and Sovacool 2014).

Availability encompasses the exploration and production of the raw materials (up-stream), transportation of the raw materials to the refineries (mid-stream), and refining, processing, and purifying the raw materials at the refineries (downstream) (Petro Online Homepage 2021). Accessibility refers to the transportation and distribution of the refined materials or fuels to be finally delivered to the tanks of the vehicles. Acceptability corresponds to the acceptance of the fuel by the society, primarily from the environmental and emission perspective. However, apart from the environmental impacts, acceptability also includes societal and economic impacts (Paravantis and Kontoulis 2020). The societal impacts encompass the human rights issues of the people involved in exploration and labour issues (health and safety, livelihood, fair wage, training and education) of those involved in production. Acceptability also encompasses the economic impacts ranging from unemployment to food security issues of the affected population (Schuchard et al. 2015). The affordability of a fuel to the end consumer depends on all three phases in the WTM journey of the fuel life cycle. All the phases in the WTM journey need investments that go into the creation and up-gradation of critical infrastructure, skilled workforce, innovation of environment-friendly technologies, maintenance of the existing system, and towards creating the governing institutions that would ensure the formulation of the required norms and guidelines and monitor the adherence to the same.

An exhaustive list of conventional and emerging transport fuels options that are either functional or awaiting deployment or under R&D is presented in Fig. 3.3. Based on the physico-chemical, ignition and safety parameters discussed in Sect. 3.2, the potential for each of the existing and emerging transport fuel options for Internal Combustion Engines (ICE) in India is assessed. Among the fuel options highlighted in Fig. 3.3, the existing fuel options in India are discussed in Sect. 3.3, and the futuristic fuel options for ICEs are discussed in Sect. 3.4. Section 3.5 discusses the pros and cons of the fuel options delineated in the Sects. 3.3 and 3.4 over their technical feasibility based on the travel segment, travel distance, constraints associated with dense urban settings and other key attributes. In the Sect. 3.6, a conclusion is drawn over the potential for the commercial deployment of the fuel options based on their technical feasibility.

3.2 Various Technical Parameters of a Fuel

The ignition characteristics and physico-chemical properties of potential fuel options for the road transport need to be analysed qualitatively as these characteristics establish the energy potential, the storage requirements, and the safety hazards (in case of an accident) of a potential energy source. The parameters defining these properties and characteristics are defined and their relevance is stated hereafter.

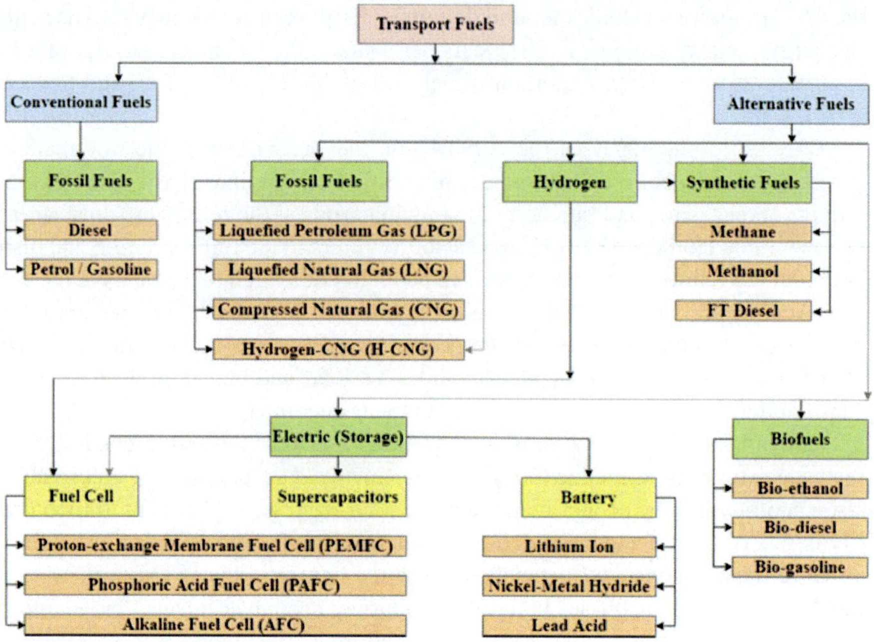

Fig. 3.3 Transport fuels options available across the world

3.2.1 Physico-chemical Parameters

- **Kinematic viscosity [mm²/s]** measures the internal resistance provided by a fluid while flowing under the gravitational forces. The thick fuels have high viscosity, while the thin ones have lower viscosity, enabling them to pour more easily at low temperatures. Therefore, thin fuels reduce friction in engines by helping them to start quickly in the cold weather. It also directly influences the fuel atomisation quality and size of the fuel droplet in the spray (Hasan and Rahman 2017). High kinematic viscosity results in slower air–fuel mix and higher injection pressure is needed in the vehicle propulsion system to perform better. At high temperatures and loads, high kinematic viscosity allows sustenance of film strength and fuel pressure.
- **Density at 288 K [kg/m³]** is the ratio of the mass to volume of a material which tells how heavy a fuel is. The heavier the fuel, the higher is quantity of the fuel injected in the combustion chamber and it results in larger fuel droplets which leads to incomplete combustion due to difficult atomization. This further results in decreased efficiency of the fuel and low Lower Heating Value (LHV) (Panchasara and Ashwath 2021). This observation is true only in the drop-in injection strategy as the new generation injectors are able to regulate the amount of fuel entering the combustion chamber by volume and not by mass. In combustion optimization, the impact of density on combustion can be minimized by adopting different injection strategies.

- **Lower heating value [MJ/kg]** is also called as net calorific value. It is the amount of heat released by combusting unit mass of the fuel which was initially at 298 K and then bringing the temperature of the combusted fuel to 423 K. It is assumed that the latent heat of vaporisation of water is not recovered from the reaction products (Hydrogen Tools Homepage 2021).
- **Higher heating value [MJ/kg]** is also known as gross calorific value. It is the amount of heat released by combusting unit mass of the fuel which was initially at 298 K and bringing the temperature of the combusted fuel back to 298 K. It is assumed that the latent heat of vaporisation of water is recovered in the reaction products (Hydrogen Tools Homepage 2021). Therefore, the numerical value of HHV is always higher than the LHV of a fuel. The difference between HHV and LHV of a fuel is important to ascertain the production of liquid water during the combustion of a fuel (Clarke-Energy Homepage 2021).
- **Latent heat of vaporisation [kJ/kg]** is the amount of energy required to change the state of a unit mass of liquid to vapour at a constant temperature. Higher latent heat of vaporisation of a fuel provides additional knock resistance due to charge cooling effect (particularly in direct injection); and this results in reduced in-cylinder temperature and lowered chances of pre-ignition of the fuel (Chupka et al. 2015). High latent heat of vaporization of a fuel leads to an increase in the volume efficiency of the fuel and it also helps in engine breathing.
- **Boiling point of a fuel [K]** is the temperature at which a fuel can change its state from liquid to a gas throughout the bulk of the liquid. It is the temperature at which the saturated vapour pressure of the fuel equals the surrounding atmospheric pressure (Purdue University Homepage 2021). A fuel's vaporization and combustion potential get affected by the boiling point (Kook and Pickett 2010). A low boiling point indicates quick evaporation and rapid mixing of the fuel with the ambient gas which can lead to pre-mixed combustion of the fuel, less visible flame, higher flame temperature, and lower particulate emissions (Sholes et al. 2002).
- **Cloud point of a fuel [K]** is the temperature at which a cloud of wax crystals first appear in a liquid fuel. It is an indicator of the performance of the fuel under cold weather conditions (Akhil et al. 2017). Cloud point is important for fuels as compared to the lubricating oils while the pour point is significant for lubricants. In colder climatic conditions, additives are added to the fuels to control the cloud point of the fuel.
- **Pour point of a fuel [K]** the lowest temperature at which the movement of a liquid fuel is observed, and the fuel can be pumped easily. It is the lowest temperature at which a fuel performs satisfactorily, and beyond this temperature, the fuel stops flowing and starts to freeze (Akhil et al. 2017). The cloud point of a fuel is always higher than the pour point of the fuel as the crystallization of the fuel starts at the cloud point and the freezing process gets completed at the pour point.

3.2.2 *Ignition Parameters*

- **Stoichiometric air–fuel ratio** is the theoretical mass of air (provides oxygen for combustion) needed to combust the unit mass of a fuel completely. The higher the stoichiometric ratio, the more air is required to mix with the fuel to burn it completely (Kosky et al. 2013). It is important to maintain the stoichiometric air–fuel ratio at the optimum level to achieve the best combustion efficiency as too much air can lead to rich combustion process which can lead to wastage of energy. Lower ratio of air–fuel mixture will lead to partial burning of the fuel and it would lead to wastage of fuel.
- **Adiabatic Flame Temperature in air [K]** is the temperature occurring from the complete combustion process of a fuel without any work, heat transfer or kinetic or potential energy changes. The higher the flame temperature, the higher is the effectiveness of heat (Hanby 2012). It is used to determine the quality of the fuel as well as the suitability of the fuel for a particular application. The adiabatic flame temperature of a fuel plays an important role in designing and scaling-up the processes where heat transfer is involved.
- **Flash Point of a fuel [K]** is the lowest temperature at which a heated liquid's vapour or air mixture can be ignited or flashed by a flame or spark or any other ignition source placed above the liquid surface. It is the temperature at which the flame is momentarily ignited (Paratherm Homepage 2021). Its knowledge is important for fire protection and in relation to the safety of the clean-up operations in case of spillage. However, the fuels with low flash point are attractive options from environmental perspective as they burn cleanly.
- **Fire point of a fuel [K]** is the lowest temperature at which a heated liquid's vapour or air mixture will burn continuously when ignition sources support combustion. It is the temperature at which the burning becomes continuous after ignition (Paratherm Homepage 2021). This parameter is important to ascertain the safety parameters need to be adopted while transporting and storing.
- **Auto-ignition temperature of a fuel [K]** at which the air mixture reaches a sufficient temperature for the fuel to self-ignite without an ignition source. The higher the auto-ignition temperature, the higher is the safety of the fuel (Paratherm Homepage 2021). It is also known as the kindling point of the fuel. It is the temperature which is high enough to provide the fuel with energy to start the chemical reaction for self-ignition. With an increase in the pressure or oxygen concentration, the auto-ignition temperature of a fuel decreases.
- **Quenching gap in air [cm]** is the distance between two flat plates at which the ignition of a flammable mixture is suppressed and the flame is just quenched. It corresponds to the smallest diameter of a tube through which a flame can propagate. It is used to establish the burning velocity of a fuel. Fast burning fluids have smaller quenching gaps (HySafe Hydrogen Fundamentals Homepage 2021).
- **Octane number of a fuel** is defined as the volume per cent of iso-octane in a blend of iso-octane and n-heptane that would have the same anti-knocking capacity as the fuel under test. It measures a fuel's resistance to detonation (engine

knocking) in spark ignition (SI) internal combustion engines. It is not a measure of energy content or better combustion. A high octane number can help increase the efficiency and performance of an engine (Allegrini and Olivieri 2011).
- **Cetane number of a fuel** is the volume per cent of cetane in a blend of cetane and methylnaphthalene that gives the same ignition delay period as the test sample. It provides information about the speed of self-ignition of fuel when injected into hot air through the fuel injector. Higher cetane numbers mean shorter ignition delay and better performance of the combustion engine (Mabanaft Homepage 2021).
- **Laminar flame speed [cm/s]** is the propagation speed of the flame cone or sphere under the laminar conditions at which the oxidation takes place. It can be understood as the speed at which an oxidation reaction occurs between the combustible constituents of a gas and oxygen at the surface of the inner core of the flame (Energy 2021). It is a key indicator of turbulent combustion. It also tells about the reactivity, diffusivity and exothermicity of a fuel (Dagaut 2019).

3.2.3 Safety Parameters

- **Diffusion coefficient [cm^2/s]** implies that the mass of the substance diffuses through a unit surface in a unit time at a concentration gradient of unity. It is also dependent on the molecule size of the fuel being diffused. The higher the diffusion coefficient, faster is the diffusion in the medium (air in case of fuel), which ensures greater safety for a fuel (Veziro and Barbir 1992).
- **Vapour Flammability limits [vol%]** are well-defined lower and upper bounds within which mixtures of the dispersed combustible materials and oxygen in the air undergo effective combustion. The vapour flammability limits stand for the entire range of concentrations of a mixture of flammable vapour or gas in the air over which a flash will occur, or a flame will travel if the mixture is ignited (Ludwig 1999). Higher the vapour flammability limits, higher is the chance to ignite the fuel in the presence of a spark. The relevant information pertaining to the various parameters associated with the different types of transport fuels for ICEs have been summarized in the Table 3.1.

3.3 Alternative Transport Energy Sources in India

3.3.1 Fossil-Based Fuels

3.3.1.1 Compressed Natural Gas (CNG)

Compressed Natural Gas (CNG) has methane as its main constituent, and it is produced in its usable form by compressing the natural gas to less than 1% of the

Table 3.1 Physico-chemical, ignition and safety parameters of ICE based fuels

Parameter	Units	Fossil-fuel (Crude)		Fossil-fuel (Natural Gas)				Biofuel	
		Gasoline	Diesel	Hydrogen	CNG	LNG	LPG	Biodiesel	Ethanol
Kinematic viscosity (313 K)	mm^2/s	0.5–0.6 (Sakthivel et al. 2018)	1.9–3.8 (Balat 2006)	125* (HyResponse Homepage 2021)	0.01–0.03	0.23	0.01–0.03	3.7–5.8 (Balat 2006)	1.2–1.5 (Sakthivel et al. 2018)
Density at 288 K	kg/m^3	750–765 (Sakthivel et al. 2018)	830–950 (Sarıkoç 2020)	0.084 (HyResponse Homepage 2021)	215 # (Smajla et al. 2019)	466 @ (Smajla et al. 2019)	520–560 (Polytechnic University of Madrid Spain Homepage 2021)	880 (Polytechnic University of Madrid Spain Homepage 2021)	785–810 (Sakthivel et al. 2018)
Specific gravity		0.68–0.74 (Fluid Switch Homepage 2021)	0.84–0.86 (Balat 2006)	0.071 (HyResponse Homepage 2021)	0.59 (Green Gas Limited Homepage 2021)	0.554 (Indian Institute of Technology Delhi Homepage 2021)	0.59 (Methanol Institute Homepage 2021)	0.87–0.89 (Balat 2006)	0.79 (Methanol Institute Homepage 2021)
Lower-higher heating value (LHV—H HV)	MJ/kg	43–47.3 (Sarıkoç 2020)	43.3–45.6 (Boretti 2020)	120–142 (HyResponse Homepage 2021)	50–55.5 (HyResponse Homepage 2021)	50–55 (Boretti 2020)	45–50 (Polytechnic University of Madrid Spain Homepage 2021)	39.3–39.8 (Polytechnic University of Madrid Spain Homepage 2021)	26.9–30 (Sakthivel et al. 2018)
Specific energy	Wh/kg	12,889	12,667	39,406	14,889	14,889	13,778	11,722	8333
Energy density	Wh/l	9500	10,722	1479	2500	6167	7028	9167	6667
Energy content (gas or liquid)	MJ/m^3 or MJ/l	42.7 (Khan et al. 2015)	36 (Khan et al. 2015)	4.5–5.3	24.6 (Khan et al. 2015)	21–24 (Smajla et al. 2019)	23.4–26.5 (Sarıkoç 2020)	33	24
Latent heat of vaporisation	kJ/kg	380–400 (Sakthivel et al. 2018)	250 (Boretti 2020)	447 (Methanol Institute Homepage 2021)	510 (Boretti 2020)	510 (Boretti 2020)	449 (Methanol Institute Homepage 2021)	230 (Raza et al. 2020)	900–920 (Sakthivel et al. 2018)

(continued)

Table 3.1 (continued)

Parameter	Units	Fossil-fuel (Crude)		Fossil-fuel (Natural Gas)				Biofuel	
		Gasoline	Diesel	Hydrogen	CNG	LNG	LPG	Biodiesel	Ethanol
Boiling point range (Initial—Final)	K-K	318–480 (Sakthivel et al. 2018)	466–566 (Cameo Chemicals NOAA Homepage 2021)	20 (HyResponse Homepage 2021)	111 (Sarıkoç 2020)	111 (Sarıkoç 2020)	231–273 (Sarıkoç 2020)	692 (Boretti 2020)	351–352 (Sakthivel et al. 2018)
Cloud point	K	255 (Kheiralla et al. 2012)	256–265 (Balat 2006)	(-)	(-)	112 (Indian Institute of Technology Delhi Homepage 2021)	(-)	262–289 (Balat 2006)	267 (Kheiralla et al. 2012)
Pour point	K	251 (Kheiralla et al. 2012)	237–243 (Balat 2006)	(-)	(-)	(-)	(-)	258–286 (Balat 2006)	265 (Kheiralla et al. 2012)
Stoichiometric air/fuel ratio	Ratio	14.7:1 (Khan et al. 2015)	14.6:1 (Khan et al. 2015)	34.3:1 (Goyal and Sharma 2014)	17.2:1 (Goyal and Sharma 2014)	17.2:1 (Goyal and Sharma 2014)	15.7:1 (Methanol Institute Homepage 2021)	13.8:1 (Methanol Institute Homepage 2021)	9:1 (Methanol Institute Homepage 2021)
Minimum ignition energy (MIE) in air	mJ	0.24 (Goyal and Sharma 2014)	(-)	0.011 (Goyal and Sharma 2014)	0.29 (Goyal and Sharma 2014)	0.29 (Goyal and Sharma 2014)	0.26 (Methanol Institute Homepage 2021)	(-)	(-)

(continued)

Table 3.1 (continued)

Parameter	Units	Fossil-fuel (Crude)		Fossil-fuel (Natural Gas)				Biofuel	
		Gasoline	Diesel	Hydrogen	CNG	LNG	LPG	Biodiesel	Ethanol
Flame temperature in air	K	2423 (Khan et al. 2015)	2327 (Khan et al. 2015)	2318 (Goyal and Sharma 2014)	2163 (Khan et al. 2015)	2339 (Jha et al. 2007)	2253 (Methanol Institute Homepage 2021)	2380 (Jha et al. 2007)	2078
Flash point	K	228–262 (HyResponse Homepage 2021)	325–350 (Balat 2006)	20 (HyResponse Homepage 2021)	89–124 (Methanol Institute Homepage 2021)	425 (Boretti 2020)	172–200 (Methanol Institute Homepage 2021)	408–423 (Balat 2006)	289 (Institute and of Technology Delhi Homepage 2021)
Auto ignition temperature	K	501–744 (Goyal and Sharma 2014)	450–603 (Cameo Chemicals NOAA Homepage 2021)	839–858 (Goyal and Sharma 2014)	813 (Goyal and Sharma 2014)	813 (Goyal and Sharma 2014)	728–783	373–443	638 (Institute and of Technology Delhi Homepage 2021)
Quenching gap in air [at NTP]	cm	0.2 (Goyal and Sharma 2014)	(-)	0.064 (Goyal and Sharma 2014)	0.203 (Goyal and Sharma 2014)	(-)	(-)	(-)	(-)
Octane number	#	91–100 (Sakthivel et al. 2018)	(-)	>130 (Goyal and Sharma 2014)	120–130 (Goyal and Sharma 2014)	92–110 (Raslavičius et al. 2014)	120 (Raslavičius et al. 2014)	(-)	108–110 (Sakthivel et al. 2018)
Cetane number	#	(-)	47–55 (Balat 2006)	(-)	(-)	(-)	(-)	46–70 (Balat 2006)	(-)

(continued)

Table 3.1 (continued)

Parameter	Units	Fossil-fuel (Crude)		Fossil-fuel (Natural Gas)				Biofuel	
		Gasoline	Diesel	Hydrogen	CNG	LNG	LPG	Biodiesel	Ethanol
Laminar flame speed (at 100 kPa, 325 K)	cm/s	33 (Sakthivel et al. 2018)	(-)	265–325 (Khan et al. 2015)	41 (Khan et al. 2015)	(-)	39 (Methanol Institute Homepage 2021)	(-)	39 (Sakthivel et al. 2018)
Burning velocity in air (at NTP)	cm/s	37–43 (Goyal and Sharma 2014)	(-)	325 (Goyal and Sharma 2014)	45 (Goyal and Sharma 2014)	(-)	(-)	(-)	(-)
Diffusion coefficient in air	cm^2/s	0.05 (Hosseini and Butler 2020)	0.04 (Raza et al. 2020)	0.61 (Hosseini and Butler 2020)	0.16 (Hosseini and Butler 2020)	(-)	(-)	0.07 (Raza et al. 2020)	(-)
Vapour flammability limits/ignition limits	vol%	0.6–8 (Sakthivel et al. 2018)	1–6 (Boretti 2020)	4–75 (Goyal and Sharma 2014)	4.3–15.2 (Khan et al. 2015)	5.3–15 (Boretti 2020)	2.2–9.5	(-)	3.5–15 (Sakthivel et al. 2018)
Equivalence ratio flammability limit in air (at NTP)		0.7–3.8 (Goyal and Sharma 2014)	0.6–2	0.1–7.1 (Goyal and Sharma 2014)	0.7–4 (Goyal and Sharma 2014)	(-)	(-)	(-)	(-)
Flame emissivity	%	34–43 (Hosseini and Butler 2020)	(-)	17–25 (Hosseini and Butler 2020)	25–33 (Hosseini and Butler 2020)	(-)	(-)	(-)	(-)
Life cycle carbon intensity	gCO_{2eq}/MJ	92–99 (Schuchard et al. 2015)	92–95 (Schuchard et al. 2015)	(-)	65–90 (Schuchard et al. 2015)	65–90 (Schuchard et al. 2015)	78 (Schuchard et al. 2015)	11–16 (Schuchard et al. 2015)	58–73 (Schuchard et al. 2015)

@ Density of LNG is at 111 K and atmospheric pressure
# Density of CNG is at 246.7 atm and room temperature
* Kinematic viscosity for Hydrogen has been calculated by dividing the absolute viscosity of hydrogen at 323 K with its density at 323 K
(-) denotes that the values either are not applicable or are not available within the search limit or cannot be calculated

volume occupied by it at the standard atmospheric pressure (1 atm) (Alternative Fuels Data Centre Homepage 2021a). CNG is mainly used in surface transport, driving locomotives and LNG transport through CNG carrier ships (Sims et al. 2014). The commercial success of CNG as a transport fuel has been proven in developing countries such as China, Iran, India, Pakistan, Uzbekistan, Thailand, Argentina, Brazil, Colombia, Egypt, and Italy. In India, it is being used as a passenger segment fuel in the nodes around Delhi and National Capital Region, Ahmadabad, and Mumbai (Ministry of Petroleum and Natural Gas, Government of India (GoI) 2020). It has been widely used in the 3/4-wheelers and buses while its use is being expanded to freight carrying vehicles such as light, medium and heavy trucks (CTCN Homepage 2021).

CNG is a lightweight fuel as its density is 215 kg/m^3, which is substantially lower as compared to that of gasoline (750 kg/m^3). Therefore, it can produce a better homogeneous air–fuel mixture ratio of 17.2:1 compared to gasoline (14.7:1). It is also evident by the lower molar mass of CNG (16 g/mol) as compared to that of gasoline (114.23 g/mol). It has a higher Lower Heating Value (LHV, net calorific value) of 50 MJ/kg, 20% more than gasoline (42.9 MJ/kg). Due to its lower energy density (2500 Wh/l), more space is required to store the fuel in the CNG-powered vehicles than gasoline-powered vehicles.

CNG is one of the safest transportation fuels in the present date due to its safe ignition parameters compared to gasoline. Its narrow range of ignition limits which range from 4.3 to 15.2% by volume in air ensures that there will not be any burning in the air beyond those flammability limits even in case of a flame or a spark. It also has a high auto-ignition temperature (813 K) in comparison of gasoline (501–744 K) which makes it safe for storage. At the distribution outlets, rigid cylindrical metallic containers (maintained at a pressure of 197–245 atm) are used to store and distribute this transport fuel (Speight 2018). As CNG is gas at room temperature and normal atmospheric conditions, it has a high level of miscibility with atmospheric air. Its higher diffusion coefficient (0.16 cm^2/s) as compared to gasoline (0.05 cm^2/s) is useful for its good combustion as well as easy dispersal in case of a leak. In contrast, the regular liquid fuels need longer time to get completely atomised and vaporised to form a homogeneous air–fuel mixture for complete combustion.

CNG has high research octane number (120–130) as compared to gasoline (91–100). This implies that the engine of the vehicle can function at a higher compression ratio (16:1) without getting knocked (Poulton 1994). This increases the ability of the CNG engines to give efficiency levels up to 35% as compared to the gasoline engine (25%) (Khan et al. 2015). However, the gasoline engines retrofitted to CNG are not able to utilise the advantage associated with the high octane number associated with CNG because the compression ratios in the vehicles are set to levels needed for gasoline (Khan et al. 2015). The density of the natural gas is very low at the atmospheric pressure (0.68 kg/m^3) in its gaseous state as against the density of air (1.202 kg/m^3). It helps the natural gas vapour to rise and dissipate into the air very quickly instead of forming pools of gas on the lower surface near the ground as it can happen in the case of gasoline and diesel (Poulton 1994). The rapid dispersal of CNG reduces the probability of a fire substantially if the tank is breached.

It is established that any leakage of CNG into the environment results in climate change irrespective of the part of its life cycle, i.e. Well to Wheels. There is a possibility of leakage of methane that can take place during its production or transpiration or even during the final delivery, which makes CNG-fuelled vehicles not a long-term mitigation strategy for climate change (Alvarez et al. 2012). The production-phase leakage of methane contributes largely to the GHG emissions that emanate in CNG's life cycle (Khan et al. 2015). Unburned methane is about 30 times more harmful than carbon dioxide in trapping the reflected heat from the earth's surface, leading to global warming (Kelly 2014).

3.3.1.2 Auto-Gas (Liquefied Petroleum Gas, LPG)

Liquefied Petroleum Gas (LPG) is generated during the refining of the crude oil, and it is primarily a mixture of propane, butane, iso-butane, butylenes, and propylene. The proportion of the constituents present in the mix varies in different geographical regions, depending upon the purpose for which it is being used (e.g., for heating, cooking, refrigeration and as the transport fuel) (Elgas Homepage 2021a). LPG is composed of liquid or gas (vapour), depending on pressure and the gas temperature. When LPG is used as a transport fuel, it is called auto-gas. It is used through cylinders across many different markets in diverse sectors such as agriculture, recreation, hospitality, industry, construction, sailing and fishing. As a transport fuel, auto-gas is usually used in cars and buses, but it has also been deployed in trucks (light and heavy) in a pilot-scale diesel-LPG dual system (Ashok et al. 2015). In India, LPG is mainly used in the 3-wheelers with 1300 filling stations spread over more than 500 cities of the Indian states (Gujarat, Maharashtra, Karnataka, Andhra Pradesh Kerala, and Tamil Nadu (India Auto-Gas Homepage 2021). LPG has been commercially successful in countries such as South Korea, Japan, the USA, Canada, Australia, Netherlands, France, Poland, Italy, Russia, Mexico, Turkey, and China (International Energy Agency (IEA) NITI-Aayog, Government of India (GoI) 2020).

The advantages of LPG as a transport fuel emerge from the simplicity of its production by de-gasolining of natural gas or crude oil stabilisation (Paczuski et al. 2016). It invites relatively low investment costs. It has a marginally higher LHV (45 MJ/kg) than gasoline (43 MJ/kg), implying that it has higher energy content for the same fuel mass. However, LPG has lower energy content per volume of fuel (25.4 MJ/m^3) than gasoline (34.2 MJ/m^3) due to its low density compared to gasoline. Therefore, auto-gas powered vehicles would require a larger volume of the storage tank in the vehicle than gasoline-powered counterparts for providing the same energy content. Unlike natural gas, LPG is heavier than air due to its higher molecular weight and thus flows along the floors, and it tends to settle in low-lying spots, such as basements (Elgas Homepage 2021b).

LPG also has a high octane number of 120, which allows the vehicle's engine to function at a higher compression ratio without getting knocked. The auto-ignition temperature of LPG lies in the range 728–783 K, which is more than that of gasoline;

however, LPG has a lower flash point range of 172–200 K as against gasoline (228–262 K). It makes LPG a safer fuel in the absence of a spark or an ignition source. LPG is designated as an environment-friendly fuel because of its several benefits ranging from social, economic, and ecological, which has been made possible by its impeccable storage parameters. The CO_2 emissions from LPG (73.6 gCO_2/MJ) are considerably low as compared to gasoline (85.8 gCO_2/MJ) and diesel (87.4 g CO_2/MJ) due to its higher calorific value and hydrogen to carbon ratio (Paczuski et al. 2016).

3.3.1.3 Hydrogen-Compressed Natural Gas (H-CNG)

Hydrogen-Compressed Natural Gas (H-CNG) is a bridge fuel that brings hydrogen into the transport system before the full-fledged introduction of hydrogen-based fuel cell vehicles (FCVs). In this way, it helps in laying out the infrastructure for the deployment of hydrogen as a mainstream transport fuel (Morrison et al. 2012). H-CNG is a blend made up of hydrogen and CNG in which the ideal concentration of hydrogen is 18% by volume. H-CNG can reduce carbon monoxide emissions up to 70% as compared to CNG, in addition to nearly 5% savings in fuel cost (Indian Oil Corporation Limited Homepage 2021). It has been piloted as a transport fuel in Norway, the USA, France, Sweden, India and Canada. In India, all the public transport buses running in Delhi on CNG are being transitioned to H-CNG. The transition from CNG to H-CNG has huge viability as the vehicle propulsion system and the infrastructure (piping network and dispensing system) built for CNG can be shared by H-CNG technology without any large investments (Energy.gov Homepage 2021).

As per an illustration by Anstrom and Collier (2016), the on-site blending processes for H-CNG are schematically depicted in Fig. 3.4. The pre-compression and post-compression blending mechanisms shown in the figure highlight the pathway in which the existing transportation and distribution infrastructure can undergo a major transition to facilitate adoption of this futuristic fuel. The pre-compression H-CNG blending process offers an additional flexibility to blend CNG and hydrogen (in the respective pure forms) in the vehicle fuel tank directly. On the other hand, in post-compression H-CNG blending process, the blending of high-pressure CNG and hydrogen occurs in the dispenser unit, and it is supplied to the vehicle tank.

The cities with large fleets of vehicles running on CNG can transition to H-CNG from CNG by bringing in some minor modifications in the vehicles' engines. The H-CNG blend used in the buses in Delhi is 18% by volume of hydrogen (2.9% by weight); it provides the twin benefits of low GHG emissions as well as the improved performance of the engines in the vehicle (Indian Oil Corporation Limited Homepage 2021). Hydrogen used in this blend is produced through gasification, where carbonaceous material containing 10–20% moisture by weight is heated to more than 973 K in an environment devoid of oxygen (Indian Oil Corporation Limited

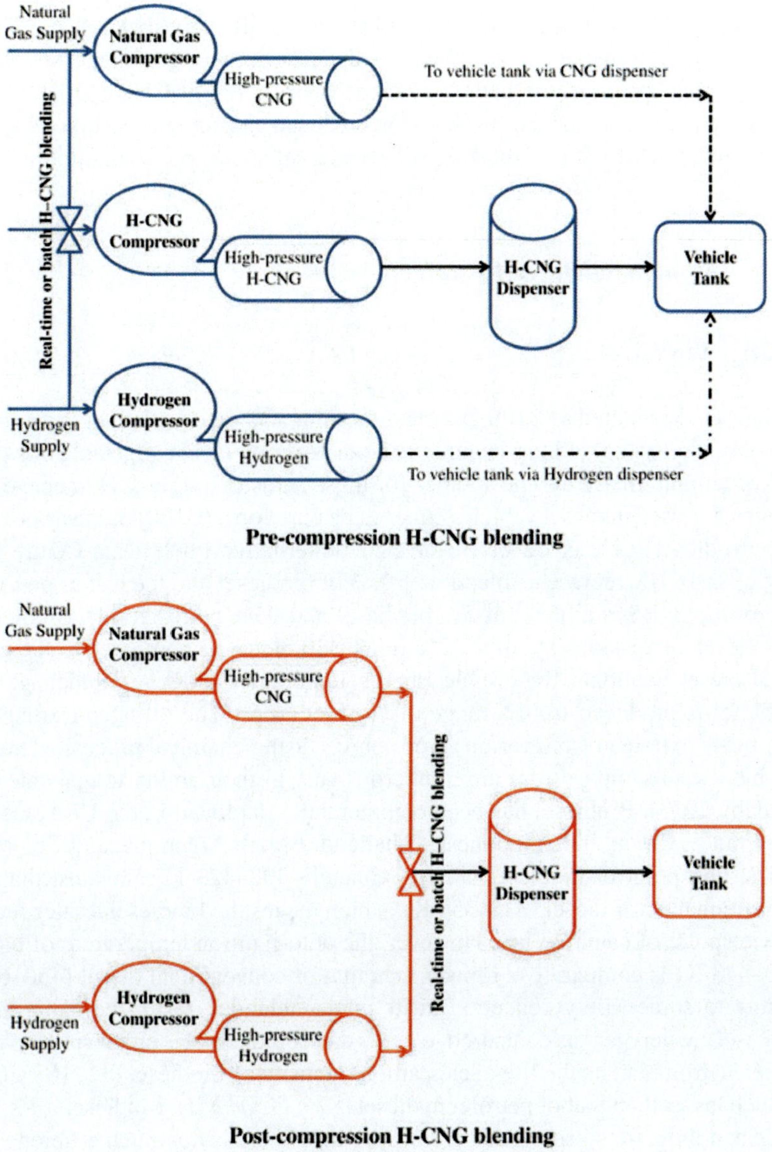

Fig. 3.4 On-site H-CNG pre-compression and post-compression blending process (Anstrom and Collier 2016)

Homepage 2021). A high daily production capacity can be built through gasification as against other hydrogen production pathways (viz. bio-photolysis, anaerobic digestion, and fermentation). It is essential to meet the large-scale demand of hydrogen while minimising the GHG emissions simultaneously (Indian Oil Corporation Limited Homepage 2021). With the blending of hydrogen with CNG, the auto-ignition temperature of the blend is increased from that of CNG (813 K), and the vapour flammability limits of the blend get reduced as compared to that of hydrogen (4–75 vol%), which makes the H-CNG blend a safer transport fuel option.

3.3.2 *Non-fossil Based Fuels*

3.3.2.1 Bio-diesel

Bio-diesel, designated as B100, is a clean-burning and renewable fuel that comprises of mono-alkyl esters of long-chain fatty acids. Animal fats or vegetable oils or recycled restaurant grease are the sources for these acids (Bio-diesel Homepage 2021). Biodiesel is used in diesel vehicles either in its pure form (B100) or in various blends of petro-diesel such as B2 or B5 or B20 (Alternative Fuels Data Centre Homepage 2021b). BX represents blend with X% of biodiesel and the rest as petro-diesel (for example, B5 is a blend of 5% biodiesel and 95% petro-diesel). Biodiesel can reduce GHG emissions by up to 15% in its B20 blend, as compared to the conventional diesel (National Renewable Energy Laboratory (NREL) Homepage 2021). Biodiesel is produced from a range of bio-fuel crops. The oil is converted to fuel after being extracted from various crop sources using chemical processes; the resultant bio-diesel samples from different crops vary in their ability to generate energy (Shalaby 2013). Biodiesel has been commercially deployed in the USA, Germany, UK, France, China, India, Indonesia, Thailand, Brazil, Argentina, and Colombia.

The flashpoint of biodiesel is approximately 408–423 K as against that of the conventional petro-diesel (325–350 K), which makes the biodiesel a safer fuel from the viewpoint of catching fire. However, the auto-ignition temperature of biodiesel (373–443 K) is comparatively lower than that of conventional diesel (450–603 K), leading to some safety concerns. In its pure unblended form, the damages from biodiesel are very less as compared to petro-diesel if the fuel is either spilt or released in the environment as the life cycle carbon intensity of biodiesel (11–16 gCO_2/MJ) is much lesser than that of petroleum diesel (87.4 gCO_2/MJ). Biodiesel is a safe fuel while handling, for storing, and for transporting. There is not much difference in the cetane number of biodiesel and diesel, and they remain in the range of 45–60 for both, which reflect a similar ignition delay for both fuels.

The LHV of biodiesel (39.3 MJ/kg) is about 10% lower than that of petro-diesel (43.3 MJ/kg), which implies that the energy content in biodiesel is less than diesel despite other physico-chemical parameters such as kinematic viscosity, specific gravity, latent heat of vaporisation, and density being nearly the same. Biodiesel tends to form gels in cold weather because of its low pour point (258–286 K), which makes it difficult for blending in diesel in colder regions as diesel has a lower pour point of 237–243 K. As the biofuels are made from animal and vegetable fat, prudent planning is required for the cultivation of the pertinent crops. Crops for biodiesel should not be cultivated without ensuring a robust state of food crop agriculture to avoid local or regional food crises. Water usage to cultivate bio-diesel crops should also be planned carefully to avoid any undue shortage of local potable water (Wilkinson et al. 2013).

3.3.2.2 Ethanol Blended Petrol (EBP)

Ethanol blended petrol is the blending of a certain percentage of ethanol in petrol so that the blend becomes cleaner than petrol and, it provides equivalent efficiency at a lower cost than the conventional petrol. The addition of ethanol to petrol in varying degrees of percentages (E5 or E10 or E20) reduces carbon monoxide emissions emanating from the petrol, and improves the octane number of the blended fuel (Anh et al. 2011). EX is the X% blend of ethanol in petrol (for example, E5 is 5% ethanol blend). The research octane number of bioethanol (100–110) is higher than petrol (91–100). India has achieved the blending of 8.5% as of FY 2020–21 (NITI-Aayog, Government of India (GoI) 2021b). Considering the high population, India needs to address the twin problem of food security and diversion of agricultural land for bio-fuel crops that will compete against each other due to the limited land availability.

3.4 Futuristic Energy Sources for India

3.4.1 Liquefied Natural Gas (LNG)

Liquefied Natural Gas (LNG) is obtained from the liquefaction of natural gas at 112 K. This reduces its volume by approximately 600 times and this process makes LNG an economical fuel for transportation (USEIA Homepage 2021). LNG has been used to fuel large freight vehicles going over large distances, in high horsepower (HP) engines involved in oil drilling, mining, locomotive running and marine operations, and as a feedstock in the manufacturing of fertilisers, plastics and other commercially important organic chemicals (SHV Energy Homepage 2021). LNG's commercial use as a long-distance transport fuel in trucks has begun in the USA, China, and Europe (the UK, Norway, and Netherlands). The top three importers of LNG are Japan, South Korea, and China. It is also an attractive option for commercial buses plying over long distances. India has recently started to use LNG fuelled buses on a pilot basis in Kerala.

The energy density of LNG (21–24 MJ/litre) is 2.4 times higher than that of CNG (9 MJ/l). It means that for the same volume of fuel, LNG can provide 2.4 times energy as compared to CNG which makes it very economical for its transportation to the terminals. This parameter also makes LNG a better fuel for larger freight vehicles travelling over long distances. However, its energy density is 60–70% of that of diesel (36 MJ/litre); thus, LNG use would necessitate a larger fuel tank in the vehicle compared to diesel to travel the same distance. Diesel-LNG dual fuel Internal Combustion Engines are currently in use in China, and the commercial use of LNG has started on the artery roads in the USA and Europe. The auto-ignition temperature of LNG (813 K) is higher than that of diesel (450–603 K) which makes LNG a relatively safer fuel. It is also colourless and odourless, non-toxic and non-corrosive. The life cycle CO_2 emissions of LNG (65–90 gCO_2/MJ) are about 5–35% lower than that of diesel (92–95 gCO_2/MJ), depending on the composition of the fuel and the efficiency of the vehicle engine.

Because of its cryogenic nature, LNG is required to be stored in a well-insulated and optimally pressurised tank (in the range of 1.36–10.21 atm) to minimise the evaporation losses and weathering. In contrast, the containers meant for diesel are low pressurised tanks, and they need not be insulated (Arefin et al. 2020). A complex fuel control system is required in an LNG-driven vehicle, consisting of a pressure management system, vaporiser, fill and vent connections, and a cryogenic tank. As there is a boil-off valve in the cryogenic tank of the vehicle, LNG vehicles should be operated continuously over long distances without long parking to minimise the evaporation losses. Being a cryogenic liquid, LNG can lead to frost burns if it comes into contact with the human skin. The LNG storage tank may also face metal embrittlement and structural failure in the case of metal cracks by LNG spill (Arefin et al. 2020).

3.4.2 Ethanol

Ethanol is a clean and renewable fuel. It is sourced from agricultural and biological feedstock. With continued research, three generations of bio-ethanol making processes have been devised. The first-generation bio-fuels are extracted from the biomass feedstock, which is edible and hence, they raise the debate of fuel versus food. The second generation of bio-fuels is produced from several types of feedstock, ranging from non-food crops (lingo-cellulosic feedstock) to municipal solid wastes. The bio-fuels from the third-generation are extracted from algal biomass and they also use CO_2 as a feedstock (Lee and Lavoie 2013). The top two contributors towards bio-fuel generation are the United States of America (USA) and Brazil, which have 84% of the global share, which is nearly 92 billion litres. They are followed by the European Union (EU), China, India, Canada, and Thailand regarding bio-fuel generation. Brazil is the leading consumer in the world. The average share of the use of 100% hydrous ethanol has increased to 46% in the transport sector of Brazil as its demand has been driven mainly by flex-fuel vehicles (NITI-Aayog, Government of India (GoI) 2021b). Ethanol has been primarily used in India for blending with gasoline as a bio-fuel additive.

Ethanol has many physico-chemical advantages over gasoline as the transport fuel. It has comparatively higher Research Octane Number (RON) of 110 against gasoline (RON—90) as well as high embedded oxygen content at 35%, which allows for high combustion efficiency, high combustion temperature, and high latent heat of vaporisation (900–920 K). These result in the reduction in loss of fuel, and higher laminar flame propagation speed of combustion (39 cm/s for ethanol as against 33 cm/s for gasoline), leading to higher thermal efficiency. Ethanol has a lower life cycle carbon intensity of 58–73 gCO_2/MJ than gasoline (92–99 gCO_2/MJ). Ethanol has a calorific value of 26.9–30 MJ/kg, which is slightly lower than gasoline (42.9–46.4 MJ/kg). This situation results in the Gasoline Gallon Equivalency (GGE) value of ethanol as 1.5; it implies that to replace one gallon of gasoline, 1.5 gallons of ethanol is needed in terms of energy content. From the safety point of view, ethanol has a higher flash point (289 K) as against gasoline (228–262 K) and lower flame temperature in air (2078 K) in comparison with gasoline (2423 K).

In order to facilitate the popularisation of ethanol as a mainstream transport fuel, flex-engines have been designed that enables the vehicles to run on either 100% gasoline or 100% ethanol in the same tank, with no need for switching back and forth from gasoline to other fuel as in the case of other bi-fuel vehicles (Pearson and Turner 2014). The advancement in the designing of modern flex-fuel engines has ensured that with a fuel composition sensor, the fuel injection and spark timing are adjusted automatically depending upon the proportion of the blend in the combustion chamber. Flex-Fuel Engine (FFE) technology is employed in over 80% of the total number of new vehicles sold in Brazil in 2019 (DBpedia Homepage 2021). The Society of Indian Automobile Manufacturers suggests that E100 can be a possible solution for India only when it is sold at 30% lower than gasoline and its availability across India is ensured (NITI-Aayog, Government of India (GoI) 2021b).

3.4.3 *Hydrogen*

Hydrogen has been used as rocket fuel for several decades, and it is being used in fuel-cell buses for many years. The potential of hydrogen to be used as an automotive fuel through both fuel cells (FCs) and internal combustion engines (ICE) makes it a suitable alternative to carbonaceous fossil fuels. Hydrogen has also been used for cooking and heating purposes across the globe. In the present day, most of the hydrogen used globally is produced from fossil fuels such as steam reforming, methane pyrolysis, and coal gasification (Kalamaras and Efstathiou 2013). Hydrogen generated through these processes results in carbon emissions into the atmosphere, and, hence, this hydrogen is called grey hydrogen. If the emissions described above are captured and stored, in that case, the produced hydrogen is called blue hydrogen. Hydrogen is also generated from the electrolysis of water, and this involves the use of electricity in this process. If the required electricity is generated from green sources, the hydrogen generated through high-temperature electrolysis (HTE) is labelled as green hydrogen (Bethoux 2020).

The most significant advantage of hydrogen is its Lower Heating Value (120 MJ/kg), which is approximately three times larger than that of gasoline (43 MJ/kg) and diesel (43.3 MJ/kg). It also has a higher stoichiometric air–fuel mixture ratio (34.3:1) compared to petroleum (14.7:1), which means comparatively a lesser amount of hydrogen is needed in the combustion chamber of the vehicle. It also creates a significantly lesser amount of zero harmful tailpipe emissions, i.e., water or its vapour, when either combusted in an ICE engine or used in a fuel cell vehicle (FCV). Hydrogen also has a higher auto-ignition temperature (839–858 K) as against gasoline (501–744 K) or diesel (450–603 K). Hydrogen is a much safer fuel than gasoline due to its higher ignition temperature and greater diffusion coefficient (0.61) than gasoline (0.05). However, hydrogen has very high vapour flammability limits of 4–75% by volume, making it highly susceptible to ignition if leaked into the air. As hydrogen burns itself, it has a very low flash point of 20 K which makes it highly combustible in the presence of a spark. Hydrogen has a very high laminar flame speed (265–325 cm/s) as compared to other conventional petro-fuels (33–41 m/s), leading to very high thermal efficiency. The octane number of hydrogen (>130) is very high as compared to gasoline (91–100), which makes hydrogen a potential competitor to gasoline to be used in internal combustion transportation engines.

Despite all these advantages, the biggest drawback associated with hydrogen as a fuel lies in the requirement of large storage volumes. When stored in the liquid form, hydrogen needs about four times the volume required for storing gasoline of the same energy content, and the number grows to approximately 19 times if hydrogen is to be stored in the gaseous form (Sharma and Ghoshal 2015). The large volume requirement for storing gaseous hydrogen is owed to its low specific gravity (0.0696). Evidently, for hydrogen to be considered as a fuel for surface transport, a large storage tank is required for the vehicle to be able to drive over a considerable range to make the travel cost-effective. In addition, a hydrogen-based transportation system needs a robust infrastructure to transport hydrogen from the production plants to refuelling stations. The distance of the delivery outlets from the point of generation and the number of refuelling stations are crucial to delivering hydrogen as the required safety measures associated with the inflammable hydrogen gas increase the cost with the increased delivery distance (Yang and Ogden 2007).

3.5 Discussion

The fuel options currently in use in India and the suggested alternatives have pros and cons as no fuel can be termed as a perfect and the ultimate solution. There are various parameters (such as energy content, weight, volume, air–fuel ratio, octane/cetane number, auto-ignition temperature, flash point, vapour flammability limits, carbon emissions etc.) that play a very important role in deciding the deployability of a fuel. At present, gasoline is mainly used for passenger transport segment in which the vehicles predominantly travel over short to medium distances (20–200 km). Diesel is mainly used for the light and heavy freight transport, with the vehicles predominantly travelling over medium to long distances (more than 100 km). India has commercially deployed relatively clean fuels such as CNG, LPG and H-CNG in the passenger segment for running the 3 and 4 wheelers and the city buses that travel over smaller distances. These get largely limited in their use because of the lack of accessibility of clean fuel beyond a region. They face huge competition from Electric Vehicles (EVs), as the EVs are also adequately equipped to tackle the problem of localisation of air pollution in the dense urban settings. However, the ***'range anxiety'*** remains with both CNG/LPG-driven conventional vehicles and battery-powered EVs owing to the lack of mid-way refuelling infrastructure across the country (Srikanth 2018).

The way forward during the transition phase of the surface transport sector in India lies in the prudent amalgamation of the suitable and viable options in fuel-mix based on the strengths of each candidate, such that the environmental, social and economic costs can be alleviated to an acceptable level. Limited success has been achieved in the deployment of alternative fuels, such as H-CNG, ethanol-blended petrol and bio-diesel, as creating a new energy system poses challenges in terms of affordability. India will need to enhance the Research and Development (R&D) to facilitate and expedite commercial deployment of the flex-engines along with the bi-fuel and hybrid propulsion systems in order to achieve a relatively cleaner surface mobility during the transition phase. High latent heat of vaporization of ethanol gives it the advantage of providing additional knock resistance in direct injection engines due to charge cooling effect. Ethanol can be used in urban conditions with heavy traffic as it provides greater volume efficiency, and it also helps in engine breathing. As the priority currently remains towards decarbonisation, the utility of green hydrogen as a potential transport fuel needs to be explored. Green hydrogen is generated through high temperature electrolysis process powered by the electricity from '*green energy*' sources. As hydrogen has multiple applications, in case of excess production of hydrogen, several industrial processes can absorb the green hydrogen to fulfil their heating requirements. Very high calorific value of hydrogen makes it a good candidate as alternative fuel but its optimal stoichiometric air–fuel ratio needs to be maintained for lowering the energy wastage during combustion.

In case of bio-fuels, the key challenges are attributed to the availability of the fuel itself as well as the cost associated with the resources required to produce bio-fuels. However, they would emerge as the potential alternatives if they are sourced from either second or third generation methods so that the debate over food versus fuel does not arise. Even though they have lesser energy content than their fossil fuel counterparts, they are comparatively cleaner and safer fuels for which there is no need for a new energy system and infrastructure to be put into place. Bio-diesel and LNG have the potential to be the long-distance transport fuel for the freight on the artery highways. They have comparatively high kinematic viscosity which is helpful in carrying heavy load by vehicles running on these fuels. It would significantly reduce the emissions as the freight-carrying vehicles are responsible for more than 50% of the GHG emissions coming from the surface transport. In order to create a favourable ecosystem for LNG, the availability of natural gas at the LNG terminals needs to be ensured as India has limited natural gas reserves. In addition, there will be a need for putting up bays where the goods containing containers can be transferred from an LNG vehicle to another vehicle (fuelled by a source other than LNG) for the dispatch of the goods to their final destinations.

The public and private sector investment aimed at creating the infrastructure to facilitate the transition of the surface transport in India is the key for on-ground implementation. As the arrival of a green ecosystem needs a lot of infrastructural as well as governance supports, the role of existing eco-system becomes even more important in shaping up the transition process while minimising the emissions. It necessitates the identification of suitable alternate fuels that can use the existing energy eco-system for transport in the short to medium term. CNG and LPG are the present-day alternatives to petrol in the passenger segment and they remain strong contenders as potential alternate fuels in the short to medium term. Biodiesel has technical feasibility to emerge as a potential alternative to diesel in the freight segment, without requiring any additional infrastructure. The alternate fuels, which can completely replace the current use of petrol and diesel for passenger and freight segment respectively in the long term, need to be separately identified in order to prioritize the initiatives. The technical parameters of ethanol make it a potential alternative to petrol for passenger segment in long term if the availability of raw material for its production is ensured. As LNG and Hydrogen need heavy infrastructure costs, they will remain long-term solutions for freight segment in road transport. However, they will face heavy competition from the electricity-dependent dedicated railway freight corridors.

3.6 Conclusion

As India is exploring a number of feasible alternatives for petrol and diesel in the fuel mix for the road transport sector, technical availability becomes the prime criterion to choose a futuristic fuel. Once the technical viability has been established, there is a need to look into the supply chain and the management of the fuel eco-system to commercially deploy a fuel in the short term, as well as, in the medium-to-long term. As the investment and, thereby, the affordability remains the primary driving factor for a futuristic fuel to be adopted, a number of steps are being taken to upgrade the existing eco-system of petrol and diesel. The fuel quality standards for petrol and diesel have been increased over the last few years. The fuel standards were raised from Bharat Stage IV to VI in April 2020. The enhancing of the fuel standards has reduced the impurities (such as lead, sulphur etc.) in the fuel without compromising on the technical parameters of these fuels. However, the concerns around their limited availability and carbon-based GHG emissions continue to remain. As the electrification in India has improved significantly, and daily availability of the electricity is now ensured for a larger duration, India envisages to achieve e-mobility or electricity-powered greener surface mobility. In the regime of EVs, the vehicles will be powered by the electric motors, which operate in a very different way as compared to the ICE-based vehicles.

The green surface mobility era (i.e., fuel cells and battery-powered vehicles) in the future would also prefer the electricity to be sourced from renewable sources. Mere a large market share of the battery-powered EVs and fuel-cell powered EVs (FCEVs) in the competitive global vehicle ecosystem would not suffice in achieving the goal of green mobility. Therefore, it needs to be ensured that the electricity charging the batteries is coming from green and renewable sources, and for that a lot of infrastructural improvement is needed including good coverage of grid connectivity across the country, power grid ability to absorb a large amount of electricity coming from the variable renewable energy sources (i.e., resilience to the intermittencies), and creation of ample number of charging stations across the country to ensure better availability and reliability. The installed electricity generation capacity from renewable energy sources is projected to increase to 52% of the electricity mix by 2030. If the installed capacities are put in place but not utilised, the energy mix in India will remain fossil-fuel heavy, and the clean transition of the Indian transport sector may get delayed. As the present-day fossil fuel-based ICE-driven transport era and the futuristic EV-driven green mobility era are diametrically opposite in their inherent attributes and infrastructural requirements, the transition needs to be well planned, so that creation of new capacity does not jeopardize the employment scenario and the overall societal welfare that is attributed to the Indian surface transport sector.

To summarize, it would be fair to mention that there can be many pathways towards achieving the goal of a cleaner and sustainable surface transport sector in India. Considering the current position, the possible technological options for cleaner transport sector encompass advanced ICE engines (such as flex-engines), which can accommodate cleaner fuels such as bio-fuel or natural gas or hydrogen, and eventually the much-awaited electric motor-based electric vehicles, which would gradually enter into the mix even though the electricity remains largely fossil-based in the near-to-medium term future (say, by 2040). The transition plan should prepare us for the transformation from fossil fuel to a renewable energy era in which the passenger and light and heavy freight vehicles will be powered completely by the electrically charged batteries powered by the renewable sources. While formulating the fuel diversification strategy for Indian surface transport sector, the ancillary uses of the candidate fuels (such as natural gas, ethanol and hydrogen) in other sectors apart from transportation need to be considered as well, making their production and distribution even more viable in the deeper pockets of India. For bringing any new system to fruition and practice, the decadal inertia associated with the development and deployment of the new energy system needs to be borne with.

References

Akhil AG, Mohammed PK, Akhilesh S, Muhammad ACA, Khan S, Kanna R (2017) Determination of cloud and pour point of various petroleum products. Int Refereed J Eng Sci (IRJES) 6(9):1–4. http://www.irjes.com/Papers/vol6-issue9/A6910104.pdf

Allegrini F, Olivieri AC (2011) A new and efficient variable selection algorithm based on ant colony optimization. Applications to near infrared spectroscopy/partial least-squares analysis. Anal Chimica Acta 699(1):18–25. https://doi.org/10.1016/j.aca.2011.04.061

Alternative Fuels Data Centre Homepage (2021a). https://bit.ly/3ymZQaB. Last accessed 30 June 2021

Alternative Fuels Data Centre Homepage (2021b). https://bit.ly/36ibxmI. Last accessed 30 June 2021

Alvarez RA, Pacala SW, Winebrake JJ, Chameides WL, Hamburg SP (2012) Greater focus needed on methane leakage from natural gas infrastructure. Proc Natl Acad Sci 109(17):6435–6440. https://doi.org/10.1073/pnas.1202407109

Ang BW, Choong WL, Ng TS (2015) Energy security: definitions, dimensions and indexes. Renew Sustain Energy Rev 42:1077–1093. https://doi.org/10.1016/j.rser.2014.10.064

Anstrom JR, Collier K (2016) Blended hydrogen–natural gas-fueled internal combustion engines and fueling infrastructure. In: Compendium of hydrogen energy. Woodhead Publishing, pp 219–232. https://doi.org/10.1016/B978-1-78242-363-8.00008-6

Arefin MA, Nabi MN, Akram MW, Islam MT, Chowdhury MW (2020) A review on liquefied natural gas as fuels for dual fuel engines: opportunities, challenges and responses. Energies 13(22):6127. https://doi.org/10.3390/en13226127

Ashok B, Ashok SD, Kumar CR (2015) LPG diesel dual fuel engine–A critical review. Alex Eng J 54(2):105–126. https://doi.org/10.1016/j.aej.2015.03.002

Balat M (2006) Fuel characteristics and the use of biodiesel as a transportation fuel. Energy Sources, Part A 28(9):855–864. https://doi.org/10.1080/009083190951474

Bethoux O (2020) Hydrogen fuel cell road vehicles and their infrastructure: an option towards an environmentally friendly energy transition. Energies 13(22):6132. https://doi.org/10.3390/en13226132

Bio-diesel Homepage (2021). https://bit.ly/3dLDRSW. Last accessed 30 June 2021

Boretti A (2020) Advances in diesel-LNG internal combustion engines. Appl Sci 10(4):1296. https://doi.org/10.3390/app10041296

Cameo Chemicals NOAA Homepage (2021). https://bit.ly/3wcHAz3. Last accessed 30 June 2021

Chupka GM, Christensen E, Fouts L, Alleman TL, Ratcliff MA, McCormick RL (2015) Heat of vaporization measurements for ethanol blends up to 50 volume percent in several hydrocarbon blendstocks and implications for knock in SI engines. SAE Int J Fuels Lubr 8(2):251–263. https://doi.org/10.4271/2015-01-0763

Clarke Energy (2021). https://bit.ly/3dI14Fm. Last accessed 30 June 2021

Clarke-Energy Homepage (2021). https://bit.ly/3dGytAs. Last accessed 30 June 2021

CTCN Homepage (2021). https://bit.ly/3hx1z6t. Last accessed 30 June 2021

Dagaut P (2019) Experiments for kinetic mechanism assessment. In: Computer aided chemical engineering, vol 45. Elsevier, pp 445–471. https://doi.org/10.1016/B978-0-444-64087-1.00008-5

DBpedia Homepage (2021). https://bit.ly/3hyJXam. Last accessed 30 June 2021

Elgas Homepage (2021a). https://bit.ly/36emgi8. Last accessed 30 June 2021

Elgas Homepage (2021b). https://bit.ly/3qPXw9E. Last accessed 30 June 2021

Energy.gov Homepage (2021). https://bit.ly/3dGPJ8s. Last accessed 30 June 2021

Fluid Switch Homepage (2021). https://bit.ly/36aAwZ9. Last accessed 30 June 2021

Goyal P, Sharma SK (2014) Review on opportunities and difficulties with HCNG as a future fuel for internal combustion engine. Adv Aerosp Sci Appl. http://www.ripublication.com/aasa-spl/aasav4n1spl_13.pdf

Green Gas Limited Homepage (2021). https://bit.ly/3jGZhEi. Last accessed 30 June 2021

Gupta S, Patil V, Himabindu M, Ravikrishna RV (2016) Life-cycle analysis of energy and greenhouse gas emissions of automotive fuels in India: part 1–tank-to-wheel analysis. Energy 96:684–698. https://doi.org/10.1016/j.energy.2015.11.031

Hanby VI (2012) Combustion and pollution control in heating systems. Springer Science & Business Media. https://doi.org/10.1007/978-1-4471-2071-1

Hänggi S, Elbert P, Bütler T, Cabalzar U, Teske S, Bach C, Onder C (2019) A review of synthetic fuels for passenger vehicles. Energy Rep 5:555–569. https://doi.org/10.1016/j.egyr.2019.04.007

Hasan MM, Rahman MM (2017) Performance and emission characteristics of biodiesel–diesel blend and environmental and economic impacts of biodiesel production: a review. Renew Sustain Energy Rev 74:938–948. https://doi.org/10.1016/j.rser.2017.03.045

Hosseini SE, Butler B (2020) An overview of development and challenges in hydrogen powered vehicles. Int J Green Energy 17(1):13–37. https://doi.org/10.1080/15435075.2019.1685999
Hydrogen Tools Homepage (2021). https://bit.ly/2Uk6EHe. Last accessed 30 June 2021
HyResponse Homepage (2021). https://bit.ly/3yt1uaR. Last accessed 30 June 2021
HySafe Hydrogen Fundamentals Homepage (2021). https://bit.ly/3qJMYJ3. Last accessed 30 June 2021
India Auto-Gas Homepage (2021). https://bit.ly/3hGfuHw. Last accessed 30 June 2021
Indian Oil Corporation Limited Homepage (2021). https://bit.ly/3hyzdsy. Last accessed 30 June 2021
Indian Institute of Technology Delhi Homepage (2021). https://bit.ly/36cotdZ. Last accessed 30 June 2021
International Energy Agency (IEA) (2021a) Data and statistics–total final consumption (TFC) by sector: India 1990–2018. https://bit.ly/3hesFjK. Last accessed 30 June 2021
International Energy Agency (IEA) (2021b) India energy outlook–2021. https://bit.ly/3jEUcMX. Last accessed 30 June 2021
International Energy Agency (IEA), NITI-Aayog, Government of India (GoI) (2020) India 2020–in-depth energy policy review. https://bit.ly/3dJeAsx. Last accessed 30 June 2021
Jha SK, Fernando SD, To SF (2007) Flame temperature analysis of biodiesel blends and components. In: 2007 ASAE annual meeting. American Society of Agricultural and Biological Engineers, p 1. https://doi.org/10.1016/j.fuel.2007.10.026
Kalamaras CM, Efstathiou AM (2013) Hydrogen production technologies: current state and future developments. In: Conference papers in science, vol 2013. Hindawi. https://doi.org/10.1155/2013/690627
Kelly M (2014) A more potent greenhouse gas than CO_2, methane emissions will leap as Earth warms (Nature). Princeton Research. https://blogs.princeton.edu/research/2014/03/26/a-more-potent-greenhouse-gas-than-co2-methane-emissions-will-leap-as-earth-warms-nature
Khan MI, Yasmin T, Shakoor A (2015) Technical overview of compressed natural gas (CNG) as a transportation fuel. Renew Sustain Energy Rev 51:785–797. https://doi.org/10.1016/j.rser.2015.06.053
Kheiralla AF, El-Awad MM, Hassan MY, Hussen MA, Osman HI (2012) Experimental determination of fuel properties of ethanol/gasoline blends as biofuel for SI engines. In: International conference on mechanical, automobile and robotics engineering, Penang, Malaysia. http://psrcentre.org/images/extraimages/24.%20212181.pdf. Last accessed 30 June 2021
Kook S, Pickett LM (2010) Effect of fuel volatility and ignition quality on combustion and soot formation at fixed premixing conditions. SAE Int J Engines 2(2):11–23. https://www.jstor.org/stable/26275403
Kosky P, Balmer R, Keat W, Wise G (2013) Exploring engineering. Academic Press, Boston, MA, USA, pp 317–337. https://doi.org/10.1016/B978-0-12-815073-3.00007-7
Le Anh T, Nguyen DK, Tuan PM (2011) Investigation of motorcycle engine's characteristics fueled with ethanol-gasoline blends. In: The 4th AUN/SEED–net regional conference on new and renewable energy
Lee RA, Lavoie JM (2013) From first-to third-generation biofuels: challenges of producing a commodity from a biomass of increasing complexity. Anim Front 3(2):6–11. https://doi.org/10.2527/af.2013-0010
Ludwig EE (1999) Chapter 7—Process safety and pressure-relieving devices. In: Applied process design for chemical and petrochemical plants, vol 1. Gulf Professional Publishing. https://doi.org/10.1016/S1874-8635(99)80008-9
Mabanaft Homepage (2021). https://bit.ly/2V82YJ9. Last accessed 30 June 2021
Methanol Institute Homepage (2021). https://bit.ly/3Atj14o. Last accessed 30 June 2021
Ministry of Petroleum and Natural Gas, Government of India (GoI) (2020) Indian petroleum and natural gas statistics 2018–19. https://bit.ly/3dI8aKh. Last accessed 30 June 2021
Ministry of Road Transport and Highways, Government of India (GoI) (2019) Road transport year book (FY 2016–17). https://bit.ly/369w0KA. Last accessed 30 June 2021

Ministry of Statistics and Program Implementation, Government of India (GoI) (2020) Sector-wise GDP of India (2020). https://bit.ly/3jF4zQG. Last accessed 30 June 2021

Morrison GM, Kumar R, Chugh S, Puri SK, Tuli DK, Malhotra RK (2012) Hydrogen transportation in Delhi? Investigating the hydrogen-compressed natural gas (H-CNG) option. Int J Hydrogen Energy 37(1):644–654. https://doi.org/10.1016/j.ijhydene.2011.09.090

National Renewable Energy Laboratory (NREL) Homepage (2021). https://bit.ly/3qJ4q0g. Last accessed 30 June 2021

NITI-Aayog, Government of India (GoI) (2021a) Fast tracking freight in India—a roadmap for clean and cost-effective goods transport. https://bit.ly/2UX8huG. Last accessed 30 June 2021

Niti-Aayog, Government of India (GoI) (2021b) Ethanol blending in India 2020–25. https://bit.ly/3hcZPQJ. Last accessed 30 June 2021

Paczuski M, Marchwiany M, Puławski R, Pankowski A, Kurpiel K, Przedlacki M (2016) Liquefied petroleum gas (LPG) as a fuel for internal combustion engines. In: Alternative fuels, technical and environmental conditions, pp 105–136. https://doi.org/10.5772/61736

Panchasara H, Ashwath N (2021) Effects of pyrolysis bio-oils on fuel atomisation—a review. Energies 14(4):794. https://doi.org/10.3390/en14040794

Paratherm Homepage (2021). https://bit.ly/3ylPCXR. Last accessed 30 June 2021

Paravantis JA, Kontoulis N (2020) Energy security and renewable energy: a geopolitical perspective. In: Renewable energy-resources, challenges and applications. IntechOpen. https://doi.org/10.5772/intechopen.91848

Pearson RJ, Turner JWG (2014) Using alternative and renewable liquid fuels to improve the environmental performance of internal combustion engines: key challenges and blending technologies. In: Alternative fuels and advanced vehicle technologies for improved environmental performance. Woodhead Publishing, pp 52–89. https://doi.org/10.1533/9780857097422.1.52

Petro Online Homepage (2021). https://bit.ly/2SNcHnn. Last accessed 30 June 2021

Polytechnic University of Madrid Spain Homepage (2021). https://bit.ly/3hgAyFl. Last accessed 30 June 2021

Poulton ML (1994) Alternative fuels for road vehicles. https://www.osti.gov/etdeweb/biblio/575115. Last accessed 30 June 2021

Purdue University Homepage (2021). https://bit.ly/3qM8RYj. Last accessed 30 June 2021

Raslavičius L, Keršys A, Mockus S, Keršienė N, Starevičius M (2014) Liquefied petroleum gas (LPG) as a medium-term option in the transition to sustainable fuels and transport. Renew Sustain Energy Rev 32:513–525. https://doi.org/10.1016/j.rser.2014.01.052

Raza A, Mehboob H, Miran S, Arif W, Rizvi SFJ (2020) Investigation on the characteristics of biodiesel droplets in the engine cylinder. Energies 13(14):3637. https://doi.org/10.3390/en13143637

Ren J, Sovacool BK (2014) Quantifying, measuring, and strategizing energy security: determining the most meaningful dimensions and metrics. Energy 76:838–849. https://doi.org/10.1016/j.energy.2014.08.083

Sakthivel P, Subramanian KA, Mathai R (2018) Indian scenario of ethanol fuel and its utilization in automotive transportation sector. Resour Conserv Recycl 132:102–120. https://doi.org/10.1016/j.resconrec.2018.01.012

Sarıkoç S (2020) Fuels of the diesel-gasoline engines and their properties. Diesel Gasoline Engines 31. https://doi.org/10.5772/intechopen.89044

Schuchard R, Springer N, Grant S (2015) The sustainability impacts of fuel. In: A BSR future of fuels initiative publication. https://bit.ly/3hz14st. Last accessed 30 June 2021

Shalaby EA (2013) Biofuel: sources, extraction and determination. In: Liquid, gaseous and solid biofuels-conversion techniques. InTech, Croatia, pp 451–478. https://doi.org/10.5772/51943

Sharma S, Ghoshal SK (2015) Hydrogen the future transportation fuel: from production to applications. Renew Sustain Energy Rev 43:1151–1158. https://doi.org/10.1016/j.rser.2014.11.093

Sholes KR, Odaka M, Goto Y, Ishii H, Suzuki H (2002) Study of the effect of boiling point on combustion and PM emissions in a compression ignition engine using two-component n-paraffin fuels (No. 2002-01-0871). SAE Technical Paper. https://doi.org/10.4271/2002-01-0871

SHV Energy Homepage (2021). https://bit.ly/3dLh8q5. Last accessed 30 June 2021

Sims R et al (2014) Transport. In: Climate change 2014: mitigation of climate change. contribution of working group III to the fifth assessment report of the IPCC. https://bit.ly/3jKmRA6. Last accessed 30 June 2021

Smajla I, Karasalihović Sedlar D, Drljača B, Jukić L (2019) Fuel switch to LNG in heavy truck traffic. Energies 12(3), 515.https://doi.org/10.3390/en12030515

Speight JG (2018) Natural gas—a basic handbook. Gulf Professional Publishing.https://doi.org/10.1016/B978-0-12-809570-6.00005-9

Srikanth R (2018) Role of electric mobility in a sustainable, and energy-secure future for India. Curr Sci 114(4):732–739. http://eprints.nias.res.in/1458/

UNSDG Homepage (2021). https://bit.ly/2UqQewA. Last accessed 30 June 2021

USEIA Homepage (2021). https://bit.ly/3jKOKIu. Last accessed 30 June 2021

Veziro TN, Barbir F (1992) Hydrogen: the wonder fuel. Int J Hydrogen Energy 17(6):391–404. https://doi.org/10.1016/0360-3199(92)90183-W

Wilkinson J et al. (2013) Biofuels and food security, A report by the high level panel of experts (HLPE Report 5), Committee on World Food Security (CFS), Rome 2013. https://www.fao.org/3/i2952e/i2952e.pdf. Last accessed 30 June 2021

Winzer (2012) Conceptualizing energy security. Energy Policy 46, 36–48.https://doi.org/10.1016/j.enpol.2012.02.067

Yang C, Ogden J (2007) Determining the lowest-cost hydrogen delivery mode. Int J Hydrogen Energy 32(2):268–286. https://doi.org/10.1016/j.ijhydene.2006.05.009

Part II
Biofuels for Sustainable Mobility

Chapter 4
Alternative Refinery Process of Fuel Catalytic Upgrade in Aqueous Media

Nikolaos C. Kokkinos

4.1 Introduction

Gasoline does not meet all desired specifications when produced in a refinery. Using gasoline as a fuel in internal combustion engines, it ought to meet specific parameters. One of the most important is the anti-knocking ability of the fuel, which refers to the fuel withstanding in self-ignition. RON (research octane number), MON (motor octane number) and anti-knock index ((MON + RON)/2) are the technical specifications that describe the fuel's aforementioned property. The required values for these specifications differentiate from region to region. For instance, in Europe according to directive EN228:2012 the gasoline's minimum RON and MON are 95 and 85 respectively. Whereas in Australia according to Fuel Quality Standards act 2000, the gasoline's minimum RON ranges from 91 to 95 and minimum MON ranges from 81 to 85 (Hart 2014).

From mid 1920s until the early 1970s, tetraethyl lead $C_8H_{20}Pb$ (TEL) has been used as an inexpensive octane booster for so-called leaded gasoline. Today, lead's adverse health effects are well known and all countries, except few worldwide, have banned TEL as an additive to gasoline (UN 2019). Gasoline ethers oxygenates (GEOs) were introduced as additives in gasoline in 1973, when the lead phase-out has begun (unleaded gasoline). However, GEOs replaced TEL as appropriate octane boosters in the blended gasoline pool of the refineries and not as environmentally friendly compounds (Bonventre et al. 2012). Therefore, GEOs disperse swiftly contaminating the environment, being also vastly persistent, due to its high volatility, its high water solubility, its low biodegradability and its growing scale of use.

N. C. Kokkinos (✉)
Department of Chemistry, School of Science, International Hellenic University, Ag. Loukas, 654 04 Kavala, Greece
e-mail: nikokkinos@mwpc.gr; nck@chem.ihu.gr

Hephaestus Advanced Laboratory, Division of Petroleum Forensic Fingerprinting, School of Science, International Hellenic University, Ag. Loukas, 654 04 Kavala, Greece

G. Di Blasio et al. (eds.), *Clean Fuels for Mobility*, Energy, Environment, and Sustainability, https://doi.org/10.1007/978-981-16-8747-1_4

In spite of the fact that there are several research proposals for gasoline upgrade mainly through additives; in the current chapter, an in situ fuel upgrade in aqueous media from refinery cuts was thoroughly examined for the first time. It could completely replace GEOs by producing in situ strong anti-knocking environmental friendly alcohol mixtures. This alternative chemical process based upon the heterogenization of homogeneous catalytic reactions and overcame the weaknesses of homogeneous catalytic systems.

4.2 Octane Boosters

4.2.1 Historical Overview—Environmental Legislation

In 1952 in the United Kingdom, the prevalent reckless burning of fossil fuels (mainly coal) was the leading cause of serious environmental pollution, which became visible even to the naked eye in the form of a very dense smog cloud, known as London's Great Smog. It was considered the worst case of air pollution in the history of the United Kingdom and lasted from 5 to 9 December 1952. During this period, more than 4000 people died prematurely, while the maximum possible visibility on public roads severely reduced to few meters (Bell et al. 2004). Therefore, the British Parliament enacted the Clean Air Acts (CAAs) of 1956 (UK-Parliament 1956) and 1968 (UK-Parliament 1968) aiming at the reduction of air pollution by setting the first barriers to the reckless burning of fossil fuels.

Almost at the same time as London's Great Smog, in the early 1950s in the USA, geologist Patterson was studying the age of the earth using isotopic lead data, when he found that the erroneous results of his calculations were due to the contamination of the examining specimens by the environment (Patterson 1956). Then, Patterson (Chow and Patterson 1962; Patterson 1965) proved that lead contamination of the environment dates back to the time when tetraethyl lead used as an effective anti-knocking additive to gasoline in internal combustion engines. A few years later, Patterson's findings led the US Congress to pass the Air Pollution Control Act (1955), followed by the US CAA (1963) and the Air Quality Act (1967), which merely announced the start of fundamental research in the development of gaseous emission thresholds (Schnelle and Brown 2001). The first really drastic and major environmental legislation was passed in 1970 by the US Congress that instituted the gradual phase-out of lead additives in motor fuels and stressed that the oil refining industry should serve the triptych: energy, economy and environment, simultaneously.

Despite the fact that the effects on human health from the use of TEL were already known to Thomas Midgley, TEL inventor, and although leaded gasoline was banned since the early 1970s, the European Union has completely withdrawn leaded gasoline on the 1st of January 2000 (European-Parliament-and-the-Council 1998). An important contribution to the faster removal of TEL from motor fuels was the installation of the lead-intolerant two-way and then three-way catalytic converters in the car exhaust system since 1975 (Petersen and Rosen

1975). Thus, the refining industry was driven to use high-octane oxygenated components, called "oxygenates", to meet the standards of the US CAA Amendments 1977 and 1990 as well as the requirements of the EU Fuel Quality Directive (98/70/EC) (European-Parliament-and-the-Council 1998) and its amendments to the EU Directive 2009/30/EC (European-Parliament-and-the-Council 2009).

4.2.2 Gasoline Ether Oxygenates (GEOs)

According to ASTM D 4814, tertiary alkyl ethers belong to fuel oxygenates. Most common ether oxygenates are: methyl tert-butyl ether (MTBE), ethyl tert-butyl ether (ETBE) and tert-amyl methyl ether (TAME). The first used and produced on an industrial scale was MTBE by the Italian company ANIC in 1973 and shortly afterwards, in 1976, by the German company Chemische Werke Hüls (Caetano et al. 1994). Since then, gasoline ether oxygenates or fuel ethers are used in certain types of gasoline, in order to rise the octane number of the final fuel and improve its combustion efficiency.

Methyl tert-butyl ether (MTBE). Methyl tert-butyl ether is the most broadly used ether oxygenate. It has been the predominant choice for octane boosting in gasoline, because of its high-octane value and low cost. Starting back in 1970s methyl tetr-butyl ether was the leading gasoline additive after TEL worldwide. MTBE is a colorless liquid with a distinct odor and low viscosity. MTBE's oxygen content is 18.2 wt% and has a RON value of 117. MTBE is produced from the reaction of methanol (produced from natural gas) with isobutene in the liquid state, catalyzed by an acidic catalyst at 100 °C (Eq. 4.1). MTBE, when blended to gasoline, boosts its octane number without adversely affecting its other properties.

$$\underset{i-Butene}{(CH_3)_2C(CH_2)} + \underset{Methanol}{(CH_3)OH} \overset{H^+}{\Leftrightarrow} \underset{MTBE}{(CH_3)_3CO(CH_3)} \tag{4.1}$$

Ethyl tert-butyl ether (ETBE). ETBE is the fourth in consumption worldwide gasoline octane booster. Ethyl tetr-butyl ether is consumed primarily in Japan and Western Europe. This is mostly due to the mandatory use of biofuels which is driving consumption for bioethanol based ETBE. Especially in Japan, following the commitment under the Kyoto protocol (greenhouse gases reduction) there has been a shift to ETBE; blending rates were increased in 2017 from 7 to 22% by volume. Ethyl tert-butyl ether appears as a pale yellow liquid with a characteristic odor. ETBE's oxygen content is 15.7 wt% and has a RON value of 118. Ethyl tert-butyl ether, similarly to MTBE, is synthesised by reacting isobutylene with ethanol (Eq. 4.2). The reaction is catalyzed using an acidic catalyst. ETBE increases the octane number (RON) of gasoline and at the same time a decrease is observed in the calorific value (Bardin et al. 2014).

$$\underset{i-Butene}{(CH_3)_2C(CH_2)} + \underset{Ethanol}{(CH_3)(CH_2)OH} \overset{H^+}{\Leftrightarrow} \underset{ETBE}{(CH_3)_3CO(CH_2)(CH_3)} \tag{4.2}$$

Tert-amyl methyl ether (TAME). Tert-amyl methyl ether began being produced at the end of 1980s both in Europe and in USA. It is currently the last in consumption by volume gasoline oxygenate worldwide. Compared to the other two ether oxygenates (MTBE and ETBE) TAME is closer to MTBE in terms of synthesis process. It has slightly lower blending octane value from MTBE and ETBE, but the RVP (Reid vapor pressure) blending value is half of ETBE's and significantly lower than MTBE's. Tert-amyl methyl ether's oxygen content is 15.7 wt% and it has a RON value of 112. In Europe, Greece, Germany, Italy, Finland and France produce and use TAME as gasoline oxygenate (Kokkinos et al. 2015a). TAME can be easily produced by isoamylenes' etherification (Eq. 4.3). Thus, TAME is synthesised from the reaction of methanol (produced from natural gas) with the two reactive isoamylenes (2-methyl-1-butene and 2-methyl-2-butene), catalyzed by an acidic catalyst (Hamid and Ali 2004). Tert amyl methyl ether's octane blending values are lower than MTBE values by roughly 5%.

$$\underset{isoamylene}{(CH_3)_2C(CH)(CH_3)} + \underset{Methanol}{(CH_3)OH} \overset{H^+}{\Leftrightarrow} \underset{TAME}{(CH_3)(CH_2)(CH_3)_2CO(CH_3)} \tag{4.3}$$

Drawbacks of GEOs' use. Gasoline ether oxygenates were selected mainly for their optimal effect on gasoline properties rather than their environmental impact, even though TEL was banned due to health and environmental issues. History revealed that the value chain of GEOs (production, distribution, storage and use) resulted in their release into the soil, groundwater and atmosphere.

MTBE. The most famous octane booster alternative to TEL became rapidly a major environmental issue. The reason was MTBE's ease to exit gasoline and enter aquifers in cases of leaking underground storage tanks or when gasoline spilled onto the ground. MTBE belongs on hydrophilic ethers with a solubility of 50 g/L in water (Kinner 2001) which cannot be easily metabolized by many organisms. MTBE is a highly hindered ether and this makes the initial step of hydrolysis to alcohols difficult (Davis and Erickson 2004).

Although methyl tetra butyl ether has a low affinity for sorption to soil and a very high vapor pressure, which make it respond well to low-temperature thermal desorption and soil vapor extraction, its rapid transfer from the soil into the groundwater adversely affects remediation. MTBE appeared to be the main contaminant in groundwater near to fuel stations, due to its chemical and biological stability (Briganti et al. 2020).

Achten et al. (2002a) overviewed MTBE's concentrations in 315 German river water samples and 82 wastewater samples, where they reported detection of approximately 250 ng/L MTBE concentrations in urban areas and approximately 50 ng/L in rural areas. In addition, they reported that MTBE quantity in river water increased at precipitation days. MTBE was also detected in drinking water, despite riverbank filtration and drinking water processing (Achten et al. 2002b). The results showed an average MTBE concentration of 88 ng/L in riverbank filtered water, raw water and recovering well water. In drinking water, MTBE concentration was between 43 and 110 ng/L, whereas tap water samples at Frankfurt revealed an average MTBE concentration of 35 ng/L with a maximum of 71 ng/L.

The possibility of MTBE concentration was measured in air among gas stations and in air of salesrooms of gas stations. MTBE was found to be present (Fembacher et al. 2017). Moreover, air samples were taken in a road traffic tunnel in Los Angeles, US and the MTBE content was measured (Fraser et al. 1998). The tests revealed that automobile exhaust gases contained 155 mL of MTBE per burnt liter of gasoline that was emitted into the atmosphere. The used gasoline contained 10% MTBE.

There are also serious health effects from exposure to MTBE. Increased frequency of respiratory, neurologic and allergic reactions have been experienced by humans exposed to gasoline containing MTBE. Chemical pneumonitis may result from MTBE aspiration into the lungs, while MTBE can induce human lymphocytes to suffer DNA double—strand breaks at 200 uM (Chauvin et al. 1990). Furthermore, studies on rodents revealed that MTBE affects primarily the kidney and liver. MTBE exposure results in reversible central nervous system effects, which include hypoactivity, sedation, ataxia and anesthesia at higher concentrations. In male rats increased incidences of kidney and testicular tumors were produced by inhalation exposure to MTBE (Chauvin et al. 1990).

ETBE. It was introduced as the promising alternative to MTBE. ETBE is currently favored in several European markets as a result of the EU biofuels directives 2003/30/EC and 2009/28/EC, mostly because it can be synthesized from both renewable and fossil-based sources, bioethanol and isobutylene, respectively. However, it seems that ETBE is at least as problematic as MTBE in terms of contamination.

ETBE follows similar paths with MTBE; both of them having poor biodegradability. Nevertheless, ETBE presents a lower solubility in water (10 g/L) compared to MTBE (40 g/L) and therefore it tends to create much smaller plumes. Moreover, it has a lower vapor pressure, which makes it easier to blend than MTBE.

In Rhone district, France a leaking gas station was studied by Fayolle-Guichard et al. (2012). They reported that although the soil which surrounded the gas station was highly heterogenated and natural barriers existed, ETBE overcame them and consequently entered and contaminated all nearby aquifers at concentrations up to 175 mg/L. (Fayolle-Guichard et al. 2012). Moreover, Stupp et al. (2012) presented an extensive distribution of ETBE in the surface water (rivers and channels) in the Netherlands. High concentrations found in the river Rhine, in the river Meuse and their channels. Examining 363 samples for MTBE and 325 samples for ETBE, the final water quality report on the Rhine showed that the average monthly values for MTBE and ETBE were 0.405 μg/L and 0.231 μg/L, respectively (Stupp et al. 2012; Wezel et al. 2009).

Ether oxygenates dissolved in water, exhibit an unpleasant and potent odor. Mansotte et al. (2012) reported that in France, part of a large local authority's drinking water distribution network was contaminated by ETBE in June 2009. Customers' continuous complaint reports of the "ether" odor in potable water to the company in charge of the drinking water distribution network, revealed for the first time in France ETBE's impact in drinking water. It must be noted that France, at the time, did not include ETBE as part of the French sanitary control regulation (Mansotte et al. 2012). The concentration of ETBE was also measured in air among gas stations and in air of salesrooms of gas stations. Fembacher et al. (2017) found ETBE to be present in the surrounding air at concentrations of milligram per cubic meter levels.

Likewise MTBE, there are also health effects from exposure to ETBE. Male volunteers exposed to ETBE at 25 and 50 ppm made them feel bad taste in mouth, irritation in airways and throat and a slightly impaired lung function (Weng et al. 2019).

TAME. Compared to MTBE, TAME has a lower octane number, it contributes less oxygen to gasoline which leads to larger volumes of TAME needed to achieve MTBE equal oxygen contents and it is the second in volume ether used as MTBE alternative. TAME's biodegradation deactivates under iron-reducing, nitrate and methanogenic conditions in surface water sediments (Stupp et al. 2012).

TAME's sorption ability to a soil surface could be attributed as a major drawback. TAME's sorption can be estimated from the Koc value. The log Koc value for TAME is 1.6 (Koc = 39 L/kg), which is the highest among all ether oxygenates and benzene. This means that TAME is the oxygenate that has the strongest physical adherence to a solid surface such as the soil/aquifer matrix (Stupp et al. 2012).

Between 2000 and 2010 selected European countries were overviewed by Concawe for the production and use of GEOs in the European Union (Stupp et al. 2012). TAME was found to have a certain presence in soil in the UK, Sweden and Finland. It is worthy of remark that Finland until 2008 was a large TAME producer/consumer. Greece and Italy also produce and use TAME, but they were not among the selected EU countries overviewed by Concawe. More specifically in the UK, soil samples from 631 sites were analyzed and TAME was detected in 19 of the samples in a range of 0.011 mg/kg up to 49 mg/kg. In addition, Vainiotalo et al. (2006) reported TAME concentrations of 1.3 mg/m^3 in the surrounding air of service stations and repair shops in Finland (Vainiotalo et al. 2006).

Landmeyer et al. (2010) examined three sites (Carmans River, Ford Pond, Tiana Bay) in the Long Island, New York and they found all of them to have underground and surface water contaminated with TAME by leaking underground gasoline storage tanks between 2003 and 2005. TAME concentrations greater than 10 mg/L have been detected in monitoring wells located within 90 m of the Carmans River, and bed sediment pore water in the hyporheic zone and surface water also contained TAME at the milligrams per liter level. The second study site, Ford Pond, receives water only from groundwater discharge and precipitation. The TAME contaminated groundwater coming from the gasoline station discharges to the pond. Investigations indicated TAME concentrations at the milligrams per liter level in bed-sediment pore water. TAME concentrations have been likewise detected in the ground water of the third study site too, Tiana Bay, coming from the gasoline station located about 1200 m up gradient.

Regarding health effects from exposure to TAME, studies in human volunteers showed that TAME's half-life in blood varies between 1.2 and 6.3 h. Chemical pneumonitis may result from aspiration into the lungs. Consciousness could low from exposure at high levels. TAME liquid can also defat the skin. Moreover, road tanker drivers in Finland, carrying TAME oxygenated gasoline, after a work shift were reported to have TAME concentrations in urine samples on average of 16 nmol/L. After 16 h the concentration dropped to 9 nmol/L. Tests in laboratory animals showed that sub-chronic or pre-chronic TAME exposure via inhalation for 6 h per day, 5 days per week for 4 weeks and high dosage of 4000 ppm, on 28 rats caused 7 of them to die. The remaining rats suffered transient central nervous system depression (ataxia, sedation and reduced activity), changes in body temperature, high limb splay, tail pinch and righting reflex.

4.3 Alternative Fuel Upgrade

In the current study an alternative fuel upgrade process is examining in order to avoid the aforementioned serious drawbacks from the use of GEOs by the refineries. The whole process based upon the heterogenization of homogeneous catalytic reactions in aqueous media applied in refinery cuts.

4.3.1 Applied Heterogenized Homogeneous Catalysis

The relatively newly formed field of heterogenized homogeneous catalysis has managed to overcome the disadvantages of homogeneous catalysis using biphasic homogeneous catalytic systems. The biphasic homogeneous catalytic systems achieved the heterogeneity of the catalyst and the product in two distinct and immiscible phases (Manassen 1973; Bailar 1974; Dror and Manassen 1977). Thus, after performing the reactions, the catalyst exists in the solution of one phase and the substrate with the products in the other phase (Fig. 4.1). The continuous agitation of the two phases in the reactor facilitates the proper interaction of the catalyst with the substrate. When the reaction is complete, the stirring stops and the mixture of the

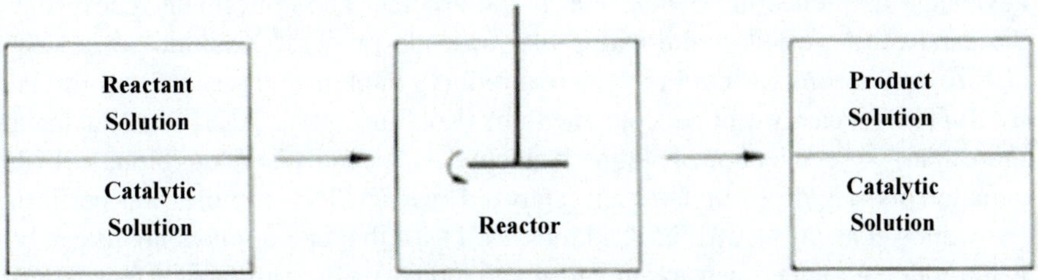

Fig. 4.1 Schematic representation of biphasic homogeneous catalytic system before and after the reaction

two phases is separated into two layers, the lighter one contains the product with part of the substrate and the heavier one the catalyst. The separation of the two phases is accomplished by decantation of the lighter phase; while the catalyst in the heavier phase is ready for reuse.

The development of the organometallic chemistry contributed considerably in the development of catalysts for biphasic homogeneous systems. Nowadays, organometallic compounds are gradually being used to catalyze organic reactions. The reason for that is the intense activity and selectivity of the organometallic catalysts, which is achieved by the appropriate selection of the active metal center and its organic ligands. However, organometallic catalysts do not cease to be homogeneous retaining the main downside of homogeneous catalysts, which is the recovery of the catalyst itself. Manassen (1973) first tried to eliminate the disadvantages of homogeneous catalysis by plainly citing the stabilization of organometallic catalysts using two immiscible liquid phases.

It all started when Manassen led a research group focused on catalyst heterogenization at the Weizmann Institute in Rehovot, Israel. The theoretical results of the above works were first presented in 1972 at the Conference of the NATO Scientific Committee. At this conference, the discovery of biphasic organometallic catalysis was presented as the most viable solution to the general disadvantage of homogeneous catalysis, regarding the easy and cost-effective separation of the catalyst from the products. Although the heterogenized homogeneous catalysis was pioneered by Manassen (1973) (Joó 2001; Cornils and Herrmann 2004), concurrently Joó was working at the same field during the elaboration of his diploma thesis, published in 1972 at Lajos Kossuth University of Debrecen, Hungary, under the supervision of Professor Beck; his first relevant articles were published in 1973 (Joó and Beck 1973). Later, Dror and Manassen (1977) managed to overcome the problems of homogeneous catalysis based on the findings of the literature review of Bailar (1974), on the heterogenization of the homogeneous catalysis, but also of Joó and Beck (1975), on the selection of a suitable ligand; thus, they developed a two-phase catalyst with organometallic complexes (Rhodium-Phosphine).

Homogeneous catalysis in refinery processes is rare, mainly due to the separation difficulties that emerge. Thus, the very many benefits of the homogeneous catalysis are still kept apart from the petroleum industry. In this chapter, a successful application of heterogenized homogeneous catalysis was implemented taking into advantage its pioneering benefit that is the efficient and convenient recovering of the catalyst. A two-step alternative fuel upgrade process (Kokkinos et al. 2013, 2015b) was accomplished. Firstly, a real refinery naphtha cut was hydroformylated by Rh/TPPTS catalyst in aqueous medium (Kokkinos et al. 2011, 2013) achieving more than 95% conversion of olefins to aldehydes. Secondly, aqueous biphasic hydrogenation of the hydroformylated cut catalyzed by Ru/TPPTS complex was performed (Kokkinos et al. 2009, 2015b; Kokkinos 2021). To that end, homogeneous catalysis is introducing to the downstream petroleum industry efficiently and effectively.

4.3.2 Materials and Methods

Materials and Instrumentation. All manipulations and reactions were performed under argon or nitrogen atmosphere using standard Schlenk techniques (Shriver and Drezdzon 1986). $RuCl_3 \cdot xH_2O$ used as catalyst precursor and was procured from Aldrich. TPPTS used as ligand and was obtained from Alfa Aesar Chemicals. Toluene was obtained from Merck. H_2 5.0 N, He 5.0 N, N_2, Ar and Zero Air, and were procured from Axarlis. Na_2SO_4 for drying purposes was obtained from Panreac. All the reactions were performed in 100 mL Autoclave Engineers reactor with a low carbon stainless steel vessel (316LSS) and a URC control unit. Both the products and the substrate were analyzed using an Agilent 7890A GC/FID and an Agilent 6890 N GC coupled with an MSD 5975B equipped with Agilent 7683B ALS; both of them were fitted with a Petrocol DH 150 capillary column (150 m long, 0.25 mm id, 1.0 μm film). Column temperature was kept initially constant at 40 °C for 5 min, then it was raised to 200 °C at a rate of 2 °C min^{-1} and held isothermally there for 45 min. Helium (99.999%) was used as a carrier gas with a flow rate 20 cm/sec at 175 °C and 65 psig. The temperatures of the ion source at GC/MS were set at 230 °C and the transfer line at 320 °C. In order to clean the column from various remaining redundant compounds the oven temperature kept at 230 °C for 180 min (the temperature of both the injector and detector was 280 °C) prior to analysis.

Hydroformylation procedure. The hydroformylated fuel was prepared according to Kokkinos et al. (2013). The whole process is illustrated in Fig. 4.2, where the contribution of heterogenized homogeneous catalysis is noticeable by successfully recycling the catalytic system in aqueous medium.

The substrate of the biphasic catalytic hydroformylation is a light cracked naphtha (LLCN) cut granted by Aspropyrgos refinery (Hellenic Petroleum S.A.). Rh-TPPTS catalyst was prepared in situ by direct addition of $RhCl_3$ to TPPTS in aqueous solutions under syn-gas ($CO/H_2 = 1/1$) pressurization and heating. The final product of the current catalytic reaction is the hydroformylated fuel (Table 4.1).

Hydrogenation procedure. After the hydroformylation procedure has taken place and the hydroformylated fuel was ready for use, the Ru-TPPTS catalyst was composed in situ by direct addition of $RuCl_3xH_2O$ catalyst precursor (10 mg; 0.04 mmol) and TPPTS ligand (111 mg; 0.15 mmol) in deoxygenated distilled demineralized water (15 mL). The above aqueous solution along with the hydroformylated fuel (7.5 mL; 38.02 mmol aldehydes) in toluene (7.5 mL) were placed in the reactor. Afterwards, the reactor was purged with hydrogen. Later, hydrogen pressurization and heating taken place to the desired conditions. After the end of the reaction, the catalyst remained at the lower aqueous phase and the products at the upper organic phase. The two immiscible phases of the reaction mixture were divided through a separation funnel.

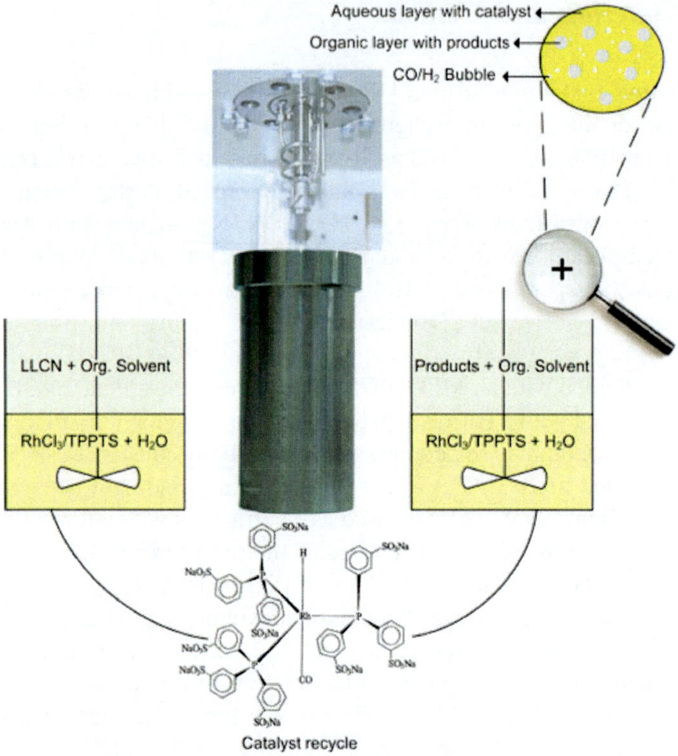

Fig. 4.2 Aqueous biphasic catalytic hydroformylation reaction (Kokkinos et al. 2013)

4.3.3 Results and Discussion

Effect of reaction temperature and pressure. The influence of reaction temperature was examined in a range of 30–120 °C (Fig. 4.3). Increase in reaction yield was observed by rising temperature. In particular, there was a sharp increase in the conversion of the reaction from 90 (22.5% mol) to 120 °C (98.9% mol). In addition, taking into account the fact that the catalyst system is a water-soluble analogue of the Wilkinson catalyst, which is active at ambient temperature, the reaction was carried out at a temperature of 30 °C. However, the conversion of the reaction under such mild conditions was only 3.6% mol, but even at 60 °C it reached only 4.2% mol (Kokkinos et al. 2009). This is in line with the literature (Hernandez and Kalck 1997a, b; Wilkinson 1988) that reveals RuHCl $(TPPTS)_3$ to be produced under conditions of 20–35 bar and 40–85 °C. At the same time, the intense difficulties of mass transfer between the two liquid phases of biphasic systems without the use of surfactants were evident. Therefore, at room temperature, the access of aldehydes to the Ru catalytic complexes was limited and the reaction rate reduced, mainly due to mass transfer difficulties. It is worthy of remark the presence of side reactions

Table 4.1 PIONA analysis of hydroformylated fuel

Compounds	Molar fraction (%mol)
PARAFINS	**5.80**
n-Butane	0.07
n-Pentane	5.73
ISOPARAFINS	**26.14**
Isopentane	26.03
2,2-Dimethyl-Butane	0.11
OLEFINS	**10.57**
trans-2-Butene	0.25
3-Methyl-1-Butene	0.05
1-Pentene	0.12
2- Methyl-1-Butene	1.07
trans-2-Pentene	1.61
cis-2-Pentene	0.54
2- Methyl-2-Butene	6.82
Cyclopentene	0.11
NAPHTHENS	**1.01**
Cyclopentane	1.01
ALDEHYDES	**55.40**
3-Methyl-Butanal	3.20
2-Methyl-Butanal	0.50
Pentanal	0.34
2,3-Dimethyl-Butanal	10.07
4-Methyl-Pentanal	0.10
2-Methyl-Pentanal	13.88
2-Ethyl-Butanal	4.57
3-Methyl-Pentanal	15.53
Hexanal	4.91
2,2-Dimethyl-Butanal	0.19
Cyclopentyl-Methanal	2.11
ALCOHOLS	**1.08**
3-Methyl-Butanol	0.09
2-Methyl-Butanol	0.25
2,3-Dimethyl-Butanol	0.03
4-Methyl-Pentanol	0.04
2-Methyl-Pentanol	0.12
3-Methyl-Pentanol	0.16

(continued)

Table 4.1 (continued)

Compounds	Molar fraction (%mol)
Hexanol	0.22
2,2-Dimethyl-Butanol	0.04
Cyclopentyl-Methanol	0.13

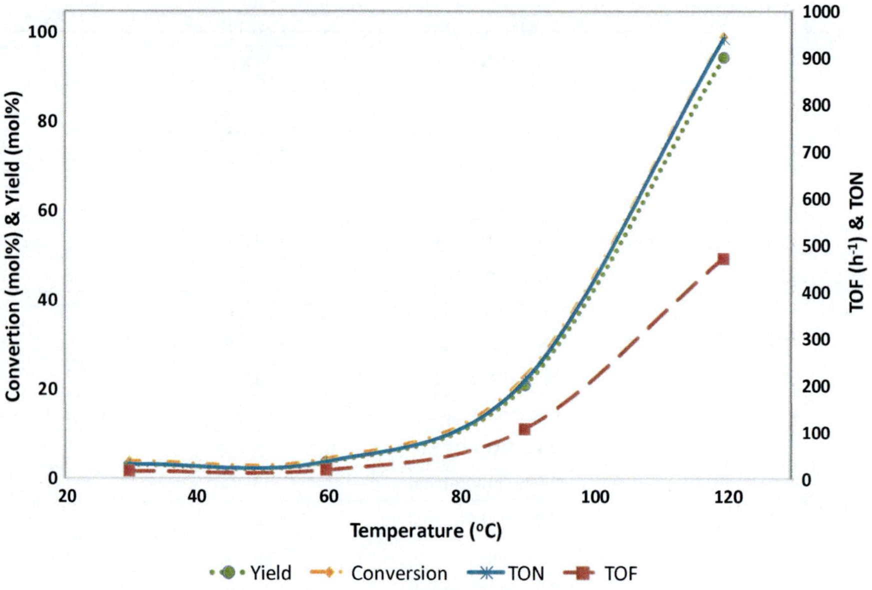

Fig. 4.3 Effect of reaction temperature in aqueous biphasic catalytic hydrogenation of the hydroformylated fuel

from 90 °C and higher, which is justified by the increase of the deviation between the conversion and the yield of the reaction.

The effect of hydrogen pressure on the catalytic biphasic hydrogenation of hydroformylated LLCN by Ru/TPPTS complexes was also tested in 50 and 75 bar. Throughout the study of the hydrogen pressure of the reaction, all other parameters of the reaction remained constant. Thus, the temperature was constant at 90 °C, the reaction time at 2 h, the ruthenium concentration at 256 ppm, the molar ratios of CH = O/Ru equal to 1000, the TPPTS/Ru equal to 4, and aqueous/organic phase volumes was 1:1. The increase of the hydrogen pressure from 50 to 75 bar favored the rise in the catalytic activity from TOF = 27 to TOF = 104. It could be explained by the fact that hydrogen dissolution in the liquid phase increases as the hydrogen pressure increases. In addition, the rise of the hydrogen concentration in the solution and in the aqueous-organic interface favors the formation and the further stabilization of $RuHCl(TPPTS)_3$ catalytic complex (Hernandez and Kalck 1997a, b; Nuithitikul and Winterbottom 2007).

Table 4.2 Comparing the ultimate upgraded fuel with the unleaded gasoline specifications

Specifications	ISO/EN	Unleaded 95, 98, 100, LRP	Ultimate upgraded fuel
Lead (mg/l max.)	EN 237:1996	5	0
Benzene (% v/v max.)	EN 12177:1998 EN 238:1996	1	0.2
Aromatics (% v/v max.)	EN 12177:1998 EN 238:1996	35	22.8
Olefins (% v/v max.)		18	1
Ethers (% v/v max.)	EN 1601:1997 prEN 13132:1998	15	0

Upgraded fuel versus unleaded gasoline specifications. The end product of the biphasic catalytic hydrogenation of the hydroformylated substrate is an upgraded fuel. Table 4.2 shows the specifications of the unleaded gasoline in accordance with the international standards (ISO) adopted by the European Union and compares them with the ultimate upgraded fuel of the current alternative refinery process. The alternative refinery process of fuel catalytic upgrade in aqueous media creates a completely free of lead and ethers product with favorable concentrations of other organic groups and compounds and within specifications.

Hydrogenation Process Modeling and simulation. PR76 and PR78 CEOS (Peng and Robinson 1976; Robinson and Peng 1978) have gained special preference of fuel industry. In such complicated mixtures, like refinery cuts, the PR78 for a pure component:

$$P = \frac{R \cdot T}{V_m - b_i} - \frac{a_i(T)}{V_m^2 + 2b_i V_m - b_i^2} \tag{4.4}$$

with:

$$R = 8.314472 \frac{J}{mol \cdot K}$$

$$b_i = 0.0777960739 \cdot \frac{R.T_{c_i}}{P_{c_i}}$$

$$a_i(T) = 0.457235529 \cdot \frac{R^2 \cdot T_{c_i}^2}{P_{c_i}} \cdot \left[1 + m_i \cdot \left(1 - \sqrt{\frac{T}{T_{c_i}}}\right)\right]^2$$

$$For\ \omega_i \leq 0.491, \quad m_i = 0.37464 + 1.54226 \cdot \omega_i - 0.26992 \cdot \omega_i^2$$

$$\text{For } \omega_i > 0.491, \quad m_i = 0.379642 + 1.48503 \cdot \omega_i - 0.164423 \cdot \omega_i^2 + 0.016666 \cdot \omega_i^3$$

where P is pressure, R the ideal gas constant, T the temperature, V_m the molar volume, T_c the critical temperature, P_c the critical pressure and ω the acentric factor, was modified and parameter α used a quadratic mixing rule as a function of mole fraction (x_i, x_j) of the components and of the adjustable $\delta_{i,j}$ binary interaction parameter:

$$a = \sum_{i=1}^{N} \sum_{j=1}^{N} x_i \cdot x_j \cdot \sqrt{a_i \cdot a_j} \cdot (1 - \delta_{i,j}) \tag{4.5}$$

and parameter b used a linear mixing rule:

$$b = \sum_{i}^{N} x_i \cdot b_i \tag{4.6}$$

Moreover, a new alpha function that avoids negative alpha values and is considerably accurate in prediction of hydrogen properties (Nasrifar 2010) was used (Twu et al. 1991, 1995):

$$\alpha(T) = T_r^{N \cdot (M-1)} \cdot e^{L \cdot \left(1 - T_r^{N \cdot M}\right)} \tag{4.7}$$

where L, M, N are component-dependent constants and T_r is the reduced temperature, T/T_c. The parameter N increased the flexibility of the initial alpha function proposed by Twu (1988) improving the vapor pressure forecast for highly polar components with high normal boiling points (NBP). For compounds that there no available data regarding vapour pressure data, the following equation was used presenting alpha function as a function of reduced temperature (T_r) and acentric factor (ω) (Twu et al. 1995):

$$\alpha = \alpha^{(0)} + \omega \cdot \left(\alpha^{(1)} - \alpha^{(0)}\right) \tag{4.8}$$

with:

$$\alpha^{(0)} = T_r^{N^{(0)} \cdot \left(M^{(0)} - 1\right)} \cdot e^{L^{(0)} \cdot \left(1 - T_r^{N^{(0)} \cdot M^{(0)}}\right)} \tag{4.9}$$

$$\alpha^{(1)} = T_r^{N^{(1)} \cdot \left(M^{(1)} - 1\right)} \cdot e^{L^{(1)} \cdot \left(1 - T_r^{N^{(1)} \cdot M^{(1)}}\right)} \tag{4.10}$$

Using PR78 CEOS embedded with Twu's alpha function, the phase behavior diagram of the organic phase (fuel) was developed (Fig. 4.4). The ultimate alternative fuel presents a critical point at $T_c = 294.2$ °C, $P_c = 41.12$ bar, its cricondentherm at 294.5 °C and its cricondenbar at 41.12 bar. Also, the organic phase of the system

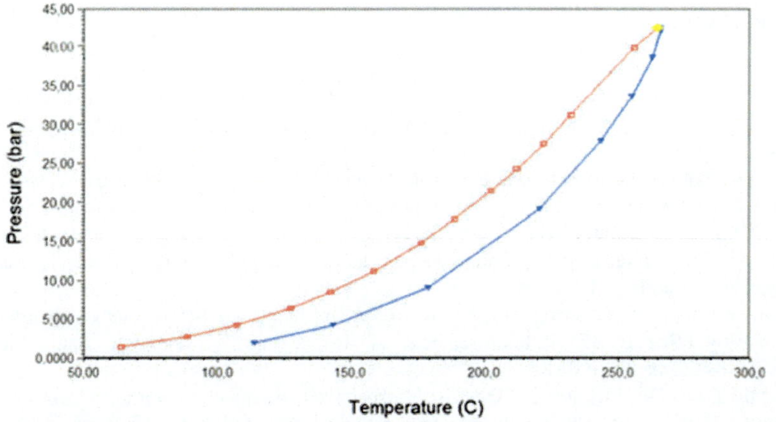

Fig. 4.4 The envelope of the ultimate upgraded fuel (red: bubble point curve, blue: dew point curve, yellow: critical point)

remains in the liquid phase in all experimental conditions. Finally, the average absolute relative deviation (AARD) of the model with the experimental outcomes from the laboratory was less than 14.5%.

4.4 Conclusions

Gasoline ethers oxygenates replaced tetraethyl lead as appropriate octane boosters in the blended gasoline pool of the refineries and not as environmentally friendly compounds. The contribution of heterogenized homogeneous catalysis on alternative fuel upgrade in aqueous media from refinery cuts proved to be significant. $RuCl_3$/TPPTS catalyst was suitable and effective for the proposed alternative fuel upgrading process achieving high conversion of the hydroformylated substance (98.9%). The ultimate upgraded fuel was totally free of GEOs and lead. The current alternative refinery process of fuel catalytic upgrade is an in situ environmentally friendly process with no byproducts, used mild operating conditions, achieved high reaction yield and conversion and succeeded catalyst and water recycle. A scale-up study of this pioneering chemical process could be implemented by using the proposed simulation model. A future use of this alternative process in refineries should fully replace the GEOs from the final gasoline blending pool with an in situ and environmentally benign way.

References

Achten C, Kolb A, Püttmann W, Seel P, Gihr R (2002a) Methyl tert-butyl ether (MTBE) in river and wastewater in Germany. 1. Environ Sci Technol 36(17):3652–3661. https://doi.org/10.1021/es011492y

Achten C, Kolb A, Püttmann W (2002b) Occurrence of methyl tert-butyl ether (MTBE) in riverbank filtered water and drinking water produced by riverbank filtration. 2. Environ Sci Technol 36(17):3662–3670. https://doi.org/10.1021/es011493q

Bailar JC (1974) Heterogenizing homogeneous catalysts. Catal Rev Sci Eng 10(1):17–36. https://doi.org/10.1080/01614947408079625

Bardin M, Hussin A, Gushchin PA, Vinokurov V, Burluka A (2014) Technical aspects of Ethyl Tert-Butyl Ether (ETBE) for large-scale use as gasoline improver. Energy Technol 2.https://doi.org/10.1002/ente.201300130

Bell ML, Davis DL, Fletcher T (2004) A retrospective assessment of mortality from the London smog episode of 1959: the role of influenza and pollution. Environ Health Perspect 112(1):6–8

Bonventre JA, White LA, Cooper KR (2012) Craniofacial abnormalities and altered WNT and MMP mRNA expression in zebrafish embryos exposed to gasoline oxygenates ETBE and TAME. Aquat Toxicol 120–121:45–53. https://doi.org/10.1016/j.aquatox.2012.04.008

Briganti A, Tuccimei P, Voltaggio M, Carusi C, Galli G, Lucchetti C (2020) Assessing methyl tertiary butyl ether residual contamination in groundwater using radon. Appl Geochem 116.https://doi.org/10.1016/j.apgeochem.2020.104583

Caetano NS, Loureiro JM, Rodrigues AE (1994) MTBE synthesis catalysed by acid ion exchange resins: kinetic studies and modeling of multiphase batch reactors. Chem Eng Sci 49(24, Part A):4589–4604. https://doi.org/10.1016/s0009-2509(05)80043-0

Chauvin Y, Gilbert B, Guibard I (1990) Catalytic dimerization of alkenes by nickel complexes in organochloroaluminate molten salts. J Chem Soc, Chem Commun 23:1715–1716

Chow TJ, Patterson CC (1962) The occurrence and significance of lead isotopes in pelagic sediments. Geochim Cosmochim Acta 26(2):263–308. https://doi.org/10.1016/0016-7037(62)90016-9

Cornils B, Herrmann WA (2004) Aqueous-phase organometallic catalysis, 2nd edn. Wiley-VCH, Weinheim

Davis LC, Erickson LE (2004) A review of bioremediation and natural attenuation of MTBE. Environ Prog 23(3):243–252. https://doi.org/10.1002/ep.10028

Dror Y, Manassen J (1977) Hydrogenation of olefins with rhodium-phosphine complexes, having substrate and catalyst in two different immiscible phases. An alternative method for the heterogenization of a homogeneous catalyst. J Mol Catal 2(3):219–222. https://doi.org/10.1016/0304-5102(77)80055-2

European-Parliament-and-the-Council (1998) Directive 98/70/EC of the European Parliament and of the Council of 13 October 1998 relating to the quality of petrol and diesel fuels and amending Council Directive 93/12/EEC. Official Journal of the European Communities 41:L 350/358–368

European-Parliament-and-the-Council (2009) Directive 2009/30/EC of the European Parliament and of the Council of 23 April 2009 amending Directive 98/70/EC as regards the specification of petrol, diesel and gas-oil and introducing a mechanism to monitor and reduce greenhouse gas emissions and amending Council Directive 1999/32/EC as regards the specification of fuel used by inland waterway vessels and repealing Directive 93/12/EEC. Official Journal of the European Communities 52:L 140/188–113

Fayolle-Guichard F, Durand J, Cheucle M, Rosell M, Michelland RJ, Tracol JP, Le Roux F, Grundman G, Atteia O, Richnow HH, Dumestre A, Benoit Y (2012) Study of an aquifer contaminated by ethyl tert-butyl ether (ETBE): site characterization and on-site bioremediation. J Hazard Mater 201–202:236–243. https://doi.org/10.1016/j.jhazmat.2011.11.074

Fembacher L, Horn M, Völkel W, Fromme H (2017) Exposure to aromatic hydrocarbons, ETBE and MTBE at gas stations and in salesrooms of gas stations. Gefahrstoffe—Reinhalt Luft 77(6):231–236

Fraser MP, Cass GR, Simoneit BRT (1998) Gas-phase and particle-phase organic compounds emitted from motor vehicle traffic in a Los Angeles roadway tunnel. Environ Sci Technol 32(14):2051–2060. https://doi.org/10.1021/es970916e

Hamid H, Ali MA (2004) Handbook of MTBE and other gasoline oxygenates. Marcel Dekker, New York

Hart (2014) International fuel quality standards and their implications for Australian standards (trans: Australian Government DotE). Hart, Energy Research & Consulting, Australia

Hernandez M, Kalck P (1997a) Water-soluble ruthenium complexes containing tris(m-sulfonatophenyl) phosphine (TPPTS). Preparation of a series of [Ru(H)(η6-arene)$(TPPTS)_2$]Cl complexes, $[Ru(H)_2(CO)(TPPTS)_3]$ and revisited procedures for previously described ruthenium-TPPTS compounds. J Mol Catal A: Chem 116(1–2):117–130. https://doi.org/10.1016/s1381-1169(96)00232-4

Hernandez M, Kalck P (1997b) Study of the hydrogenation of α, β-unsaturated carbonyl compounds catalyzed by water-soluble ruthenium-TPPTS complexes. J Mol Catal a: Chem 116(1–2):131–146. https://doi.org/10.1016/s1381-1169(96)00233-6

Joó F (2001) Aqueous organometallic catalysis. Kluwer Academic Publishers, Dordrecht

Joó F, Beck MT (1973) Vízoldható foszfinkomplexek redoxireakciói—új lehetőségek a homogén oldatkatalízisben. Magyar Kémiai Folyóirat 79:189–191

Joó F, Beck MT (1975) Formation and catalytic properties of water-soluble phosphin complexes. React Kinet Catal Lett 2(3):257–263

Kinner NE (2001) Fate, transport and remediation of MTBE. The University of New Hampshire. https://archive.epa.gov/oust/mtbe-a/web/pdf/kinner.pdf. Accessed 7 July 2020

Kokkinos NC (2021) Modeling and simulation of biphasic catalytic hydrogenation of a hydroformylated fuel. Int J Hydrogen Energy 46(37):19731–19736. https://doi.org/10.1016/j.ijhydene.2020.09.082

Kokkinos NC, Lazaridou A, Nikolaou N, Papadogianakis G, Psaroudakis N, Chatzigakis AK, Papadopoulos CE (2009) Hydrogenation of a hydroformylated naphtha model (mixture of specific aldehydes) catalysed by Ru/TPPTS complex in aqueous media. Appl Catal A 363(1–2):129–134. https://doi.org/10.1016/j.apcata.2009.05.002

Kokkinos NC, Lazaridou A, Papadopoulos CE, Nikolaou N, Psaroudakis N, Mertis K (2011) An environmentally friendly potential refinery process of light-light cracked naphtha olefins conversion to valuable oxygenated products. In: International conference on chemical engineering (CET/CEN2011), Shanghai, China, 28–30 Oct 2011, pp 146–149

Kokkinos NC, Kazou E, Lazaridou A, Papadopoulos CE, Psaroudakis N, Mertis K, Nikolaou N (2013) A potential refinery process of light–light naphtha olefins conversion to valuable oxygenated products in aqueous media—Part 1: Biphasic hydroformylation. Fuel 104:275–283. https://doi.org/10.1016/j.fuel.2012.08.040

Kokkinos NC, Mitropoulos AC, Nikolaou NA (2015a) An environmentally benign catalytic process enhances in situ the quality of gasoline. Paper presented at the Abu Dhabi international petroleum exhibition and conference, Abu Dhabi, UAE, 9 Nov 2015

Kokkinos NC, Nikolaou N, Psaroudakis N, Mertis K, Mitkidou S, Mitropoulos AC (2015b) Two-step conversion of LLCN olefins to strong anti-knocking alcohol mixtures catalysed by Rh, Ru/TPPTS complexes in aqueous media. Catal Today 247:132–138. https://doi.org/10.1016/j.cattod.2014.07.058

Landmeyer JE, Bradley PM, Trego DA, Hale KG, Haas Ii JE (2010) MTBE, TBA, and TAME attenuation in diverse hyporheic zones. Groundwater 48(1):30–41. https://doi.org/10.1111/j.1745-6584.2009.00608.x

Manassen J (1973) In: Bassolo F, Burwell RL (eds) Catalysis: progress in research. Plenum Press, London, pp 177, 183

Mansotte F, Dejean G, Renault C (2012) Feedback on an accidental pollution of a drinking water network by ETBE. Tech Sci Methodes 10:42–47. https://doi.org/10.1051/tsm/201210042

Nasrifar K (2010) Comparative study of eleven equations of state in predicting the thermodynamic properties of hydrogen. Int J Hydrogen Energy 35(8):3802–3811. https://doi.org/10.1016/j.ijhydene.2010.01.032

Nuithitikul K, Winterbottom M (2007) Crucial parameters in the selective biphasic hydrogenation of cinnamaldehyde by biphasic Ru-TPPTS and $RhCl(TPPTS)_3$ catalysts. Catal Today 128(1–2):74–79. https://doi.org/10.1016/j.cattod.2007.05.019

Patterson CC (1956) Age of meteorites and the earth. Geochim Cosmochim Acta 10(4):230–237. https://doi.org/10.1016/0016-7037(56)90036-9

Patterson CC (1965) Contaminated and natural environments of man. Arch Environ Health 11:344–360

Peng D-Y, Robinson DB (1976) A New Two-Constant Equation of State. Ind Eng Chem Fundam 15(1):59–64. https://doi.org/10.1021/i160057a011

Petersen PC, Rosen EM (1975) The petersen automotive troubleshooting & repair manual. Grosset & Dunlap, New York, p 493

Robinson DB, Peng D-Y (1978) The characterization of the Heptanes and Heavier fractions for the GPA Peng–Robinson programs. Gas Processors Association (RR-28), pp 1–36

Schnelle KB, Brown CA (2001) Air pollution control technology handbook. CRC Press/Taylor & Francis Group, Boca Raton

Shriver DF, Drezdzon MA (1986) The manipulation of air-sensitive compounds, 2nd edn. John Wiley and Sons, New York, p 326

Stupp D, Gass M, Leiteritz H, Pijls C, Thornton S, Smith J, Dunk M, Grosjean T, Den Haan K (2012) Gasoline ether oxygenate occurrence in Europe, and a review of their fate and transport characteristics in the environment. CONCAWE Reports (4)

Twu CH (1988) A modified Redlich-Kwong equation of state for highly polar, supercritical systems. In: International symposium on thermodynamicsin chemical engineering and industry, 30 May–2 June 2, 1988, pp 148–169

Twu CH, Bluck D, Cunningham JR, Coon JE (1991) A cubic equation of state with a new alpha function and a new mixing rule. Fluid Phase Equilib 69:33–50. https://doi.org/10.1016/0378-3812(91)90024-2

Twu CH, Coon JE, Cunningham JR (1995) A new generalized alpha function for a cubic equation of state Part 1.Peng-Robinson equation. Fluid Phase Equilib 105(1):49–59. https://doi.org/10.1016/0378-3812(94)02601-V

UK-Parliament (1956). Clean Air Act, 1956, vols 4 & 5 Eliz 2. Her Majesty's Stationery Office, London

UK-Parliament (1968). Clean Air Act, 1968. Her Majesty's Stationery Office, London

UN (2019) Just two countries away from global elimination of lead in petrol. https://www.unenvironment.org/news-and-stories/story/just-two-countries-away-global-elimination-lead-petrol. Accessed 11 June 2021

Vainiotalo S, Kuusimäki L, Pekari K (2006) Exposure to MTBE, TAME and aromatic hydrocarbons during gasoline pump maintenance, repair and inspection. J Occup Health 48:347–357. https://doi.org/10.1539/joh.48.347

van Wezel A, Puijker L, Vink C, Versteegh A, de Voogt P (2009) Odour and flavour thresholds of gasoline additives (MTBE, ETBE and TAME) and their occurrence in Dutch drinking water collection areas. Chemosphere 76(5):672–676. https://doi.org/10.1016/j.chemosphere.2009.03.073

Weng Z, Shi Y, Suda M, Yanagiba Y, Kawamoto T, Nakajima T, Wang RS (2019) Inhalation exposure to low levels of ethyl tertiary butyl ether: its genetic effects were significantly modified by ALDH2 activity. Environ Mol Mutag 60(2):145–153. https://doi.org/10.1002/em.22256

Wilkinson G (1988) Advanced inorganic chemistry, 5th edn. John Wiley & Sons, New York

Chapter 5
Ethanol Derived from Municipal Solid Waste for Sustainable Mobility

Mohd Mubashshir Naved, Amaanuddin M. Azad, Roshan Wathore, Hemant Bherwani, and Nitin Labhasetwar

5.1 Introduction

This section briefly describes municipal solid waste (MSW) generation, health and environmental impacts of unscientific landfilling, energy recovery potential, waste to energy (WtE) approach and the role of MSW derived fuel in achieving sustainable mobility for a better tomorrow.

5.1.1 Municipal Solid Waste (MSW) Generation and Disposal

Developing countries witnessed an acceleration in MSW generation due to population growth, urbanization, economic development and a rise in the standard of living (Nanda and Berruti 2021). The voluminous MSW generation calls for the development of management infrastructure, including collection, landfilling and incineration facilities which is challenging (Das et al. 2019). MSW cycle starts from production/generation, followed by temporary storage at place of origin, collection, transfer and transportation, processing and ultimately disposal/dumping (Sharholy et al. 2008). However, a debilitated waste management network leads to poor segregation, unscientific landfilling, incineration, air/water contamination and inhibits the attainment of sustainable development goals (SDG), at least 12 SDGs and their

M. M. Naved · A. M. Azad · R. Wathore · H. Bherwani (✉) · N. Labhasetwar (✉)
CSIR-National Environmental Engineering Research Institute, CSIR-NEERI, Nagpur, Maharashtra 440020, India
e-mail: h.bherwani@neeri.res.in

N. Labhasetwar
e-mail: nk_labhsetwar@neeri.res.in

M. M. Naved · A. M. Azad · R. Wathore · H. Bherwani
Academy of Scientific and Innovative Research (AcSIR), Ghaziabad, Uttar Pradesh 201 002, India

G. Di Blasio et al. (eds.), *Clean Fuels for Mobility*, Energy, Environment, and Sustainability, https://doi.org/10.1007/978-981-16-8747-1_5

targets are directly linked to waste management (Usmani et al. 2020; Rodić and Wilson 2017). The generation of MSW is inevitable; however, sustainable management with the least environmental footprints is essential and is the research focus for scientific communities across the globe.

It is estimated that 2.01 billion tons of MSW is generated annually; an absolute increment of 40% (3.4 billion tons) is expected till 2050 (Kaza et al. 2018). India has also seen manifold times growth in MSW generation: 277 million tonnes per year with 0.26–0.85 kg kg/capita/day daily waste generation (CPCB 2021; Kumar and Agrawal 2020). A significant fraction of this generated waste is disposed of on landfills without following the recommended measures creating unhygienic conditions for nearby inhabitants (Kumar and Agrawal 2020). Waste landfilling demonstrates a temporary waste management strategy; however, prolonged dumping causes a toll on the environment and human health (Liu et al. 2017).

Landfills are a major source of air, water and soil pollution. Recent studies on landfill sites show increased concentrations of heavy metals such as chromium and lead in the nearby groundwater and cadmium and copper in the nearby soil, which exceeded recommended levels, thus posing a threat to the environment (Vongdala et al. 2019; Dutta et al. 2021) Additionally, practices such as open burning to reduce waste volume emits pollutants like particulate matter (PM), carbon monoxide (CO), nitrogen oxides, sulfur dioxide and carbon dioxide (CO_2), whose exposure is also a concern to both the health and environment (Okedere et al. 2019). Long term exposure to the odour from nearby MSW landfills can lead to increased risk of respiratory illnesses and negative mental health effects (Feo et al. 2013; Vinti et al. 2021). Biodegradation of organic matter from landfill waste produces raw landfill gases having global warming potential (Duan et al. 2021).

Figure 5.1 portrays the annual waste generation by the regions. East Asia and Pacific regions generate 468 million tonnes of waste per year, 23% of entire waste generation. On the other hand, the Middle East and North African regions have the lowest annual waste generation (6%). In the coming decades, the annual MSW generation of developing countries will match developed countries (Fazeli et al. 2016). Also, with advancement and changing consumption patterns, the future MSW

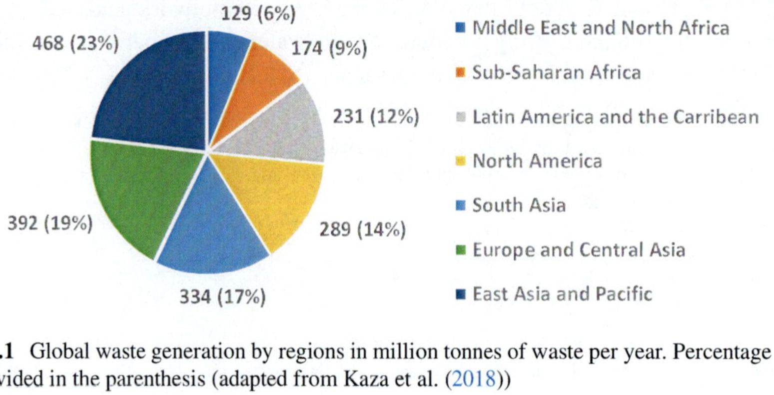

Fig. 5.1 Global waste generation by regions in million tonnes of waste per year. Percentage share is provided in the parenthesis (adapted from Kaza et al. (2018))

will be a complex mix that will demand better treatment technologies. Therefore comprehensive utilization of MSW will become the national priority to decongest landfills and achieve the climate commitments and SDGs (Jamas and Nepal 2010).

5.1.2 Waste to Energy and Mobility Fuel

Earlier incineration technologies were not aligned for energy recovery; the goal was MSW volume reduction by direct combustion (Brunner and Rechberger 2015). Acknowledging the huge challenge of MSW disposal, the waste to energy (WtE) impression came into existence with an objective of energy extraction and resource conservation. A WtE facility is one of the components of waste management where waste materials are scientifically utilized for energy generation (Cucchiella et al. 2014). The associated WtE technologies identified as incineration, pyrolysis and gasification (Stehlík 2009) are used for synthetic fuel production (Kumar and Samadder 2017).

The benefits of WtE are remarkable, including reduced demand for landfills, energy security and reduced greenhouse gas (GHG) emissions, leading to a positive impact on economic growth (Psomopoulos and Themelis 2015; Psomopoulos et al. 2014; Consonni and Viganò 2011). It is estimated that more than 80% of the global primary energy is met by conventional fossil fuels (Looney 2020); therefore, the WtE approach can play a critical role in tackling increasing energy demand and enhancing renewable energy stakes (Makarichi et al. 2018). The energy generation potential of WtE plants is found to be comparable to the energy produced from biomass and leads to significant primary energy savings in conventional utility systems (Pavlas et al. 2010).

WtE concept finds applications in utility thermal power stations, cement kilns, brick making facilities, goods transport sector and district heating. The conversion of waste into solid fuel (pellets/briquets), liquid fuel (ethanol) or gaseous (syngas) offers a great substitution opportunity with conventional mix offsetting the huge carbon footprints. However, the WtE concept has not gained enough momentum because of the poor fuel value, inconsistent supply and economic unviability. Poor segregation, variable composition, improper handling and separation and variable moisture content of MSW make the fuel quality highly inconsistent and unpredictable (Kalyani and Pandey 2014). Furthermore, the debilitated network of door-to-door collection and transportation to the point of application, i.e. WtE plants (which are generally situated far from urban sites) of MSW accompanied with high transportation charges making energy harnessing costly and intermittent (Sierzchula et al. 2014). Owing to such challenges, the WtE facilities are demonstrated only for pilot-based projects and small-scale plants. Also, in India, due to enervated source separation of MSW, developing technology, lack of proper financial and logistical planning, and policy implementation challenges are some reasons for the slow growth of WtE technology (Chand Malav et al. 2020).

There are certain sectors/applications where the WtE approach is practised, but conclusive evidence of field performance are not yet drawn under different techno-economic conditions. The biggest sector identified which lacks to take advantage of the WtE is transport mobility. The projection of the mobility sector reflects a massive growth in passenger and goods movement across the globe (Noussan et al. 2020). This massive growth needs exploration of alternative energy harvesting methods to synchronize with the advancement in passenger/goods mobility, which is likely to continue its dependency on liquid fuels in years to come.

Sustainable Mobility

Transportation networks, working in tandem with increasing populations, evolved to fulfil the growing economic and social demands for better global connectivity (Tuite et al. 2020). Transport mobility brought countries together for trade and other benefits; however, the environmental impacts (GHG emissions), social costs (accidents) and dependence on conventional fuels are serious concerns. The conventional unsustainable transport sector is considered energy, material intensive, polluting and sometimes unequipped with modern safety tools (Proost and Van Dender 2012). It is estimated that passenger mobility will consume 200 Mtoe of energy with consequent emissions of 600 million tonnes CO_2 per year till 2050 (NITI Aayog 2017). These projections are worrisome, especially when the warnings of climate tipping are hard to ignore. Table 5.1. summarizes the consequences of the traditional mode of transport.

Sustainable mobility and its impact on society are recognized worldwide. The concept of sustainable mobility arises from sustainable development, which acknowledges sustainability in environment, economic and social development (Zhou 2012). The main aim of sustainable mobility is to conserve resource depletion, address and solve environmental problems and promote social and economic welfare (Zhao et al. 2020). Sustainable mobility is considered a priority agenda and can be seen in the transport policies of many nations.

Increasing population, mindset for personal mobility and affluence may burden conventional processes/methods for achieving sustainable mobility (Vergragt and Brown 2007). Therefore, to fulfill the aim of sustainable mobility targets, there is a need for innovative alternative fuels that can support/replace the traditional fuel mix. The recent growth of low emission mobility like electric vehicles (EV) have resulted in a move from traditional internal combustion engines, however, such electrification of vehicles leads to challenges such as high operating costs, lack of charging infrastructure, poor purchasing power, difficult battery fault detection, reliability of electrical/electronic software and hardware (Kumar et al. 2020; Hidrue et al. 2011; Xiong et al. 2020; Gandoman et al. 2019; Bonsu 2020). In line with the above recent advancements in vehicle fuel, hydrogen-powered vehicles have also gained momentum; still, challenges like lower power density and nascent technology are some fundamental barriers (Hosseini and Butler 2020; Mori and Hirose 2009). The progress to decarbonize the world automotive fleet by advancing bioethanol technology is phenomenal and estimated to provide 45 exajoules in the global fuel market by 2050 (Pacheco and Silva 2019). With little or no modifications in conventional engines, ethanol +

Table 5.1 The dire consequences of unsustainable transport

Sr. No.	Consumption/Effects	Consequences	Reference
1	Energy	• ~31.6% of the world's total energy consumption is used for transport • 41.8 MJ/kg of energy is required for unit vehicle production	Holden et al. (2020), Sato and Nakata (2020)
2	Metal	• Vehicle production requires a large amount of ferrous and nonferrous materials. Production of 1 kg of vehicle requires at least 5.23 kg of raw materials. The copper metal requirement is 2.29 kg/kg vehicle	Sato and Nakata (2020)
3	Pollution	• A source of local and global pollution, responsible for 24% of direct CO_2 emissions from fuel combustion • Road vehicles, cars, trucks, buses, and two and three-wheelers account for nearly three-quarters of transport CO_2 emissions	IEA (2020), Van den Berg and Langen (2017)
4	Accidents	• Approximatively 1.25 million people worldwide lose their lives in road traffic accidents each year • Road accidents are expected to increase by 65% in 2030 to become the fifth largest cause of fatalities	Sangare et al. (2021)
5	Accesses to mobility services	• Uneven access to safe and sustainable mobility due to policy glitches and lack of financing mechanisms	Verlinghieri and Schwanen (2020), Canitez et al. (2020)

gasoline blends can be directly substituted (Lapuerta et al. 2008). The environmental footprints of ethanol production are accessed by various researchers by using the life cycle assessment (LCA) technique. However, the studies are local and consider only biomass, agricultural residues and sugar cane as precursor material.

5.1.3 *Objectives*

The objectives of the current study include.

- Estimation of the environmental impacts of ethanol production from segregated organic fraction of MSW using the LCA tool.
- Encourage researchers for working on process and pathway optimization for increasing yield of MSW to ethanol conversion.
- To stimulate policymakers, R&D institutes and oil-producing companies to explore the possibility of blending MSW derived ethanol in established fuels on a commercial scale.
- To highlight the grey areas to reduce the environmental footprints of ethanol generation from MSW.

5.2 MSW to Ethanol Conversion

Depleting crude oil reserves and volatile prices have shifted the focus of many countries from conventional to renewable and sustainable power-producing techniques. Bioethanol is a potential alternative source to replace fossil fuels, helping GHG mitigation and resource preservation. The attention towards ethanol recovery picked the pace for various reasons—global warming and climate change being the prime motives, especially in the transport sector. Ethanol is an oxygenated fuel that contains 35% oxygen which helps reduce particulate matter emissions from combustion (Bajpai 2020). Other benefits of ethanol as mobility fuels are summarized in Table 5.2.

Figure 5.2 shows biofuel consumption (particularly ethanol) and its share in transport energy demand. Brazil consumes the highest amount of produced biofuel for transport mobility, followed by China, France and Indonesia. India produces and consumes comparatively lower amounts of biofuels and utilizes 0.7% for transport fuel. However, due to various national missions of biofuel blending, the production and share of biofuel for sustainable transport mobility is projected to increase significantly.

Table 5.2 Benefits of ethanol combustion (Bajpai 2020)

Sr. No.	Markers	Merits
1	CO_2 emissions	Reduces CO_2 emissions by 30–40% with today's combustion technology
2	CO emissions	Reduces tailpipe CO emissions up to 30%
3	Particulate matter	Reduces tailpipe fine particulate matter emissions by 50%
4	Smog	Reduces smog significantly
5	Biodegradability	Rapidly biodegrades in surface water, groundwater and soil, no spillage hazard

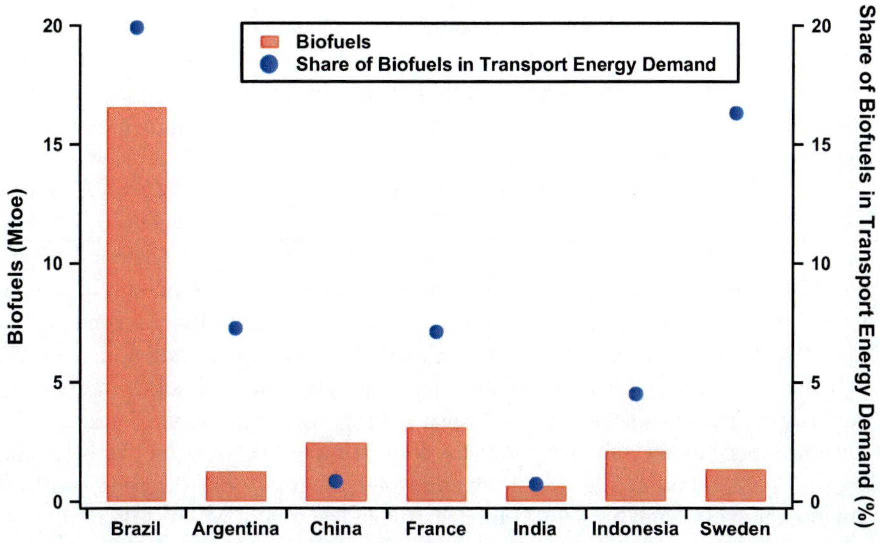

Fig. 5.2 Share of biofuel energy consumption and transport energy demand for the year 2019 (https://www.iea.org/data-and-statistics/charts/biofuel-energy-consumption-and-transport-energy-demand-shares-in-selected-countries-2016) (copyright granted)

The global ethanol production is concentrated in the US and Brazil, with a combined ~20.4 billion litres produced in 2013–18 (collectively, ~80% of global ethanol generation) (Table 5.3). India stands fourth in annual ethanol generation with 1.3 billion litres; this is expected to increase by 70% to 2.2 billion litres by 2024, in line with the increasing vehicle population (NITI-Aayog 2020). The basic precursor to obtaining ethanol is lignocellulosic biomass, specifically agricultural residue. This makes ethanol conversion favourable as biomass is abundantly available in India without geographical limitations (Liu et al. 2019). However, other sources need to be identified to reduce the burden on biomass-derived ethanol. On the same note, In India, it is proposed to blend 20% ethanol with petrol, which is quiet significant (NITI-Ayog 2020). Fulfilment of such a high demand by focusing

Table 5.3 Country-wise ethanol production and expected growth (https://www.iea.org/data-and-statistics/charts/biofuel-production-growth-in-key-markets-2019-2024)

Country	Ethanol production and expected growth by country (billion litres)	
	2013–18	2019–2024
United States	11	1.1
Brazil	9.4	6.4
China	0.5	8.1
India	1.3	2.2
ASEAN*	1.1	1.1

*Association of Southeast Asian Nations

on conventional ethanol obtaining routes may create stress on the resources. The demand–supply gap can be managed by ethanol import; however, this may be a short-term measure. This situation calls for substituting traditional raw materials (like maize, sugarcane etc.) for ethanol generation to ease the burden and support the national blending initiatives.

As discussed in Sect. 1.2, the energy extraction from MSW enables more efficient management of this waste. Therefore, the utilization of organic contents of MSW for biofuel generation is promising, especially in the Indian context. The conversion of MSW to ethanol by thermochemical and biochemical has been practised at industrial levels for decades. Thermochemical convesrion is found more promising than any other ethanol conversion methods (Bajpai 2020). The biochemical method uses cellulosic fractions, which are hydrolyzed using concentrated acids; the resulting acid sugar solution is separated, and sugar is fermented to ethanol (Li et al. 2012a). Several experimental studies reported the conversion of MSW-ethanol with a distinct process. Farmanbordar et al. (2018) reported 83.9 g butanol, 36.6 g acetone, and 20.8 g ethanol recovery from each kg of biodegradable fraction of MSW. Likewise, 194 g ethanol production per kg of dry organic fraction MSW was achieved by Mahmoodi et al. (2018). Several other studies have done experimental procedures for converting MSW to ethanol; however, studies focusing on life cycle impacts and cradle to grave analysis are scarce. In India waste composition exhibits significant variation across regions.

LCA is an effective tool to qualitatively evaluate the ecological loads of the selected fuel. In broader terms, LCA analysis includes raw material extraction, involved energy resources, conversion of resources into the product, utilization of the product and final reuse/recycle or disposal of product (Nanaki and Koroneos 2012). The objective of conducting LCA is to determine the effect of anthropogenic activities on the environment for better planning and policy inputs (Singh et al. 2010). For this work, the general methodological framework of LCA is adapted from the earlier studies (Kumar et al. 2021).

5.3 Methods

This study looks into the process of obtaining ethanol from organic fraction segregated MSW feedstock. The overall process converts MSW into ethanol by pretreatment, enzymatic hydrolysis, fermentation and recovery. The process parameters are taken from Meng et al. (2019) and tabulated in Table 5.4. The enzymatic hydrolysis process details are taken from Nielsen et al. (ISO 2006). GaBi 8.2 (2017) software and Ecoinvent 3.3 inventory databases were used for the LCA analysis. Four environmental impacts categories were selected as (i) global warming potential (GWP) quantifying in terms of CO_2 equivalents, (ii) human toxicity potential (HTP), (iii) eutrophication potential (EUP) and (iv) acidification potential (AP).

Table 5.4 Process values considered for LCA of MSW to ethanol conversion

Sr. No.	Input	Parameters	Values	Unit
1	Organic fraction MSW (Moisture content = 40%)		1	tonne
2	Mechanical shredding	Electricity	43	MJ/t
3	Enzymatic hydrolysis	Glucose slurry	1125.5	kg/t
		Electricity	0.41	MJ/t
		Heat	214.63	MJ/t
		Enzyme loading	15.90	kg/t
		OFMSW <75um	850	kg/t
		Sulfuric acid	8.59	kg/t
		Clean water usage	5366.25	kg/t
4	Fermentation	Electricity	0.684	MJ/t
		Water	4480	kg/t
		Starch	65.5	kg/t
		Yeast	2.73	kg/t
5	Distillation/Recovery	Heat	171	MJ/t
		Bioethanol recovery	44.6	kg/t
		Electricity	3.89	MJ/t

5.3.1 Scope and Functional Unit

The LCA (MSW-ethanol) model is developed in-line with the ISO Standards 14,040 and 14,044 (ISO 2006). The functional unit was identified as 1 tonne of segregated organic fraction MSW with a gate-to-gate system boundary. A schematic process flow diagram defining the process is shown in Fig. 5.3. The system boundary commences with already sorted and separated organic fraction MSW and ends with the final production of ethanol. The main product of the entire process is ethanol, while co-products (butanol, acetone and others) are assumed to allocate their impacts to the main product.

5.3.2 Waste Composition

The waste composition (only organic) used is representative of India's MSW fraction obtained from a recent report (Cpheeo 2018) consists of food (30%), paper (9%), wood (2%), plastic (7%), and cloth (3%).

Fig. 5.3 Representation of different steps for ethanol production

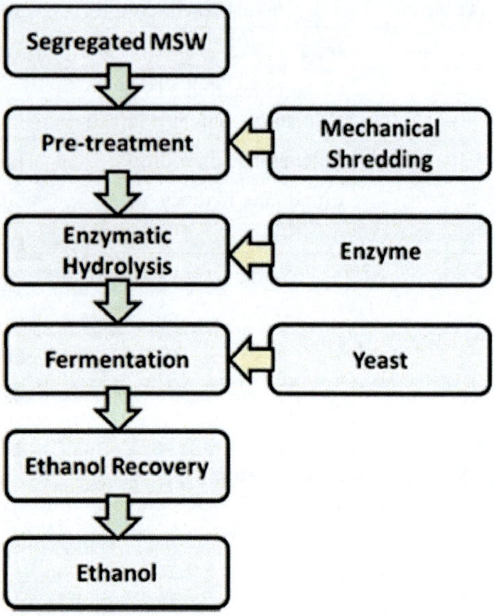

5.3.3 *Life Cycle Inventory*

Size Reduction—Mechanical shredding

The first step for ethanol conversion is to have a size reduction operation to improve the biodegradability of organic MSW (Barampouti et al. 2019). An independent study revealed particle size of < 75 μm for better hydrolysis and maximum yield (Li et al. 2012a). The energy required for size reduction is taken from the GaBi database (Fig. 5.4).

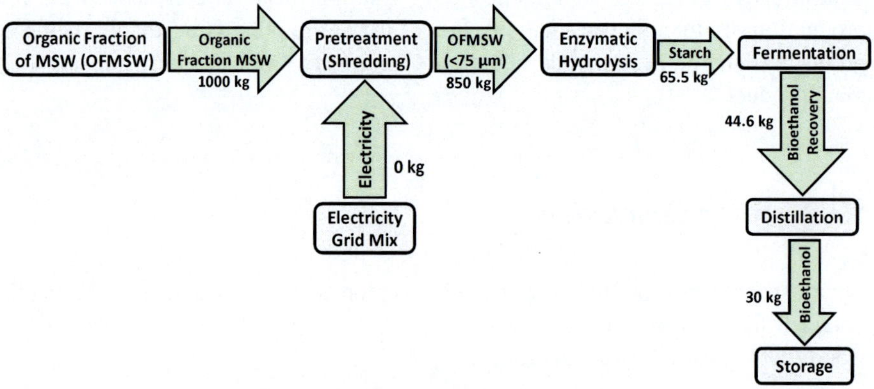

Fig. 5.4 Inputs and outputs for the process of ethanol production from MSW used for LCA

Enzymatic Hydrolysis

Enzymatic hydrolysis is a critical step in ethanol production, where complex carbohydrates are converted into simple monomers (Gupta and Verma 2015). Due to the heterogeneous nature of the organic fraction of MSW, the effect of enzymatic hydrolysis is not well investigated (Bonsu 2020). However, in the present article, an attempt has been made to replace or discuss acid hydrolysis with enzymatic hydrolysis. The selection of enzyme for the study was based on the availability in the Indian market; Novozymes Cellic CTec2 enzyme is picked because of its availability and proven performance (Meng et al. 2019). The shredded lot of MSW is loaded into the vessel, where water (20% by weight) is added to dilute the solution. The selected enzyme is introduced into the formed slurry (5% by weight) (kept in a separate tank), initiating the hydrolysis of the biogenic fibres to a fermentable sugar solution.

Fermentation

The next step after hydrolysis is fermentation, where the sugary solution is adjusted with other nutrients. The yeast extracts are selected based on the available data (Meng et al. 2019). The fermentation is conducted at a set temperature of 37 °C for 48 h (per batch). At the final step, ethanol, acetone and butanol are produced with a total yield of 44.6 kg.

Distillation/Recovery

After condensation, the acetone, butanol and ethanol are transferred to the rectification column for separation. The solution of acetone, butanol, ethanol is distilled based upon their relative volatilities into pure streams of each component.

5.4 Results and Discussion

Ethanol production acts as an important factor in terms of market value and vehicle population. This study helps frame the methodology and environmental impact associated with ethanol production chain with the help of the organic fraction of MSW. The inventory analysis results are based on CML 2001—Jan 2016 (Centrum voor Milieuwetenschappen), which strongly follows the method preferred by Ecoindicator99 developed by PRé Consultants B.V (ISO 2006). The environmental impact incorporates four categories—GWP, HTP, AP and EP based on gate-to-gate system boundaries with 1 tonne of functional unit of organic segregated MSW fraction for ethanol production. The yield of ethanol production from MSW is taken as 30 L from Barampouti et al. (2019).

Figure 5.5a–d shows the results of LCA of organic fraction MSW to ethanol production process. The assessment revealed total GWP impact with 85.03 kg CO_2 eq. which can be compared to 99.64 kg to 137.91 kg CO_2 eq. showed by Kalogo et al. (2007). The reason for such discrepancy lies in the approaches, later study being cradle to gate assessment. The total GWP (Fig. 5.5a) is the sum of electricity required

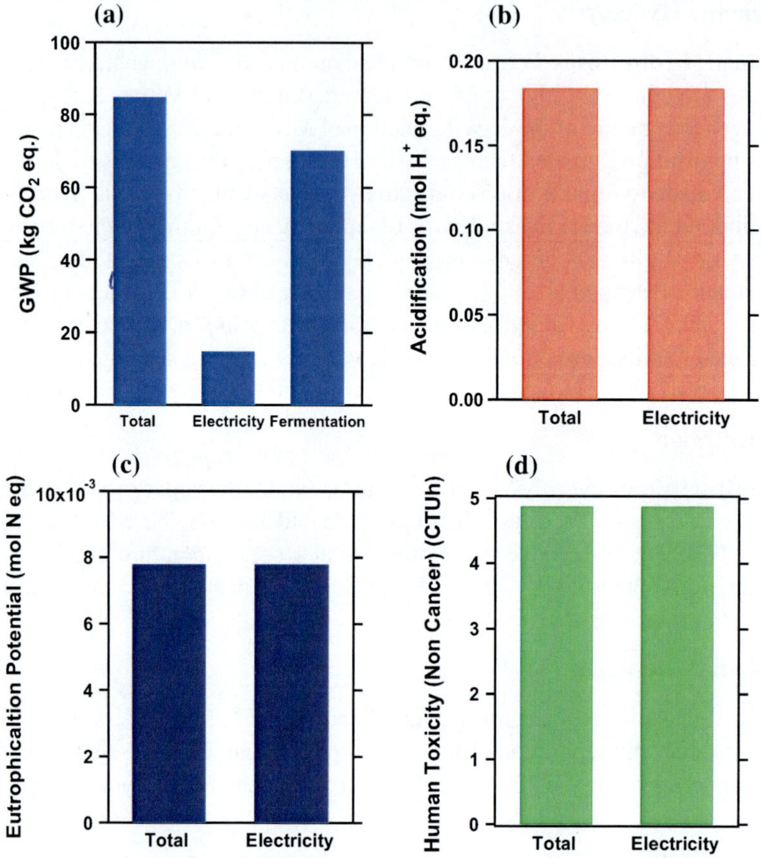

Fig. 5.5 **a** Global warming potential and **b** acidification potential **c** eutrophication potential **d** human toxicity potential of MSW to ethanol conversion

in the entire conversion process i.e., from pretreatment to distillation (14.78 kg CO_2 eq.) and footprints associated with the fermentation (70.25 kg CO_2 eq.). The high GWP associated with the fermentation process is due to (a) footprints attributed to enzyme culture (b) the duration and temperature sensitivity of the process (for the present study, the fermentation process is carried out at constant 37 °C temperature for 48 h duration) (c) the poor conversion efficiency of MSW-ethanol results in CO_2 emissions (d) addition of yeast devoid of oxygen (Singh and Bishnoi 2012; Guerfali et al. 2015).

Similarly, electricity consumption is a major factor for the high value of acidification. The AP is evaluated as 0.184 mol of H^+ eq. (Fig. 5.5b). AP represents the number of H^+ ions that can be formed theoretically to per unit mass of the released pollutant compared to SO_2 (Barnthouse 2018). Majority of the literature reported similar results with a high value of AP; impacts can be reduced by minimum application of chemicals during the process. EU, which represents the impacts on the

Table 5.5 Summary of total environmental impacts of MSW to ethanol production (values are based on a unit tonne of segregated organic MSW feedstock)

Sr. No.	Impact category	Unit	Total environmental impact
1	Global warming potential	kg CO_2 eq.	85.03
2	Acidification potential	mol of H^+ eq.	0.184
3	Eutrophication potential	mol of N eq.	7.794E−3
4	Human toxicity	CTUh	4.873

environment (terrestrial) due to over-fertilization, is already recognized for causing air pollution and forest degradation (Kim and Chae 2016). In this present study, the impact of eutrophication potential shows 7.794E−3 which is comparable to Schmitt et al. (2012). The reason for a high EU value is attributed to MSW being used as a fertilizer. Using MSW as a fertilizer leads to the accumulation of nutrients, a high concentration of which causes higher ground eutrophication. However, MSW-ethanol production comes with minimum EU impacts (Li et al. 2012b). The value of non-cancer human toxicity impact in this study is 4.873 CTUh (comparative toxicity unit for humans), which again comprises of the net electricity requirement.

The findings of the study are summarized in Table 5.5; the largest source having significant ecological impacts is the electricity consumed in the conversion process. The electricity is considered a useful inventory required for all processes like shredding of organic fraction segregated MSW, enzymatic hydrolysis process, fermentation and distillation process. The relative performance of organic fraction segregated MSW derived ethanol mainly depends on the production of enzyme and type of electricity use (Liu et al. 2019). The energy and electricity recovery may be obtained from other methods, including incineration, gasification and anaerobic digestion of MSW to offset the fossil fuel powered grid electricity footprints (Alberto et al. 2017). The onsite, decentralized production of electricity from the unutilized fraction of MSW can be useful for energy or electricity required for all processes from pretreatment to final ethanol distillation, which will directly affect or minimize the environmental impacts. The variation in the results of the present study and reported literature might be due to (i) different waste fraction (ii) shredding efficacy (iii) size of shredded fraction (iv) enzyme selection (v) approaches (like cradle to grave, well to wheel etc.).

5.5 Conclusion

The energy-security and sustainability concerns of fossil fuels are demanding smooth tuning with alternative fuels. There is a huge mismatch between the discovery rate of fossil fuels and the utilization rate, posing a fast resource depletion challenge.

The anthropogenic perturbations associated with fossil fuel extraction is at alarming levels. Transport and passenger mobility using fossil fuels add to ambient air pollution and threaten sustainable development. On the other hand, raising awareness for utilizing waste as a resource is getting appreciation. In the scenario mentioned above, a sustainable all-round approach is required to mitigate the mobility-related footprints using alternative sources, which are abundant and cheap. However, the mobility sector needs a reliable and consistent supply of alternative fuel to meet the demand. Such demand and consistent supply related tasks can be accomplished with WtE method. The waste resource recovery to power vehicles can prove a suitable option for sustainable mobility.

The present study investigates the environmental footprints of ethanol generation from MSW to support the national programmes of conventional fuel + ethanol blending. The study can be summarised as:

- The motivation of the study came from the conventional ethanol production process from lignocellulosic waste, which is complex, accompanied with unwanted by-products (weak acids, furan derivatives and phenolic compounds) (Asghar et al. 2015; Balan 2014), and has significant GHG impacts (Li et al. 2012b). Therefore, a substitute is essential which has least impacts compared to earlier processes.
- The current study presents an ethanol production method with which is less energy and resource-intensive. However, direct conversion of MSW for ethanol recovery without acknowledging the process impacts makes the process impractical for commercialization due to necessary compliance with environmental standards. The study directs to investigate the concentration of noxious pollutants (particulate matter, carbon monoxide, carbon dioxide, nitrogen dioxide, sulphur dioxide etc.) emitting from the conversion process and the safer limits suggested by local regulatory bodies to avert the cost of non-compliance in future. Therefore, to access the environmental impacts of the MSW-ethanol process, evaluation of life cycle impacts is prioritized rather than resource/yield generated.
- The findings suggest a lower yield of MSW-ethanol conversion and moderate environmental impacts. However, in an optimistic scenario, the advancement in the conversion technology with support from renewable energy sources can close the gap between high environmental footprints and low output, supporting and strengthening the national fuel blending mission.
- The LCA method in the present work has clearly indicates that the tool can be used as an important source in decision making and aligning the same for environmentally sound and sustainable strategies. Furthermore, endpoint product recovery, energy recovery, wastewater treatment and utilities are not included in the analysis, which can be considered as limitations of the study. Also, further work on other impact categories, techno-economic, waste supply, ethanol recovery facility design optimization is required.
- For a future task, reducing the uneven access of sustainable mobility can be undertaken, which is a great blockade in switching to a decarbonized fleet.

However, overcoming these barriers requires robust policy support, tax incentives, public/private infrastructure investment, positive sentiments of technology adoptions and lifestyle changes. The financing mechanism based on the production facilities, level of production (production-based incentives) together with funds collected from old vehicles' emissions-based taxes may be put forth as a solution. Furthermore, on the basis of LCA done in this study, an assessment of the benefits generated in monetary terms and environmental damage cost averted can be estimated as future task.

Acknowledgements The authors thankfully acknowledge the Council of Scientific and Industrial Research (CSIR), India and the Director, National Environmental Engineering Research Institute (NEERI) for supporting the research. Also acknowledged is funding support under the project CNP-2-2469. The manuscript is checked for plagiarism using licensed iThenticate software through CSIR-NEERI Knowledge Resource Centre wide KRC No. CSIR-NEERI/KRC/2021/JULY/ERMD-DRC/1.

References

Alberto I, Liu D, Zhao J (2017) LCA studies comparing alkaline and immobilized enzyme catalyst processes for biodiesel production under Brazilian conditions. Resour Conserv Recycling 119:117–127.https://doi.org/10.1016/j.resconrec.2016.05.009

Asghar U, Irfan M, Nadeem M (2015) Challenges in bioethanol production from lignocellulosic waste

Bajpai P (2020) Developments in bioethanol. Springer Nature

Balan V (2014) Current challenges in commercially producing biofuels from lignocellulosic biomass

Barampouti EM, Mai S, Malamis D, Moustakas K, Loizidou M (2019) Liquid biofuels from the organic fraction of municipal solid waste: a review. Renew Sustain Energy Rev 110:298–314. https://doi.org/10.1016/j.rser.2019.04.005

Barnthouse A (2018) Life-cycle impact assessment

Bonsu NO (2020) Towards a circular and low-carbon economy: insights from the transitioning to electric vehicles and net zero economy. J Clean Prod 256:120659. https://doi.org/10.1016/j.jclepro.2020.120659

Brunner PH, Rechberger H (2015) Waste to energy—key element for sustainable waste management. Waste Manag 37:3–12. https://doi.org/10.1016/j.wasman.2014.02.003

Canitez F, Alpkokin P, Kiremitci ST (2020) Sustainable urban mobility in Istanbul: challenges and prospects. Case Stud Transp Policy 8:1148–1157. https://doi.org/10.1016/j.cstp.2020.07.005

Chand Malav L, Yadav KK, Gupta N, Kumar S, Sharma GK, Krishnan S, Rezania S, Kamyab H, Pham QB, Yadav S, Bhattacharyya S, Yadav VK, Bach QV (2020) A review on municipal solid waste as a renewable source for waste-to-energy project in India: current practices, challenges, and future opportunities. J Clean Prod 277:123227. https://doi.org/10.1016/j.jclepro.2020.123227

Choudhury M, Jyethi DS, Dutta J, Purkayastha SP, Deb D, Das R, Roy G, Sen T, Bhattacharyya KG (2021) Investigation of groundwater and soil quality near to a municipal waste disposal site in Silchar, Assam, India. Int J Energy Water Resour. https://doi.org/10.1007/s42108-021-00117-5

Consonni S, Viganò F (2011) Material and energy recovery in integrated waste management systems: the potential for energy recovery. Waste Manag 31:2074–2084. https://doi.org/10.1016/j.wasman.2011.05.013

Cucchiella F, D'Adamo I, Gastaldi M (2014) Sustainable management of waste-to-energy facilities. Renew Sustain Energy Rev 33:719–728. https://doi.org/10.1016/j.rser.2014.02.015

Das S, Lee SH, Kumar P, Kim KH, Lee SS, Bhattacharya SS (2019) Solid waste management: scope and the challenge of sustainability. J Clean Prod 228:658–678. https://doi.org/10.1016/j.jclepro.2019.04.323

Duan Z, Scheutz C, Kjeldsen P (2021) Trace gas emissions from municipal solid waste landfills: a review. Waste Manag 119:39–62. https://doi.org/10.1016/j.wasman.2020.09.015

Farmanbordar S, Karimi K, Amiri H (2018) Municipal solid waste as a suitable substrate for butanol production as an advanced biofuel. Energy Convers Manag 157:396–408. https://doi.org/10.1016/j.enconman.2017.12.020

Fazeli A, Bakhtvar F, Jahanshaloo L, Che Sidik NA, Bayat AE (2016) Malaysia's stand on municipal solid waste conversion to energy: a review. Renew Sustain Energy Rev 58:1007–1016. https://doi.org/10.1016/j.rser.2015.12.270

Feo GD, Gisi SD, Williams ID (2013) Public perception of odour and environmental pollution attributed to MSW treatment and disposal facilities: a case study. Waste Manag 33:974–987. https://doi.org/10.1016/j.wasman.2012.12.016

Gandoman FH, Ahmadi A, Van den Bossche P, Van Mierlo J, Omar N, Nezhad AE, Mavalizadeh H, Mayet C (2019) Status and future perspectives of reliability assessment for electric vehicles. Reliab Eng Syst Saf 183:1–16.https://doi.org/10.1016/j.ress.2018.11.013

Guerfali M, Saidi A, Gargouri A (2015) Enhanced enzymatic hydrolysis of waste paper for ethanol production using separate saccharification and fermentation. 25–42. https://doi.org/10.1007/s12010-014-1243-1

Gupta A, Verma JP (2015) Sustainable bio-ethanol production from agro-residues: a review. Renew Sustain Energy Rev 41:550–567. https://doi.org/10.1016/j.rser.2014.08.032

Hidrue MK, Parsons GR, Kempton W, Gardner MP (2011) Willingness to pay for electric vehicles and their attributes. Resour Energy Econ 33:686–705. https://doi.org/10.1016/j.reseneeco.2011.02.002

Holden E, Banister D, Gössling S, Gilpin G, Linnerud K (2020) Grand narratives for sustainable mobility: a conceptual review. Energy Res Soc Sci 65:101454. https://doi.org/10.1016/j.erss.2020.101454

Hosseini SE, Butler B (2020) An overview of development and challenges in hydrogen powered vehicles. Int J Green Energy 17:13–37. https://doi.org/10.1080/15435075.2019.1685999

CPCB (2021) waste generation composition. https://cpcb.nic.in/uploads/MSW/Waste_generation_Composition.pdf. Last accessed 10 May 2021

Cpheeo (2018) Guidelines on usage of refuse derived fuel in various industries, Ministry of Housing and Urban Affairs

https://www.iea.org/data-and-statistics/charts/biofuel-energy-consumption-and-transport-energy-demand-shares-in-selected-countries-2016. Last accessed 2 May 2021

https://www.iea.org/data-and-statistics/charts/biofuel-production-growth-in-key-markets-2019-2024. Last accessed 2 May 2021

IEA (2020) Tracking transport 2020

ISO (2006) ISO: Environmental management—life cycle assessment—principles and framework (ISO 14040). ISO, The International Organization for Standardization, Geneva

Jamas T, Nepal R (2010) Issues and options in waste management: a social cost-benefit analysis of waste-to-energy in the UK. Resour Conserv Recycl 54:1341–1352. https://doi.org/10.1016/j.resconrec.2010.05.004

Kalogo Y, Habibi S, Maclean HL, Joshi SV (2007) Environmental implications of municipal solid waste-derived ethanol. Environ Sci Technol 41:35–41. https://doi.org/10.1021/es061117b

Kalyani KA, Pandey KK (2014) Waste to energy status in India: a short review. Renew Sustain Energy Rev 31:113–120. https://doi.org/10.1016/j.rser.2013.11.020

Kaza S, Yao L, Bhada-Tata P, Van Woerden F (2018) What a waste 2.0: a global snapshot of solid waste management to 2050. World Bank Publications

Kim TH, Chae CU (2016) Environmental impact analysis of acidification and eutrophication due to emissions from the production of concrete. Sustain 8:1–20. https://doi.org/10.3390/su8060578

Kumar A, Agrawal A (2020) Recent trends in solid waste management status, challenges, and potential for the future Indian cities—a review. Curr Res Environ Sustain 2:100011. https://doi.org/10.1016/j.crsust.2020.100011

Kumar A, Samadder SR (2017) A review on technological options of waste to energy for effective management of municipal solid waste. Waste Manag 69:407–422. https://doi.org/10.1016/j.wasman.2017.08.046

Kumar R, Jha A, Damodaran A, Bangwal D, Dwivedi A (2020) Addressing the challenges to electric vehicle adoption via sharing economy: an Indian perspective. Manag Environ Qual Int J 32:82–99. https://doi.org/10.1108/MEQ-03-2020-0058

Kumar H, Azad A, Gupta A, Sharma J, Bherwani H, Labhsetwar NK, Kumar R (2021) COVID-19 Creating another problem? Sustainable solution for PPE disposal through LCA approach. Environ Dev Sustain 23:9418–9432. https://doi.org/10.1007/s10668-020-01033-0

Lapuerta M, Armas O, Herreros JM (2008) Emissions from a diesel-bioethanol blend in an automotive diesel engine. Fuel 87:25–31. https://doi.org/10.1016/j.fuel.2007.04.007

Li S, Zhang X, Andresen JM (2012a) Production of fermentable sugars from enzymatic hydrolysis of pretreated municipal solid waste after autoclave process. Fuel 92:84–88. https://doi.org/10.1016/j.fuel.2011.07.012

Li A, Mcmanus MC, Hammond GP (2012b) Environmental life cycle assessment of lignocellulosic conversion to ethanol: a review. Renew Sustain Energy Rev 16:4638–4650. https://doi.org/10.1016/j.rser.2012.04.016

Liu Y, Kong F, Santibanez Gonzalez EDR (2017) Dumping, waste management and ecological security: evidence from England. J Clean Prod 167:1425–1437. https://doi.org/10.1016/j.jclepro.2016.12.097

Liu CG, Xiao Y, Xia XX, Zhao XQ, Peng L, Srinophakun P, Bai FW (2019) Cellulosic ethanol production: progress, challenges and strategies for solutions. Biotechnol Adv 37:491–504. https://doi.org/10.1016/j.biotechadv.2019.03.002

Looney B (2020) Statistical review of world energy, BP

Mahmoodi P, Karimi K, Taherzadeh MJ (2018) Efficient conversion of municipal solid waste to biofuel by simultaneous dilute-acid hydrolysis of starch and pretreatment of lignocelluloses. Energy Convers Manag 166:569–578. https://doi.org/10.1016/j.enconman.2018.04.067

Makarichi L, Jutidamrongphan W, Techato KA (2018) The evolution of waste-to-energy incineration: a review. Renew Sustain Energy Rev 91:812–821. https://doi.org/10.1016/j.rser.2018.04.088

Meng F, Ibbett R, de Vrije T, Metcalf P, Tucker G, McKechnie J (2019) Process simulation and life cycle assessment of converting autoclaved municipal solid waste into butanol and ethanol as transport fuels. Waste Manag 89:177–189. https://doi.org/10.1016/j.wasman.2019.04.003

Mori D, Hirose K (2009) Recent challenges of hydrogen storage technologies for fuel cell vehicles. Int J Hydrogen Energy 34:4569–4574. https://doi.org/10.1016/j.ijhydene.2008.07.115

Nanaki EA, Koroneos CJ (2012) Comparative LCA of the use of biodiesel, diesel and gasoline for transportation. J Clean Prod 20:14–19. https://doi.org/10.1016/j.jclepro.2011.07.026

Nanda S, Berruti F (2021) Municipal solid waste management and landfilling technologies: a review. Environ Chem Lett 19:1433–1456. https://doi.org/10.1007/s10311-020-01100-y

NITI Aayog (2017) India leaps ahead: transformative mobility solutions for all

NITI-Aayog (2020) Ethanol blending in India 2020–25

Noussan M, Hafner M, Tagliapietra S (2020) The evolution of transport across world regions. In: The future of transport between digitalization and decarbonization: trends, strategies and effects on energy consumption. Springer International Publishing, Cham, pp 1–28

Okedere OB, Olalekan AP, Fakinle BS, Elehinafe FB, Odunlami OA, Sonibare JA (2019) Urban air pollution from the open burning of municipal solid waste. Environ Qual Manag:67–74. https://doi.org/10.1002/tqem.21633

Pacheco R, Silva C (2019) Global warming potential of biomass-to-ethanol: review and sensitivity analysis through a case study. Energies 12. https://doi.org/10.3390/en12132535

Pavlas M, Tou M, Bébar L, Stehlík P (2010) Waste to energy—an evaluation of the environmental impact. Appl Therm Eng 30:2326–2332. https://doi.org/10.1016/j.applthermaleng.2009.10.019

Proost S, Van Dender K (2012) Energy and environment challenges in the transport sector. Econ Transp 1:77–87. https://doi.org/10.1016/j.ecotra.2012.11.001

Psomopoulos CS, Themelis NJ (2015) Environmental effects the combustion of as-received and pre-processed (RDF/SRF) municipal solid wastes as fuel for the power sector. Energy Sources, Part A Recover Util Environ Eff 37:1813–1820. https://doi.org/10.1080/15567036.2011.639845

Psomopoulos CS, Venetis I, Themelis NJ (2014) The impact from the implementation of "Waste to Energy" to the economy: a macroeconomic approach for the trade balance of Greece. Fresenius Environ Bull 23:2735–2741

Rodić L, Wilson DC (2017) Resolving governance issues to achieve priority sustainable development goals related to solid waste management in developing countries. Sustain 9. https://doi.org/10.3390/su9030404

Sangare M, Gupta S, Bouzefrane S, Banerjee S, Muhlethaler P (2021) Exploring the forecasting approach for road accidents: analytical measures with hybrid machine learning. Expert Syst Appl 167:113855. https://doi.org/10.1016/j.eswa.2020.113855

Sato FEK, Nakata T (2020) Energy consumption analysis for vehicle production through a material flow approach. Energies 13. https://doi.org/10.3390/en13092396

Schmitt E, Bura R, Gustafson R, Cooper J, Vajzovic A (2012) Converting lignocellulosic solid waste into ethanol for the State of Washington: an investigation of treatment technologies and environmental impacts. Bioresour Technol 104:400–409. https://doi.org/10.1016/j.biortech.2011.10.094

Sharholy M, Ahmad K, Mahmood G, Trivedi RC (2008) Municipal solid waste management in Indian cities—a review. Waste Manag 28:459–467. https://doi.org/10.1016/j.wasman.2007.02.008

Sierzchula W, Bakker S, Maat K, Van Wee B (2014) The influence of financial incentives and other socio-economic factors on electric vehicle adoption. Energy Policy 68:183–194. https://doi.org/10.1016/j.enpol.2014.01.043

Singh A, Bishnoi NR (2012) Bioresource technology enzymatic hydrolysis optimization of microwave alkali pretreated wheat straw and ethanol production by yeast. Bioresour Technol 108:94–101. https://doi.org/10.1016/j.biortech.2011.12.084

Singh A, Pant D, Korres NE, Nizami AS, Prasad S, Murphy JD (2010) Key issues in life cycle assessment of ethanol production from lignocellulosic biomass: challenges and perspectives. Bioresour Technol 101:5003–5012. https://doi.org/10.1016/j.biortech.2009.11.062

Stehlík P (2009) Contribution to advances in waste-to-energy technologies. J Clean Prod 17:919–931. https://doi.org/10.1016/j.jclepro.2009.02.011

Tuite AR, Bhatia D, Moineddin R, Bogoch II, Watts AG, Khan K (2020) Global trends in air travel: implications for connectivity and resilience to infectious disease threats. J Travel Med 27:1–8. https://doi.org/10.1093/JTM/TAAA070

Usmani Z, Kumar V, Varjani S, Gupta P, Rani R, Chandra A (2020) Municipal solid waste to clean energy system: a contribution toward sustainable development. Elsevier

Van den Berg R, De Langen PW (2017) Environmental sustainability in container transport: the attitudes of shippers and forwarders. Int J Logist Res Appl 20:146–162. https://doi.org/10.1080/13675567.2016.1164838

Vergragt PJ, Brown HS (2007) Sustainable mobility: from technological innovation to societal learning. J Clean Prod 15:1104–1115. https://doi.org/10.1016/j.jclepro.2006.05.020

Verlinghieri E, Schwanen T (2020) Transport and mobility justice: evolving discussions. J Transp Geogr 87. https://doi.org/10.1016/j.jtrangeo.2020.102798

Vinti G, Bauza V, Clasen T, Medlicott K, Tudor T, Zurbrügg C, Vaccari M (2021) Municipal solid waste management and adverse health outcomes: a systematic review. 1–26

Vongdala N, Tran HD, Xuan TD, Teschke R, Khanh TD (2019) Heavy metal accumulation in water, soil, and plants of municipal solid waste landfill in Vientiane, Laos. Int J Environ Res Public Health 16:1–13. https://doi.org/10.3390/ijerph16010022

Xiong R, Sun W, Yu Q, Sun F (2020) Research progress, challenges and prospects of fault diagnosis on battery system of electric vehicles. Appl Energy 279:115855. https://doi.org/10.1016/j.apenergy.2020.115855

Zhao X, Ke Y, Zuo J, Xiong W, Wu P (2020) Evaluation of sustainable transport research in 2000–2019. J Clean Prod 256:120404. https://doi.org/10.1016/j.jclepro.2020.120404

Zhou J (2012) Sustainable transportation in the US: a review of proposals, policies, and programs since 2000. Front Archit Res 1:150–165. https://doi.org/10.1016/j.foar.2012.02.012

Chapter 6
Bioethanol from Wastes for Mobility: Europe on the Road to Sustainability

Mariana S. T. Amândio, Jorge M. S. Rocha, Luísa S. Serafim, and Ana M. R. B. Xavier

Abbreviations

ACEA	European Automobile Manufacturers' Association
AEBIOM	Bioenergy Europe
BtL	Biomass-to-liquids
EIA	Energy Information Administration
ePure	European Renewable Ethanol
ETBE	Ethyl tertiary-butyl ether
EU	European Union
FAME	Fatty acid methyl ester
FFVs	Flexi-fuel vehicles
GHG	Greenhouse gas
HEFA	Hydrotreated esters and fatty acids
HVO	Hydrotreated vegetable oil
IEA	International Energy Agency
ILUC	Indirect land-use change
IRENA	International Renewable Energy Agency
RED-II	Renewable Energy Directive 2018/2001
RISE	Research Institutes of Sweden
SSL	Spent sulphite liquor
Toe	Tonne of oil equivalent

M. S. T. Amândio · L. S. Serafim · A. M. R. B. Xavier (✉)
CICECO—Aveiro Institute of Materials, Department of Chemistry, University of Aveiro, Campus Universitário de Santiago, 3810-193 Aveiro, Portugal
e-mail: abx@ua.pt

M. S. T. Amândio · J. M. S. Rocha
CIEPQPF, Department of Chemical Engineering, Faculty of Sciences and Technology, University of Coimbra, Pólo II, Rua Sílvio Lima, 3030-790 Coimbra, Portugal

G. Di Blasio et al. (eds.), *Clean Fuels for Mobility*, Energy, Environment, and Sustainability, https://doi.org/10.1007/978-981-16-8747-1_6

6.1 Introduction

Presently, the European energy sector is facing a deep transition, with governments pushing for a switch from fossil to bio-based fuels, targeting sustainable mobility (Panoutsou et al. 2021; Ebadian et al. 2020). The transport sector in the European Union (EU) accounted for about one-third of the total energy consumption and is still one of the most dependent on fossil fuels (Danielis et al. 2021; European Commission Statistics—Eurostat 2020). Greenhouse gas (GHG) emissions increased in the transport sector over the last years, contrary to the declining trend that has been observed in the remaining sectors (Danielis et al. 2021; European Environment Agency 2020). Emissions derived from the transport sector represented about 27% of total EU-28 GHG emissions in 2017, of which road transport accounted for almost 72% (Danielis et al. 2021; European Environment Agency 2020). The projections for 2030 pointed out that GHG emissions will account for about 17% from 1990 levels. Therefore, given the transport sector's impact, the EU has directed efforts to increase the share of renewable energy as a key to a successful low-carbon transition (Danielis et al. 2021; Dahal et al. 2018). Moreover, intensive work has been dedicated to developing innovative and highly efficient combustion technologies that can reduce emissions combined with alternative fuels (Ianniello et al. 2021).

Advanced biofuels are a promising solution for the short-medium term, despite the strong focus of the European framework for decarbonising the transport sector on electrification (Panoutsou et al. 2021; República Portuguesa, Fundo Ambiental, Agência Portuguesa do Ambiente 2019). Biofuels can rapidly improve energy security and accelerate the replacement of fossil fuels by taking advantage of existing fuel distribution infrastructures with minimal investment costs (Panoutsou et al. 2021; IRENA 2019). The RED II Directive changed the paradigm by introducing renewables to all forms of transport (road, rail, maritime and aviation) by strengthening energy efficiency and strongly encouraging a sharp fall in GHG emissions. Furthermore, the focus on non-food competing feedstock, namely wastes, was reinforced (Lászlok et al. 2020; Cadillo-Benalcazar et al. 2021; Paris et al. 2021).

Within the EU, about 225 million tons of wastes per year are available to be converted into value-added products within the biorefinery concept. These wastes are mainly agricultural residues (54%), municipal waste (28%), and forestry residues (18%). The identification, quantification, and characterisation of these residues is crucial for a successful switch towards a bioeconomy. All of these wastes could be a promising feedstock for cellulosic ethanol production (Romaní et al. 2018).

6.2 Biofuels: Current and Prospective Status

6.2.1 The Biofuel Concept

Biofuel refers to any liquid and gaseous fuel derived from biomass. Biomass is any type of renewable organic matter, undoubtedly recognized as the most important renewable energy source (IEA 2011).

Conventional biofuels, most of them classified as first-generation, involve well-established technologies with industrial production. This category includes ethanol and biodiesel. The most used feedstocks for bioethanol production include corn, sugarcane, sugar beet, and sweet sorghum. For biodiesel, animal fats and vegetable oils derived from soybean, canola, oil palm, or sunflower are some examples of the typical raw materials. However, the extensive use of first-generation raw materials leads to food-fuel competition, negatively impacting food prices. Also, some environmental problems may arise, namely food insecurity, soil degradation, water scarcity, among others (Bin et al. 2021; Zabed et al. 2017).

Recently, the EU established that advanced biofuels are liquid or gaseous biofuels derived from raw materials that do not compete with food or feed crops and must comply with minimum GHG emissions (Panoutsou et al. 2021). The detailed list published in Part A of Annex IX of the Renewable Energy Directive 2018/2001 (commonly referred to as RED-II) includes algae, animal manure, forestry residues, agricultural residues (cereal straw, cobs, husks, nutshells, grape marc), industrial and municipal solid waste, and so on (Espí et al. 2020). The feedstocks used in this category are generally abundant, worldwide available, low-cost, and do not contribute to the competition with the food chain (Arevalo-Gallegos et al. 2017; Tian et al. 2018). The biofuels of this category are commonly classified as second- or third-generation (Espí et al. 2020). Cellulosic ethanol production requires typically four main stages: pretreatment, hydrolysis, fermentation and recovery. An additional detoxification step is often recommended. Advanced biodiesel includes hydrotreated vegetable oil (HVO) and biomass-to-liquids (BtL) diesel. Various new conversion routes have been researched in the last decade, namely synthetic fuels production (IEA 2011).

6.2.2 Biofuels as Main Driving Forces for a Bio-based Economy

Biofuels have emerged as a renewable and sustainable alternative to meet the rising global energy consumption due to population growth besides the health and environmental impacts of fossil fuels utilisation and their finite nature (EIA Independent Analysis and Statistics 2018; Jansen et al. 2017; Zabed et al. 2016). Compared to fossil fuels, biofuels have many advantages, including reducing GHG emissions and, consequently, atmospheric pollution. Moreover, biofuels are biodegradable and have low toxicity (Sarris and Papanikolaou 2016). In addition, they can reduce the external energy dependence of fossil fuels importing countries, providing them independence and security of energy supply. In fact, unlike fossil fuels, renewable resources typically used are available worldwide (Chandel et al. 2019; Jansen et al. 2017; Nigam and Singh 2011).

Climate change mitigation, energy security, and rural growth are the fundamental principles of guiding policies related to biofuels (Linares and Pérez-Arriaga 2013). Biofuels, particularly advanced biofuels, can significantly reduce GHG emissions (above 80%) compared to fossil alternatives. Moreover, considering the high dependence on fossil fuel imports, particularly for the transport sector, it is crucial to ensure the security of energy supply, promoting some energy self-sufficiency (Londo et al. 2010). Biofuels can also support the agriculture sector by job creation and boost competitiveness and innovation through a new economic model (Callegari et al. 2020; Londo et al. 2010).

Advanced biofuels are considered the most promising and ready-available alternative to allow the transport sector to meet the GHG emissions target (REN21 2020). Electrification and, even more, hydrogen depend on the investment level in restructuring the automotive industry and creating infrastructures along the value chain (production, storage, and supply) (Danielis et al. 2021; REN21 2020). Electrification is only a sustainable alternative if electricity derives from renewable resources (REN21 2020). Otherwise, no positive contribution to GHG savings can be observed. Furthermore, an efficient value chain is required, particularly for the lithium and other rare minerals extraction for batteries manufacturing (Danielis et al. 2021). Finally, hydrogen is still challenging due to technology maturity, costs, and safety issues (Panoutsou et al. 2021).

6.2.3 The Role of Biofuels in the Transport Sector

Currently, biofuels have the most important renewable energy contribution to the transport sector, representing more than 90%. However, biofuels have a minimal share of 3% regarding global transport energy (REN21 2020). Biodiesel and bioethanol are the biofuels with the highest market share on road transport (Puricelli et al. 2021), but the production of HVO also has been growing over the last years (EurObserv'ER 2020; Puricelli et al. 2021).

Biodiesel

Fatty acid methyl esters (FAME), typically designated as conventional biodiesel, are derived from lipid raw materials, namely vegetable oils, animal fats and even waste cooking oils (Puricelli et al. 2021). Several technologies are available for producing biofuels from vegetable oil, namely transesterification, fermentation, thermal cracking, catalytic cracking, and hydroprocessing (Zhao et al. 2017). For the production of biodiesel, transesterification is the most widely used process. Esters and glycerol are produced through the transesterification reaction between triglycerides and alcohol, commonly in the presence of a catalyst (Bin et al. 2021; Singh et al. 2020).

Biodiesel brings environmental benefits since it results in lower carbon dioxide and particulate emissions than fossil-based alternatives (Callegari et al. 2020). According to some estimations, biodiesel can reduce CO_2 emissions by up to 78% compared to diesel (Mishra and Goswami 2018). In contrast, nitrogen oxides and particulate matter emissions increase due to incomplete combustion (Bin et al. 2021). Moreover, the lower energy content of biodiesel leads to a rise in fuel consumption of up to 10% (Mishra and Goswami 2018). Another issue regarding biodiesel production on an industrial scale is the production cost, mainly dependent on feedstock and catalyst selection (Callegari et al. 2020).

The fraction of biodiesel blended with conventional diesel varied typically from 7 to 10% (v/v) (B7-B10). Engine modifications are required for higher blend percentages (Puricelli et al. 2021).

More detailed information about the production of biodiesel, its properties, advantages and drawbacks was revised by several authors, namely Agarwal et al. (2017), Bozbas (2008), Knothe and Razon (2017), Kralova and Sjöblom (2010), Mishra and Goswami (2018), Mohiddin et al. (2021) and Singh et al. (2020).

Hydrotreated vegetable oil (HVO)

HVO is a paraffinic diesel derived from hydroprocessing, typically known as green diesel (Puricelli et al. 2021). Hydroprocessing consists of several catalytic reactions in the presence of hydrogen to convert unsaturated compounds into paraffins and cycloalkanes. The conversion involves two main stages: hydrotreatment and hydrocracking/isomerization (Karatzos et al. 2014). A wide range of waste fats and oils can be used as raw material, with no consequences on the quality of the final product (Aatola et al. 2009).

HVO is free of sulphur, aromatics and oxygen (Puricelli et al. 2021). Moreover, it presents some advantages over FAME: better storage stability, no deposit formation, and decreased nitrogen oxide emissions. HVO is recognized as "drop-in fuel" and can be used directly in diesel vehicles engines due to similarities with fossil fuel in terms of properties. Furthermore, it is a promising alternative fuel for the aviation sector (Mittelbach 2015; Panoutsou et al. 2021).

The production process of HVO and its properties is described in more detail by Knothe (2010), Lapuerta et al. (2011), Mittelbach (2015), and Vásquez et al. (2017).

Bioethanol

Bioethanol is a liquid fuel typically obtained through the alcoholic fermentation of sugars (Puricelli et al. 2021; Zabed et al. 2017). It presents several advantages over gasoline since the higher octane number enhances the combustion performance and leads to a lower engine knock (Zabed et al. 2017). Unlike gasoline, ethanol is an oxygenated fuel, leading to about 15% higher combustion efficiency and lower emissions (Callegari et al. 2020; Zabed et al. 2017).

One of the drawbacks of bioethanol over gasoline is the lower vapour pressure, which makes it challenging to start a cold engine. Moreover, one litre of ethanol resulted only in about two-thirds of the energy provided by the same amount of gasoline, leading to an increase in fuel consumption (Zabed et al. 2017).

In Europe, the blend with conventional gasoline is restricted to 10% (v/v) by EN 228 quality standard (E10), which may be limiting the expansion of the bioethanol market. Flexi-fuel vehicles (FFVs) can run with ethanol up to 85% (E85). However, in contrast to some bioethanol producing leader countries, such as Brazil, this category accounts for an insignificant percentage of the EU passenger car fleet (Puricelli et al. 2021).

More details about production processes and properties of bioethanol were revised by some authors, namely Aditiya et al. (2016), Balat et al. (2008), Branco et al. (2019), and Manzetti and Andersen (2015).

6.2.4 EU Policies and Directives on Biofuels

Over the last years, the EU has been focused on promoting an energy transition from fossil to renewable resources through new legislation (Panoutsou et al. 2021; Romaní et al. 2018). The switch to a low-carbon economy and climate change mitigation have been the two main drivers for developing policies related to incorporating biofuels in EU mobility (Ebadian et al. 2020).

The Renewable Energy Directive 2009/28/EC (RED I) established a goal of 20% for total energy consumption by 2020, of which 10% was targeted for transportation (Lászlok et al. 2020; Panoutsou et al. 2021). The directive was updated in 2015 by the EU Indirect Land Use Change (ILUC) 2015/1513, which restricted the contribution of conventional biofuels to 7% (Lászlok et al. 2020; Panoutsou et al. 2021; Romaní et al. 2019). Moreover, a non-binding 0.5% sub-target for advanced biofuels was defined in 2020 (Chiaramonti et al. 2021; Panoutsou et al. 2021). These policies boosted research on advanced biofuels production from agro-industrial residues, wastes, and by-products (Buchspies et al. 2020; Romaní et al. 2019). In 2018, the revised Renewable Energy Directive 2018/2001—commonly referred to as RED II—targeted a quota of 32% of renewable energy by 2030, while the transport sector should achieve the incorporation of 14% minimum (Cadillo-Benalcazar et al. 2021; Lászlok et al. 2020). Furthermore, RED II established that advanced biofuels and biogas should account for at least 0.2% in 2022, 1% in 2025 and 3.5% in 2030 of total energy consumption in the transport sector (Cadillo-Benalcazar et al. 2021; Paris et al. 2021). New sustainability criteria and GHG savings targets were also introduced (EurObserv'ER 2020).

The utilisation of biofuels in the EU differs depending on the adopted national strategy of each Member State to comply with the goals of integrating renewable energy in the mobility sector. Currently, the maximum ethanol content in gasoline is 10% (E10 fuel) to fulfil the quality standard EN 228. This standard ensures the compatibility of these fuels without requirements for engine modifications (European

Renewable Ethanol (ePure) 2018; Panoutsou et al. 2021). While world markets are shifting to higher ethanol content, the EU strives to introduce E10 fuel in all Member States. About 75% of the EU gasoline market already contains up to 5% ethanol (E5 fuel). However, the incorporation of up to 10% accounted for a minimum fraction of 9.5% in 2016 (European Renewable Ethanol (ePure) 2018). Benefits from using higher ethanol blends include GHG reduction and air quality improvement (European Renewable Ethanol (ePure) 2018). However, some investments are required to foster access to infrastructure for high ethanol blends, such as E85 fuel for compatible engines and ED95 (95% ethanol, 5% ignition improvers) for buses and trucks (Panoutsou et al. 2021; REN21 2020). Brazil is a well-known example, where slightly modified FFVs are allowed using higher blend levels, usually up to 85% ethanol (REN21 2020). This technology was first introduced in 2003, and since 2006 FFVs have dominated Brazil's new-vehicle market (Brito et al. 2019; Santos et al. 2018).

Besides the directives, the European Commission approved "The European Green Deal" in 2019, targeting climate neutrality by 2050 (Chiaramonti et al. 2021). To comply with the ambitious goals is crucial to rethink existing and planned legislation on energy, environment, and transport. However, policies promoting the incorporation of renewable energy in the mobility sector are mainly focused on road transport (Chiaramonti et al. 2021; Ebadian et al. 2020), with no commitments set in RED II for long-distance transport, namely aviation and shipping (Chiaramonti et al. 2021). In 2018, biofuels had a negligible contribution of 0.1% on aviation fuel (REN21 2020). Even though some Member States have started to propose some goals in this respect (Chiaramonti et al. 2021).

According to the description above, blending mandates evidently continue to be the main instrument used to support the incorporation of biofuels in the transport sector. However, although blending mandates contributed to establishing the biofuels market in many countries, they are not strong enough to ensure long-term growth. Some reasons pointed out include the lack of feedstock, high feedstock costs due to competing uses, lack of infrastructures, food security and sustainability concerns (Ebadian et al. 2020).

Fuel excise tax reductions, exemptions and incentives were successfully implemented, directed primarily to support biofuels suppliers and reduce the cost gap of advanced over fossil fuels (Ebadian et al. 2020; Panoutsou et al. 2021). However, these incentives are generally reduced or even abolished as biofuels technologies become more mature (Ebadian et al. 2020).

In the particular case of Portugal, its long-term strategy for energy transition targeting carbon neutrality for 2050, a commitment that is in line with the Paris Agreement, was already published (República Portuguesa, Fundo Ambiental, Agência Portuguesa do Ambiente 2019). Road transport accounts for about 96% of transport emissions, from which cars represent 60%. One of the main issues is related to low usage efficiency, even below the European average. Private vehicles are typically occupied by 1.2 passengers, while the average public transport occupancy ranges between 17 and 24%. Therefore, some key actions for the decarbonisation of the mobility sector include: greater efficiency and reinforcement of public transport systems; support shared mobility and autonomous vehicles; use of biofuels as a complement of electrification, which will dominate the transport sector until 2050. Also, hydrogen technology will be a strong bet (República Portuguesa, Fundo Ambiental, Agência Portuguesa do Ambiente 2019). Advanced biofuels will be particularly important in the heavy long-distance and aviation sector (República Portuguesa, Fundo Ambiental, Agência Portuguesa do Ambiente 2019).

Despite the driving forces supporting the production and incorporation of biofuels aforementioned, Ebadian et al. (2020) presented some concerns hindering the success of biofuels policies: the low fossil fuels prices; issues regarding food security, ILUC, and sustainability; uncertainties about funding programs and future policies. Nonetheless, sustainability criteria are progressively being incorporated, aiming GHG emissions reduction goals and pushing the production and use of advanced biofuels. Future guidelines should boost the transition to the low-carbon economy, focusing on the decarbonisation roadmap, leading to more stable markets, mainly in sectors such as aviation and navy, where the share of biofuels remains negligible (Ebadian et al. 2020; Panoutsou et al. 2021). Panoutsou et al. (2021) reviewed the challenges, current policy gaps and future interventions that should comply with the Sustainable Development Goals defined by United Nations to be achieved by 2030. These policies were categorised for the several stages involved along the value chain, namely feedstock production, conversion pathways and end-use (Panoutsou et al. 2021).

6.2.5 The Evolution of the Biofuels Market

The worldwide production of liquid biofuels grew around 5% in 2019, achieving 161 billion litres. The leading biofuel-producers included United States (41%), Brazil (26%), Indonesia (4.5%), China (2.9%), and Germany (2.8%) (REN21 2020).

In 2019, bioethanol accounted for about 59% of global biofuel production, biodiesel for 35%, and HVO and hydrotreated esters and fatty acids (HEFA) for 6%. Despite efforts made, the production of other biofuels, including biomethane and advanced biofuels, represented less than 1% (REN21 2020).

The European biofuels market is mainly based on bioethanol and biodiesel (Espí et al. 2020). Bioethanol is essentially produced from cereal grains and sugar beet within the EU context. Corn is primarily used in central Europe (particularly in Hungary), The Netherlands and Spain. Wheat is dominating in northern Europe, namely Germany and France. Sugar beet is also used in France, Germany, Czechia and Belgium (Espí et al. 2020; Flach et al. 2020). The feedstock used for bioethanol is highly dependent on the volatility of prices (Kapustová et al. 2020). In the EU, 82.3% of the renewable ethanol produced is employed as fuel (European Renewable Ethanol (ePure) 2020). Regarding biodiesel production, rapeseed oil is still the most used raw material, accounting for about 43% of total output in 2019 (Flach et al. 2020). However, its share has been progressively diminishing due to the intensive use of palm oil, recycled vegetable oil and used cooking oil (Espí et al. 2020; Flach et al. 2020; Kapustová et al. 2020).

According to the EurObserv'Er 2020 data, biodiesel consumption represented 81.2% of biofuels consumption for transport in EU-27 in 2019 (13,000.6 ktoe), whereas bioethanol accounted for 17.2% (2,761.9 ktoe) and the remainder was biogas, 1.6% (255.7 ktoe) (EurObserv'ER 2020). Overall biofuel consumption by EU-27 increased by 4.2% between 2018 and 2019. Growth was primarily driven by the biodiesel sector, namely by FAME and HVO fuels. Several EU countries implemented legal increases in their biofuel incorporation rates. Figure 6.1 shows the consumption of bioethanol, biodiesel and biogas for transport, in ktoe, for each state member of EU-27 in 2019 according to data published in EurObserv'ER 2020 (EurObserv'ER 2020). The top-five of EU-27 biofuels consumption leaders in 2019 were France (20.0%), Germany (17.1%), Spain (10.7%), Sweden (9.5%) and Italy (8.2%). Biodiesel consumption data refers to FAME and HVO, while bioethanol refers to bioethanol directly blended with gasoline or previously converted into ethyl tertiary-butyl ether (ETBE, a blend of bioethanol with an oil refining by-product isobutene) (EurObserv'ER 2020).

From the EU market point of view, there are two primary motivations for supporting biodiesel production. First, the EU is a diesel importer and gasoline exporter (Callegari et al. 2020). Second, data from 2019 shows that most light commercial vehicles circulating across the EU were still dependent on diesel (89.5%) (ACEA 2021). However, this trend has been reversed in recent years, and the diesel market share has declined (ACEA 2021). According to "The automobile industry pocket guide 2020/2021", more than half of all new passenger cars in 2019 in the EU run on petrol (58.9%), while diesel accounts only for 30.5% of sales. Remainder 10.6% corresponds to hybrid (5.9%), electrically chargeable (3.0%) and alternative fuels, namely liquefied petroleum gas, natural gas and ethanol vehicles (1.7%). Data from 2019 demonstrated a reduction of 18.7% on sales of new passenger cars by diesel compared to 2016, whereas on petrol passenger it was registered a growth

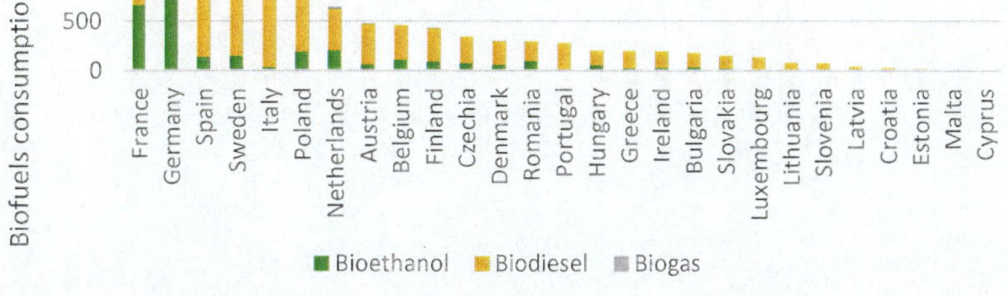

Fig. 6.1 Biofuels consumption for transport (ktoe)

of 12.3% cars for the same period (ACEA 2021). This trend could potentially be favourable for the bioethanol market.

Figure 6.2 shows the consumption of advanced biofuels and biofuels derived from used cooking oil and animal fats, in toe, for each state member of EU-27 in 2018 according to data published in EurObserv'ER 2020. "Advanced biofuels" designation refers to the biofuels that are produced from the feedstock listed in Part A of Annex IX of the RED II, while the category mentioned as "used cooking oil and animal fats" means biofuels that are produced from the feedstock listed in Part B of Annex IX of the RED II. According to the data, advanced biofuels accounted for 20.1%, while used cooking oil and animal fats represent about 70.9%. Regarding advanced biofuels consumption, the leaders are Sweden (36.4%) and Finland (31.5%), followed by Italy with a significantly lower share (10.6%). Concerning consumption of biofuels derived from used cooking oil and animal fats, Germany (33.8%), Italy (22.4%) and the Netherlands (13.8%) are the three foremost leaders, followed by Portugal (7.1%) and France (6.0%) (EurObserv'ER 2020).

More recently, the Covid-19 pandemic negatively impacted the biofuels market, leading to an inversion in the expected evolution. Due to the harsh lockdown measures, a high reduction in demand for transport fuel occurred (IEA 2020). Total biofuel demand decayed 8% from 2019 to 2020, resulting in the first reduction in annual production in the last two decades (IEA 2020, 2021). Consequently, a decrease in biofuels consumption was noticed, particularly in countries where fossil fuels' blending mandates were imposed. Moreover, fossil fuel prices have declined caused by the reduced demand, making biofuels even less economically attractive (IEA 2020).

Despite the pandemic, some key markets introduced higher blending mandates in 2020. Consequently, a less extent impact on biodiesel and HVO markets was noticed compared to ethanol. EU is the leading producer of biodiesel and HVO.

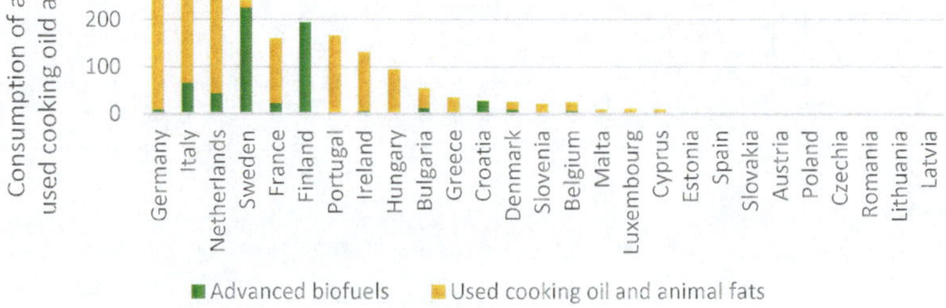

Fig. 6.2 Consumption of advanced biofuels and used cooking oil & animal fats (ktoe)

France, Germany, Spain, and The Netherlands account for more than 70% of production. However, fewer diesel car registrations coupled with the growing efficiency of passenger and road freight vehicle fleets can lead to a decline of 10% in diesel consumption over the forecast period (2020–2025), reducing biodiesel and HVO demand imposed by blending mandates (IEA 2020).

6.3 Bioethanol from Wastes

Besides minimising the dependence on fossil fuels, the management and reduction of waste generation is another key challenge of modern society (Liguori and Faraco 2016). The circular economy model considers residues and industrial wastes with high carbohydrates content as renewable feedstocks that can be converted into bio-based compounds (Liguori and Faraco 2016; Romaní et al. 2018). These renewable resources are low-cost, worldwide available and do not contribute to the competition with the food chain (Arevalo-Gallegos et al. 2017; Tian et al. 2018). Furthermore, the residues valorisation represents new and innovative market opportunities (Romaní et al. 2018). For the particular case of the EU, which is striving to move away from high ILUC-risk raw materials, the use of waste-based feedstock should be reinforced (IRENA 2019). Currently, in the EU, biomass sources mainly used for bioethanol production include wood from forests, crops and residues, by-products from wood and agricultural industries, herbaceous and woody energy crops, municipal organic wastes, and manure. Moreover, algae and marine biomass can also be integrated in the future. In 2017, in the EU, about 70% of the biomass-to-bioenergy was derived from forestry and wood industry residues, followed by agriculture biomass, namely crops and residues (18%), biowaste (12%), and municipal solid waste (7.3%) (Espí et al. 2020). The main challenges associated with the feedstocks mentioned above include their heterogeneous and complex composition (IRENA 2019; Romaní et al. 2018; Zabed et al. 2017). Furthermore, before establishing a process for wastes valorisation, it is crucial to proceed with its identification and quantification. Additionally, a deep knowledge of their structure and chemical characterisation is critical for finding successful deconstruction processes (Romaní et al. 2018).

Regarding forestry residues, they are mainly derived from the wood handling stage of the pulp and paper industries or wood processing sector, which are key sectors contributing to the economic development of Europe (Romaní et al. 2018). Forests cover about 42% of the European area and, for this reason, its bioenergy sector established synergies for decades to use sawdust, woodchips and low-quality wood (Cerruti et al. 2020; Romaní et al. 2018). Other forestry residues include leaves, small branches and stumps. A study conducted by Searle and Malin (Searle and Malins 2013) estimated that about 80 million tonnes of forestry residues were generated in the EU in 2011, from which only 8% were used for energy purposes. These authors assumed that 50% of forestry residues may be removed sustainably without compromising soil fertility. Therefore, still 40 million tonnes remained for further valorisation (Searle and Malins 2013). Recently, Cerruti et al. (2020) indicated that the sustainable availability of forestry residues is about 7.35 million tonnes, highlighting Sweden and Finland as the European countries with the highest availability.

Agricultural residues are recognised as one of the most promising feedstock for bioenergy use (AEBIOM 2019), since the EU is one of the leading producers of agricultural products (Hortsch and Corvo 2020). About 40% of the European area is dedicated to agriculture (Cerruti et al. 2020). This category can be classified as primary (residues left in the field after harvest or pruning and manure, such as straw, corn stover) or secondary residues (the portion discarded during the processing step, including olive shells/husks). Seasonality is an inherent challenge of this category of residues, which impacts availability and price and depends on several factors, such as profitability, incentives, logistics, and weather conditions (IRENA 2019).

Moreover, a substantial part of residues needs to remain on the field to safeguard soil fertility acting as a source of carbon and minerals. In contrast, another fraction already has other potential applications, such as straw use in animal bedding and horticulture (AEBIOM 2019). Searle and Malins (2013) estimated crop residues availability in the EU in 2011 of about 122 million tonnes, considering that about 66% of total residues production remains in fields for use in livestock and horticulture. Agricultural residues in the EU are mainly derived from wheat, barley, maise, sugar beet and rapeseed (Searle and Malins 2013). A recent study estimated that 146 million tonnes of residues could be sustainably collected annually in the EU (AEBIOM 2019). Recently, Cerruti et al. (2020) calculated the sustainable availability of agricultural residues, accounting for about 84.6 million tonnes. The discrepancies in estimations can be derived from differences in data sources or assumptions. Moreover, assessing global wastes availability and potential is not a straightforward task (IEA 2011).

According to data published in the Statista platform, in the EU-28 in 2018, about 252 million metric tons of municipal waste were generated (Statista 2021). Regrettably, more than a third of this waste ended up in landfills (European Commission Statistics—Eurostat 2020).

Regardless of the source of waste, ensuring a steady and year-round feedstock supply is crucial. Typically, these wastes are degraded under inadequate storage conditions. Furthermore, pre-processing could be required to control its size and moisture content (IRENA 2019).

Since wastes are used as feedstocks, advanced biofuels are estimated to have a greater potential than conventional GHG emissions savings. However, the quantification of GHG emissions through life cycle analysis is a time-consuming and complex task. Several variables need to be considered, including land use (such as deforestation, fertilizers), impact on biodiversity, energy and water consumption, transportation, distribution and end-use, among others (Jeswani et al. 2020; Sydney et al. 2019). Moreover, different raw materials are submitted to further processing technologies, leading to significant variations among estimates. In general, these are energy-intensive processes, which could compromise the GHG emissions savings (Jeswani et al. 2020). For all of these reasons, careful selection of feedstock, technologies, and even strategic location is crucial to ensure that advanced bioethanol represents a more sustainable alternative able to fulfil the GHG emissions savings required by the RED II.

6.3.1 Cellulosic Ethanol from Wastes: The Current Scenario

A biorefinery approach is emerging to efficiently convert biomass-based sources into a broad portfolio of value-added bioproducts, including sugars, biochemicals, and biofuels (Arevalo-Gallegos et al. 2017). Some authors argue that a multi-product biorefinery can improve the commercial competitiveness of bio-based products, namely cellulosic ethanol (Clauser et al. 2021; Rosales-Calderon and Arantes 2019). Rosales-Calderon and Arantes (2019) reviewed some chemicals and materials that are already commercially available and that could be integrated into a cellulosic ethanol production process within a biorefinery concept. Some examples of these bio-based products include acetic acid, furfural, lactic acid, micro fibrillated cellulose, sorbitol, succinic acid, and xylitol, among others (Rosales-Calderon and Arantes 2019).

The production of ethanol from lignocellulosic sugar via fermentation is one of the most researched routes for producing advanced biofuels (Brown et al. 2020). Table 6.1 provides an overview of the current cellulosic ethanol companies and several new plants currently being constructed and/or planned in the EU based on data published in International Energy Agency Bioenergy Task 39 (https://demoplants.bioenergy2020.eu/) and EurObserv'ER (2020). According to the data, most facilities are employed in the main EU biofuels' global leaders, namely France, Germany, Italy, Finland and Sweden (EurObserv'ER 2020; https://demoplants.bioenergy2020.eu/). There is a large variety of raw materials that are being used, depending mainly on local availability. These include agricultural residues (wheat straw, corn stover, other cereal straw), forest residues, industrial side-streams (sugarcane bagasse, spent sulphite liquor), food industry waste, energy crops (miscanthus), among others (Brown et al. 2020; EurObserv'ER 2020; https://demoplants.bioenergy2020.eu/).

One of the biggest cellulosic ethanol production sites, the Beta Renewables plant in Crescentino (Italy) operated since 2013 using wheat straw, rice straw and husks, and *Arundo donax*, an energy crop grown on marginal land, as feedstock (Flach et al. 2020; Padella et al. 2019). Extensive pretreatment issues were found during the plant operation, mainly due to rocks and dirt in the feedstock system. For this reason, a review of feedstock processing procedures was required. Moreover, the facility failed to come close to the level of production announced by the project (over 75 million litres). However, it was shut down in November 2017 due to its parent company filing for bankruptcy (Uslu et al. 2019). Later, in November 2018, the plant was acquired by Versalis (Eni's chemical subsidiary), and was restructured with an investment higher than €15 million. The plant has already been restarted and uses the PROESA® technology, converting organic residues and waste streams (Flach et al. 2020; https://demoplants.bioenergy2020.eu/). Besides ethanol, energy from biomass and biogas from wastewater treatment are also produced (Flach et al.

Table 6.1 Overview of the commercial-scale cellulosic ethanol plants (in operation, commissioning, under construction or planning based on data published in International Energy Agency Bioenergy Task 39 (https://demoplants.bioenergy2020.eu/) and EurObserv'ER (2020))

Company	Project	Status	City	Country	Start-up year	Installed capacity (tonnes/year)	Raw material
ARD	Futurol Pilot	Operational	Pomacle	France	2011	N.M	Woody and agricultural by-products, residues, energy crops
AustroCel Hallein	AustroCel Hallein Biorefinery	Operational	Hallein	Austria	2020–2021	30,000	Sulphite spent liquor (SSL, 33% dry content) from spruce wood pulping
Chempolis Ltd.	Chempolis Biorefining plant	Operational	Oulu	Finland	2008	5000	Straw, reed, empty fruit bunch, bagasse, corn stalks, wood residues
Clariant	Clariant Sunliquid	Operational	Straubing	Germany	2012	1000	Wheat straw, corn stover, miscanthus, sugarcane bagasse
	Clariant Romania	Under construction	Podari	Romania	2021	50,000	Wheat and other cereal straw
Domsjoe Fabriker	Domsjoe Fabriker	Operational	Ornskoldsvik	Sweden	1940	19,000	Sugars from pulping of lignocellulose
Energochemica	Bioskoh	Under construction	N.M	Slovakia	N.M	55,000	Dedicated crops and agro-food industrial side streams
Enviral Leopoldov site	Enviral Leopoldov site	Planned	Leopoldov	Slovakia	N.M	50,000	Lignocellulosic

(continued)

Table 6.1 (continued)

Company	Project	Status	City	Country	Start-up year	Installed capacity (tonnes/year)	Raw material
Eta Bio	Cellulosic Ethanol Plant Clariant Technology	Planned	N.M	Bulgaria	N.M	50,000	Wheat straw
IFPEN	Futurol	Operational	Bucy-Le-Long	France	2016	350	Agricultural residues
INA	INA Ethanol	Planned	Sisak	Croatia	N.M	55,000	Miscanthus and wheat straw
Kanteleen Voima	Nordfuel biorefinery	Planned	Haapavesi	Finland	2022–2023	65,000	Forest residues
ORLEN Poludnie (part of ORLEN GROUP)	Jedlicze site	Planned	Jedlicze	Poland	N.M	25,000	Wheat straw
RE Energy	RE Energy	Planned	Kalundborg	Denmark	2022	N.M	Lignocellulosics
Research Institutes of Sweden (RISE)	Biorefinery Demo plant	Operational	Ornskoldsvik	Sweden	2004	160	Sugarcane bagasse, wheat, corn stover, energy grass and recycled waste
Sainc Energy Limited	Cordoba	Planned	Cordoba	Spain	2022	25,000	Olive tree waste
St1 Biofuels Oy in cooperation withNorth European Bio Tech Oy	Cellulonix Kajaani	Operational	Kajaani	Finland	2017	8000	Sawdust
	Cellulonix Kajaani 2	Planned	Kajaani	Finland	2024	40,000	Sawdust
	Cellulonix Pietarsaari	Planned	Pietarsaari	Finland	2024	40,000	Sawdust
	Etanolix Lahti	Operational	Lahti	Finland	2009	1000	Food industry waste

(continued)

Table 6.1 (continued)

Company	Project	Status	City	Country	Start-up year	Installed capacity (tonnes/year)	Raw material
	Etanolix Gothenburg	Operational	Gothenburg	Sweden	2015	4000	Food industry waste
	Etanolix Hamina	Operational	Hamina	Finland	2008	1000	Bakery waste and process residues, bread waste
	Etanolix Vantaa	Operational	Vantaa	Finland	2009	1000	Bakery waste and process residues, bread waste
	Etanolix Jokioinen	Operational	Jokioinen	Finland	2011	7000	Bakery waste and process residues, bread waste
Versalis group Eni	Crescentino Biorefinery	Operational	Crescentino	Italy	2020 (restart)	40,000	Organic residues and waste streams

N.M. Not mentioned

2020). Due to the Covid-19 pandemic situation, biorefinery was adapted temporarily for bioethanol-based disinfectant production using corn glucose syrup (Eni 2020).

St1 Biofuels Oy manages the Cellulonix® project in collaboration with North European Bio Tech Oy. An ethanol plant started production in 2017 in Kajaani (Finland), and two similar facilities are planned to be built in this country, using sawdust as feedstock (Flach et al. 2020). Under a similar project, Etalonix® operates in some plants in Finland (Lathi, Hamina, Vantaa, Jokioinen) and Sweden (Gothenburg), using the food industry waste as feedstock (https://demoplants.bioenergy2020.eu/).

Recently Austrocel Hallein Biorefinery (Austria) started to produce cellulosic ethanol from spent sulphite liquor (SSL, 33% dry content) from spruce wood pulping. The advanced bioethanol will be used to meet the 0.5% advanced biofuels mandate (https://demoplants.bioenergy2020.eu/).

In Germany, Clariant also operates a demonstration plant using Sunliquid® technology to produce cellulosic ethanol from agricultural wastes, namely wheat straw, corn stover, miscanthus, and sugarcane bagasse. In Romania, a first-of-a-kind commercial plant using the same technology to produce cellulosic ethanol is also under construction (https://demoplants.bioenergy2020.eu/).

Research Institutes of Sweden (RISE) owns a biorefinery demo plant project, operational since 2004 using SEKAB technology. Several lignocellulosic raw materials were tested, particularly sugarcane bagasse, wheat straw, corn stover, energy grass, and recycled waste. Domsjoe Fabriker is a pulping plant that integrated another biorefinery at a commercial scale in Sweden. The raw material is organic residues and waste streams, in particular, sugars from pulping of lignocellulose. A by-product of cellulose production is used for the bioethanol plant (https://demoplants.bioenergy2020.eu/).

IFPEN is a French public research player running the Futurol project with two multi-feedstock biorefinery plants operating in France. One of them is a pilot-scale (since 2011), and the other is on a demonstration-scale phase (since 2016) (https://demoplants.bioenergy2020.eu/).

Chempolis uses formico® technology to valorise non-wood and non-food lignocellulosic biomass. This biorefining plant is located in Finland and has been operating since 2008 (https://demoplants.bioenergy2020.eu/).

Multiple biorefinery projects are planned for the next few years for the production of cellulosic ethanol in the EU, namely in Slovakia (Energochemica, Enviral Leopoldov site), Bulgaria (Eta Bio), Croatia (INA), Finland (Kanteleen Voima), Spain (Sainc Energy Limited), Denmark (RE Energy) and Poland (ORLEN Poludnie) (https://demoplants.bioenergy2020.eu/).

6.3.2 *Other Unexploited Wastes: The Particular Case of Portugal*

Portugal needs to import biofuels to comply with blending mandates (Guerrero 2017; Hassan et al. 2019), since it is one of the few European countries that still does not produce bioethanol (Flach et al. 2018; Guerrero 2015).

Portugal has a strong industrial activity in the primary sectors, namely agro-food, forestry, and fishery (Bio-based Industries Consortium 2018). Regarding bio-based sectors, the pulp and paper sector is the most abundant source of residual biomass, followed by the food and beverage industries (Bio-based Industries Consortium 2021). Data from 2018 shows that although 82% of residual streams are recovered, a significant fraction of 18% is still disposed of, which account for 1 million tonnes of residual streams that potentially could be valorised (Bio-based Industries Consortium 2021). Another relevant source of biomass derived from the organic fraction of municipal solid wastes is about 2.7 million tonnes in 2018, from which about 38% is landfilled. Once landfilling is being progressively abolished, it is crucial to find other solutions representing new market opportunities and strengths for the bio-based economy (Bio-based Industries Consortium 2021). Other relevant biomass sources are derived from animal manure and wastewaters (Ferreira et al. 2017). Therefore, it is crucial to identify and characterise readily available residues, evaluating their potential for producing bio-based products, reducing pollution, contributing to sustainability, and fostering a circular economy implementation (Romaní et al. 2018). For that purpose, it is crucial to establish industrial synergies for the valorisation of wastes derived from the main industrial sectors of the Portuguese economy, benefiting from logistics and infrastructures already existing.

Forest residues

Above 30% of the national territory is covered by forests, accounting for a total area of 3.26 million hectares. Eucalyptus, pine, and cork trees are the dominant species (Bio-based Industries Consortium 2018). Forest and the related value chains, namely pulp and paper, wood and cork, are critical segments of the primary sector (Bio-based Industries Consortium 2018). According to Ferreira et al. (2017), about 2.2 million tonnes of forest residues are generated per year in Portugal.

The pulp and paper industry is a key sector contributing to national economic development (Espí et al. 2020; Amândio et al. 2021). The Portuguese pulp and paper industry processes about 7 million cubic meters of wood per year, mainly eucalyptus (Bio-based Industries Consortium 2018). The wastes generated by this sector include wood processing residues (bark, sawdust, wood chips), pulping liquors (black liquor or spent sulphite liquor (SSL), depending on the pulping process), sludge derived from wastewater treatment and rejects (Gaspar et al. 2019). Currently, these wastes are mainly burned for energy generation (Ferreira et al. 2017). Some works reported the potential of these residues as a substrate for bioethanol production, namely from sawdust (Asada et al. 2019; Guigou et al. 2019), eucalyptus bark (Amândio et al. 2021; Gomes et al. 2021; Romaní et al. 2019), spent sulphite liquor (Henriques et al.

2018; Pereira et al. 2012, 2015; Xavier et al. 2010), and primary sludge (Kang et al. 2010, 2011; Mendes et al. 2016, 2014).

Agricultural residues

In Portugal, the area utilised for agriculture accounts for 3.6 million hectares, of which 50% is occupied by permanent grassland, 30% by temporary crops and 20% by permanent crops (Bio-based Industries Consortium 2018).

The agricultural residues are mainly derived from agricultural production by-products and the agro-food industry (Ferreira et al. 2017). Some agricultural residues derived from national territory include straw cereals (wheat, barley and rice), dried fruits shells, olive pits/stones, pomace derived from olive oil extraction and pruning residues from fruit trees, olives and vineyards (leaves and small branches) (Ferreira et al. 2009, 2017). Some of the wastes mentioned above are currently used for heat production by direct combustion (Ferreira et al. 2017). There are already several works published regarding the potential of these agricultural residues to be converted into bioethanol (Cara et al. 2008; Fernandes-Klajn et al. 2018; Francisco et al. 2020; Gaspar et al. 2019; Mateo et al. 2015). Besides the residues above referred, some perennial plants have been studied as feedstock for second-generation bioethanol, namely *Cynara cardunculus* (cardoon) and *Cistus ladanifer* (rockrose) (Fernandes et al. 2015, 2018).

Historically, the production of wine plays a significant role in Portugal's agricultural sector (Bio-based Industries Consortium 2018). The harvest of the vine and the vinification process generated a considerable amount of residues, accounting for about an annual production of 1.2–3.5 tonnes per hectare (Jesus et al. 2017). The main residues include vine pruning residue, grape stalks or stems and grape pomace or marc (Romaní et al. 2018). Romaní et al. (2018) described the chemical composition of each fraction and the respective valorisation route. There are already some published works regarding the production of bioethanol, namely from vine pruning residue (Jesus et al. 2017; Kaae Kirk et al. 2021; Senila et al. 2020), grape stalks (Egüés et al. 2013) and grape pomace (Mendes et al. 2013; Williams et al. 2019).

6.3.3 Main Challenges Related to the Conversion of Wastes into Cellulosic Ethanol

The three main challenges hindering the commercialisation of advanced biofuels, particularly cellulosic ethanol, are flexibility, reliability, and process cost (; Brown et al. 2020; Panoutsou et al. 2021). Regardless of the technology used for the operation, selecting a flexible process to handle various raw materials is crucial, mainly due to the seasonality (Panoutsou et al. 2021; Romaní et al. 2018). In many cases, feedstock availability is not enough to continuously supply a production plant (Padella et al. 2019). Moreover, it is essential to ensure the reliability related to consistency and fuel quality from an industrial perspective regardless of any variations in the process feed (Panoutsou et al. 2021). The variations in the quality and composition can be caused by the inherent heterogeneity of this type of raw material (Padella et al. 2019; Panoutsou et al. 2021; Romaní et al. 2018). Finally, the most difficult challenge to overcome is the cost, which makes these approaches highly dependent on government policies and investment incentives to reduce the financial risk (Panoutsou et al. 2021).

Implementing a process on an industrial scale must be economically competitive with the existing alternatives (Lászlok et al. 2020; Panoutsou et al. 2021). The attractiveness of biofuels is continuously changing depending mainly on oil price volatility and instability of feedstock prices (Ajanovic and Haas 2010). Furthermore, there are still some technical issues to overcome due to the complexity of the process. However, there is potential for cost reduction (Brown et al. 2020). An approach proposed to reduce capital costs is related to co-processing and co-location by taking advantage of know-how, existing infrastructures, and efficient supply chain operations (Brown et al. 2020; Panoutsou et al. 2021; Romaní et al. 2018). In general, feedstock costs associated with the collection/harvesting of biomass and its transportation are significant (IRENA 2019; Sandesh and Ujwal 2021). Some examples include integrating advanced biofuels production into power plants, pulp and paper mills and the wood processing industry (Panoutsou et al. 2021). The switch of the pulp and paper mill industries to biorefineries is already a common practice, mainly in Sweden and Finland (Berntsson et al. 2008; González-García et al. 2011; Goldfarb et al. 2017; Temmes and Peck 2020). An integrated biorefinery entails fully exploiting incoming biomass and other raw materials for simultaneous and economically optimized materials, chemicals, and energy production. Some examples of products include a) chemicals and materials, such as phenols, adhesives, activated carbon, antioxidants; b) biofuels including ethanol, pellets, lignin fuel, among others; c) electricity (Berntsson et al. 2008; González-García et al. 2011). This concept has been extended to other sectors, namely wine (Ahmad et al. 2020) and olive production (Goldfarb et al. 2017; Kourmentza et al. 2017; Schievano et al. 2015).

Integration of process steps, like, e.g. hydrolysis and fermentation, can contribute to costs reduction. Moreover, it is also crucial to focus on continuous improvement through research and development activities (Brown et al. 2020). Furthermore, the valorisation of each fraction derived from biomass and wastes generated during the production process could expand the portfolio of products and increase revenues (Brown et al. 2020; Romaní et al. 2018). Padella et al. (2019) also suggested some alternative routes for post-fermentation broth and solid residues derived from ethanol fermentation, namely recycling as a supplement for subsequent fermentation, used as feed for animals or even added anaerobic digestion plant for biogas production.

Regarding technological barriers, these mainly include the cost of enzymes, accounting for up 30–50% of the total production costs of ethanol, the simultaneous consumption of hexoses and pentose by highly efficient and robust industrial yeast strains and the downstream processing steps (Padella et al. 2019). By overcoming these challenges, operation and maintenance costs certainly should be reduced, and capital will be available on more favourable terms since investor risk perception will be lower (Brown et al. 2020).

6.4 Conclusions

Over the last years, the EU focused on promoting an energy transition from fossil fuels to renewable resources. Currently, biofuels are the most considerable contribution of renewable energy to the transport sector, representing more than 90%. Decarbonisation of the transport sector, climate change mitigation and energy security are the three main driving forces guiding biofuels policies in the EU. Despite advances in low-carbon technologies, the transition to zero emissions has proven quite challenging.

Long-term policies towards carbon neutrality are focused on electrification and hydrogen production. Nevertheless, advanced biofuels are considered one of the most promising and ready-available alternatives for enabling the transport sector to comply with GHG emission targets. Furthermore, this alternative does not require a significant restructuring of the transport sector since the existing infrastructures can be used.

Advanced bioethanol production is mainly derived from agricultural residues, forest residues, industrial side-streams and food industry waste, depending on local availability. Flexibility, reliability and costs are the main challenges hindering the commercialisation of cellulosic ethanol. Despite these barriers, cellulosic ethanol production is already a reality in the EU. In Portugal, large-scale cellulosic ethanol deployment needs to be boosted by policy decision and investment incentives. It is urgent to implement integrated biorefineries in existing pulping mills and agro-industries since they already have well-established logistics, processing infrastructures, and wastewater treatment. R&D and technological efforts should continue reducing production costs and making ethanol economically more competitive and attractive to investors. Investing in this area will support more technology implementation in the existing industrial plants. This is the only way to attain the EU targets regarding carbon neutrality for 2050, contributing to sustainability and circular economy.

Acknowledgements This work was carried out under the Project InPaCTus—Innovative Products and Technologies from Eucalyptus, Project No. 21874 funded by Portugal 2020 through European Regional Development Fund (ERDF) in the frame of COMPETE 2020 no. 246/AXIS II/2017. This work was developed within the scope of the project CICECO-Aveiro Institute of Materials, UIDB/50011/2020 and UIDP/50011/2020, financed by national funds through the Portuguese Foundation for Science and Technology/MCTES. Authors would also like to thank the CIEPQPF—Strategic Research Centre Project UIDB/00102/2020, funded by the Fundação para a Ciência e Tecnologia (FCT).

References

Aatola H, Larmi M, Sarjovaara T, Mikkonen S (2009) Hydrotreated vegetable oil (HVO) as a renewable diesel fuel: trade-off between NO_x, particulate emission, and fuel consumption of a heavy duty engine. SAE Int J Engines 1:1251–1262. https://doi.org/10.4271/2008-01-2500

ACEA (2021) The automobile industry pocket guide 2020/2021. https://www.acea.auto/publication/automobile-industry-pocket-guide-2020-2021/

Aditiya HB, Mahlia TMI, Chong WT et al (2016) Second generation bioethanol production: a critical review. Renew Sustain Energy Rev 66:631–653. https://doi.org/10.1016/j.rser.2016.07.015

AEBIOM (2019) Bionergy europe factsheet: biomass for energy—agricultural residues and energy crops. In: Biomass for energy. https://bioenergyeurope.org/article/204-bioenergy-explained-biomass-for-energy-agricultural-residues-energy-crops.html

Agarwal AK, Gupta JG, Dhar A (2017) Potential and challenges for large-scale application of biodiesel in automotive sector. Prog Energy Combust Sci 61:113–149. https://doi.org/10.1016/j.pecs.2017.03.002

Ahmad B, Yadav V, Yadav A et al (2020) Integrated biorefinery approach to valorize winery waste: a review from waste to energy perspectives. Sci Total Environ 719:1–14. https://doi.org/10.1016/j.scitotenv.2020.137315

Ajanovic A, Haas R (2010) Economic challenges for the future relevance of biofuels in transport in EU countries. Energy 35:3340–3348. https://doi.org/10.1016/j.energy.2010.04.020

Amândio MST, Rocha JMS, Serafim LS, Xavier AMRB (2021) Cellulosic bioethanol from industrial eucalyptus globulus bark residues using kraft pulping as a pretreatment. Energies 14:1–19. https://doi.org/10.3390/en14082185

Arevalo-Gallegos A, Ahmad Z, Asgher M et al (2017) Lignocellulose: a sustainable material to produce value-added products with a zero waste approach—a review. Int J Biol Macromol 99:308–318. https://doi.org/10.1016/j.ijbiomac.2017.02.097

Asada C, Sasaki C, Nakamura Y (2019) High concentration ethanol production from mixed softwood sawdust waste. Waste Biomass Valorization 10:433–439. https://doi.org/10.1007/s12649-017-0073-0

Balat M, Balat H, Öz C (2008) Progress in bioethanol processing. Prog Energy Combust Sci 34:551–573. https://doi.org/10.1016/j.pecs.2007.11.001

Berntsson T, Axegard P, Backlund B, et al (2008) Swedish pulp mill biorefineries—a vision of future possibilities. https://www.osti.gov/etdeweb/servlets/purl/951488

Bio-based Industries Consortium (2018) Mapping the potential of Portugal for the bio-based industry. https://biconsortium.eu/portugal_country_report

Bio-based Industries Consortium (2021) Mapping Portugal's bio-based potential—country report. https://biconsortium.eu/sites/biconsortium.eu/files/downloads/BIC.pdf

Bozbas K (2008) Biodiesel as an alternative motor fuel: production and policies in the European Union. Renew Sustain Energy Rev 12:542–552. https://doi.org/10.1016/j.rser.2005.06.001

Branco R, Serafim L, Xavier A (2019) Second generation bioethanol production: on the use of pulp and paper industry wastes as feedstock. Fermentation 5:1–30. https://doi.org/10.3390/fermentation5010004

Brito TLF, Islam T, Stettler M et al (2019) Transitions between technological generations of alternative fuel vehicles in Brazil. Energy Policy 134:1–10. https://doi.org/10.1016/j.enpol.2019.110915

Brown A, Waldheim L, Landälv I, et al (2020) Advanced biofuels—potential for cost reduction. In: IEA bionenergy task 41 202101. https://www.ieabioenergy.com/wp-content/uploads/2020/02/T41_CostReductionBiofuels-11_02_19-final.pdf

Buchspies B, Kaltschmitt M, Neuling U (2020) Potential changes in GHG emissions arising from the introduction of biorefineries combining biofuel and electrofuel production within the European Union—a location specific assessment. Renew Sustain Energy Rev 134:1–14. https://doi.org/10.1016/j.rser.2020.110395

Cadillo-Benalcazar JJ, Bukkens SGF, Ripa M, Giampietro M (2021) Why does the European Union produce biofuels? Examining consistency and plausibility in prevailing narratives with quantitative storytelling. Energy Res Soc Sci 71:1–14. https://doi.org/10.1016/j.erss.2020.101810

Callegari A, Bolognesi S, Cecconet D, Capodaglio AG (2020) Production technologies, current role, and future prospects of biofuels feedstocks: a state-of-the-art review. Crit Rev Environ Sci Technol 50:384–436. https://doi.org/10.1080/10643389.2019.1629801

Cara C, Ruiz E, Ballesteros M et al (2008) Production of fuel ethanol from steam-explosion pretreated olive tree pruning. Fuel 87:692–700. https://doi.org/10.1016/j.fuel.2007.05.008

Cerruti E, Di Gruttola F, Lauro G, et al (2020) Assessment of feedstocks and technologies for advanced biofuel production. In: 75° National ATI congress. E3S web of conferences, Rome, pp 1–10. https://doi.org/10.1051/e3sconf/202019705002

Chandel AK, Albarelli JQ, Santos DT et al (2019) Comparative analysis of key technologies for cellulosic ethanol production from Brazilian sugarcane bagasse at a commercial scale. Biofuels, Bioprod Biorefining 13:1–21. https://doi.org/10.1002/bbb.1990

Chiaramonti D, Talluri G, Scarlat N, Prussi M (2021) The challenge of forecasting the role of biofuel in EU transport decarbonisation at 2050: a meta-analysis review of published scenarios. Renew Sustain Energy Rev 139:1–12. https://doi.org/10.1016/j.rser.2021.110715

Clauser NM, Felissia FE, Area MC, Vallejos ME (2021) A framework for the design and analysis of integrated multi-product biorefineries from agricultural and forestry wastes. Renew Sustain Energy Rev 139:1–17. https://doi.org/10.1016/j.rser.2020.110687

Dahal K, Juhola S, Niemelä J (2018) The role of renewable energy policies for carbon neutrality in Helsinki Metropolitan area. Sustain Cities Soc 40:222–232. https://doi.org/10.1016/j.scs.2018.04.015

Danielis R, Scorrano M, Giansoldati M (2021) Decarbonising transport in Europe: trends, goals, policies and passenger car scenarios. Res Transp Econ 101068.https://doi.org/10.1016/j.retrec.2021.101068

Ebadian M, van Dyk S, McMillan JD, Saddler J (2020) Biofuels policies that have encouraged their production and use: an international perspective. Energy Policy 147:1–17. https://doi.org/10.1016/j.enpol.2020.111906

Egüés I, Serrano L, Amendola D et al (2013) Fermentable sugars recovery from grape stalks for bioethanol production. Renew Energy 60:553–558. https://doi.org/10.1016/j.renene.2013.06.006

EIA Independent Analysis and Statistics (2018) Monthly energy review, Aug 2018. U.S. Energy Information Administration (EIA) Independent Statistics & Analysis, Washington, United States of America. https://www.eia.gov/totalenergy/data/monthly/archive/00351811.pdf

Eni (2020) Versalis (Eni) starts the production of Invix, bioethanol-based disinfectant at Crescentino. In: Eni Media. https://www.eni.com/en-IT/media/press-release/2020/07/pr-versalis-invix-21-july-2020.html

Espí E, Ribas Í, Diáz C, Sastrón Ó (2020) Feedstocks for advanced biofuels. In: Llamas B, Romero MFO, Sillero E (eds) Sustainable mobility. IntechOpen, London, UK, pp 1–22. https://www.intechopen.com/chapters/70860

EurObserv'ER (2020) Biofuels barometer 2020. https://www.eurobserv-er.org/biofuels-barometer-2020/

European Commission Statistics—Eurostat (2020) Energy, transport and environment statistics. https://ec.europa.eu/eurostat/en/web/products-statistical-books/-/ks-dk-20-001

European Environment Agency (2020) Greenhouse gas emissions from transport in Europe. https://www.eea.europa.eu/data-and-maps/indicators/transport-emissions-of-greenhouse-gases-7/assessment

European Renewable Ethanol (ePURE) (2020) European renewable ethanol—key figures 2019. https://www.epure.org/wp-content/uploads/2020/11/200813-DEF-PR-ePURE-infographic-European-renewable-ethanol-key-figures-2019_web.pdf

European Renewable Ethanol (ePure) (2018) An ethanol blend to fuel Europe's clean mobility… E10. https://www.epure.org/wp-content/uploads/2020/11/190509-def-pr-revised-epure-e10-leaflet_en_web.pdf

Fernandes MC, Ferro MD, Paulino AFC et al (2015) Enzymatic saccharification and bioethanol production from *Cynara cardunculus* pretreated by steam explosion. Bioresour Technol 186:309–315. https://doi.org/10.1016/j.biortech.2015.03.037

Fernandes MC, Ferro MD, Paulino AFC et al (2018) Comparative study on hydrolysis and bioethanol production from cardoon and rockrose pretreated by dilute acid hydrolysis. Ind Crops Prod 111:633–641. https://doi.org/10.1016/j.indcrop.2017.11.037

Fernandes-Klajn F, Romero-García JM, Díaz MJ, Castro E (2018) Comparison of fermentation strategies for ethanol production from olive tree pruning biomass. Ind Crops Prod 122:98–106. https://doi.org/10.1016/j.indcrop.2018.05.063

Ferreira S, Moreira NA, Monteiro E (2009) Bioenergy overview for Portugal. Biomass Bioenerg 33:1567–1576. https://doi.org/10.1016/j.biombioe.2009.07.020

Ferreira S, Monteiro E, Brito P, Vilarinho C (2017) Biomass resources in Portugal: current status and prospects. Renew Sustain Energy Rev 78:1221–1235. https://doi.org/10.1016/j.rser.2017.03.140

Flach B, Lieberz S, Rossetti A (2018) European Union biofuels annual 2017. https://apps.fas.usda.gov/newgainapi/api/report/downloadreportbyfilename?filename=Biofuels%20Annual_The%20Hague_EU-28_6-19-2017.pdf

Flach B, Lieberz S, Bolla S (2020) Biofuels annual—European Union. https://apps.fas.usda.gov/newgainapi/api/Report/DownloadReportByFileName?fileName=BiofuelsAnnual_TheHague_European Union_06-29-2020

Francisco J, Cuevas M, Feng C et al (2020) Energetic valorisation of olive biomass: olive-tree pruning, olive stones and pomaces. Processes 8:1–38. https://doi.org/10.3390/pr8050511

Gaspar MC, Mendes CVT, Pinela SR et al (2019) Assessment of agroforestry residues: their potential within the biorefinery context. ACS Sustain Chem Eng 7:17154–17165. https://doi.org/10.1021/acssuschemeng.9b03532

Goldfarb JL, Buessing L, Gunn E et al (2017) Novel integrated biorefinery for olive mill waste management: utilization of secondary waste for water treatment. ACS Sustain Chem Eng 5:876–884. https://doi.org/10.1021/acssuschemeng.6b02202

Gomes DG, Michelin M, Romaní A et al (2021) Co-production of biofuels and value-added compounds from industrial *Eucalyptus globulus* bark residues using hydrothermal treatment. Fuel 285:1–11. https://doi.org/10.1016/j.fuel.2020.119265

González-García S, Hospido A, Agnemo R et al (2011) Environmental life cycle assessment of a swedish dissolving pulp mill integrated biorefinery. J Ind Ecol 15:568–583. https://doi.org/10.1111/j.1530-9290.2011.00354.x

Guerrero M (2015) Portugal biofuels standing report 2015. https://apps.fas.usda.gov/newgainapi/api/report/downloadreportbyfilename?filename=PortugalBiofuelsStandingReport2015_Madrid_Portugal_7-27-2015.pdf

Guerrero M (2017) Portugal biofuel market outlook 2017. https://gain.fas.usda.gov/RecentGAINPublications/PortugalBiofuelMarketOutlook_Madrid_Portugal_6-21-2017.pdf

Guigou M, Cabrera MN, Vique M et al (2019) Combined pretreatments of eucalyptus sawdust for ethanol production within a biorefinery approach. Biomass Convers Biorefinery 9:293–304. https://doi.org/10.1007/s13399-018-0353-3

Hassan SS, Williams GA, Jaiswal AK (2019) Lignocellulosic biorefineries in europe: current state and prospects. Trends Biotechnol 37:231–234. https://doi.org/10.1016/j.tibtech.2018.07.002

Henriques TM, Pereira SR, Serafim LS, Xavier AMRB (2018) Two-stage aeration fermentation strategy to improve bioethanol production by *Scheffersomyces stipitis*. Fermentation 4:1–13. https://doi.org/10.3390/fermentation4040097

Hortsch R, Corvo P (2020) The biorefinery concept: producing cellulosic ethanol from agricultural residues. Chem-Ing-Tech 92:1803–1809. https://doi.org/10.1002/cite.202000203

Ianniello R, Belgiorno G, Di Luca G et al (2021) Ethanol in dual-fuel and blend fueling modes for advanced combustion in compression ignition engines. Energy Environ Sustain 5–27.https://doi.org/10.1007/978-981-16-0931-2_2

IEA (2011) Technology roadmap biofuels for transport. https://iea.blob.core.windows.net/assets/0905b11e-53d4-417a-a061-453934009476/Biofuels_Roadmap_WEB.pdf

IEA (2020) Renewables 2020 analysis and forecast to 2025. https://www.iea.org/reports/renewables-2020

IEA (2021) Renewable energy market update - outlook for 2021 and 2022. https://www.iea.org/reports/renewable-energy-market-update-2021

International Energy Agency Bioenergy Task 39 Database on facilities for the production of advanced liquid and gaseous biofuels for transport. https://demoplants.bioenergy2020.eu/

IRENA (2019) Advanced biofuels—What holds them back? https://www.irena.org/publications/2019/Nov/Advanced-biofuels-What-holds-them-back

Jansen MLA, Bracher JM, Papapetridis I et al (2017) *Saccharomyces cerevisiae* strains for second-generation ethanol production: from academic exploration to industrial implementation. FEMS Yeast Res 17:1–20. https://doi.org/10.1093/femsyr/fox044

Jesus MS, Romaní A, Genisheva Z et al (2017) Integral valorization of vine pruning residue by sequential autohydrolysis stages. J Clean Prod 168:74–86. https://doi.org/10.1016/j.jclepro.2017.08.230

Jeswani HK, Chilvers A, Azapagic A (2020) Environmental sustainability of biofuels: a review. Proc R Soc A Math Phys Eng Sci 476:1–37. https://doi.org/10.1098/rspa.2020.0351

Kaae Kirk N, Navarrete C, Ellegaard Juhl J et al (2021) The "Zero Miles Product" concept applied to biofuel production: a case study. Energies 14:1–19. https://doi.org/10.3390/en14030565

Kang L, Wang W, Lee YY (2010) Bioconversion of kraft paper mill sludges to ethanol by SSF and SSCF. Appl Biochem Biotechnol 161:53–66. https://doi.org/10.1007/s12010-009-8893-4

Kang L, Wang W, Pallapolu VR, Lee YY (2011) Enhanced ethanol production from de-ashed paper sludge by simultaneous saccharification and fermentation and simultaneous saccharification and Co-Fermentation. BioResources 6:3791–3808. https://doi.org/10.15376/biores.6.4.3791-3808

Kapustová Z, Kapusta J, Boháčiková A, Bielik P (2020) Development status in EU biofuels market. Visegr J Bioecon Sustain Dev 9:67–71. https://doi.org/10.2478/vjbsd-2020-0013

Karatzos S, Mcmillan J, Saddler J (2014) The potential and challenges of "drop in" biofuels

Knothe G (2010) Biodiesel and renewable diesel: a comparison. Prog Energy Combust Sci 36:364–373. https://doi.org/10.1016/j.pecs.2009.11.004

Knothe G, Razon LF (2017) Biodiesel fuels. Prog Energy Combust Sci 58:36–59. https://doi.org/10.1016/j.pecs.2016.08.001

Kourmentza C, Koutra E, Venetsaneas N, Kornaros M (2017) Integrated biorefinery approach for the valorization of olive mill waste streams towards sustainable biofuels and bio-based products. In: Kalia VC, Kumar P (eds) Microbial applications. Springer International Publishing AG, pp 221–238. https://doi.org/10.1007/978-3-319-52666-9_10

Kralova I, Sjöblom J (2010) Biofuels-renewable energy sources: a review. J Dispers Sci Technol 31:409–425. https://doi.org/10.1080/01932690903119674

Lapuerta M, Villajos M, Agudelo JR, Boehman AL (2011) Key properties and blending strategies of hydrotreated vegetable oil as biofuel for diesel engines. Fuel Process Technol 92:2406–2411. https://doi.org/10.1016/j.fuproc.2011.09.003

Lászlok A, Takács-györgy K, Takács I (2020) Examination of first generation biofuel production in some selected biofuel producing countries in Europe: a case study. Agric Econ 66:469–476. https://doi.org/10.17221/237/2020-AGRICECON

Liguori R, Faraco V (2016) Biological processes for advancing lignocellulosic waste biorefinery by advocating circular economy. Bioresour Technol 215:13–20. https://doi.org/10.1016/j.biortech.2016.04.054

Linares P, Pérez-Arriaga IJ (2013) A sustainable framework for biofuels in Europe. Energy Policy 52:166–169. https://doi.org/10.1016/j.enpol.2012.10.008

Londo M, Lensink S, Wakker A et al (2010) The REFUEL EU road map for biofuels in transport: application of the project's tools to some short-term policy issues. Biomass Bioenerg 34:244–250. https://doi.org/10.1016/j.biombioe.2009.07.005

Manzetti S, Andersen O (2015) A review of emission products from bioethanol and its blends with gasoline. Background for new guidelines for emission control. Fuel 140:293–301. https://doi.org/10.1016/j.fuel.2014.09.101

Mateo S, Puentes JG, Moya AJ, Sánchez S (2015) Ethanol and xylitol production by fermentation of acid hydrolysate from olive pruning with *Candida tropicalis* NBRC 0618. Bioresour Technol 190:1–6. https://doi.org/10.1016/j.biortech.2015.04.045

Mendes JAS, Xavier AMRB, Evtuguin DV, Lopes LPC (2013) Integrated utilization of grape skins from white grape pomaces. Ind Crops Prod 49:286–291. https://doi.org/10.1016/j.indcrop.2013.05.003

Mendes CVT, Rocha JMS, Carvalho MGVS (2014) Valorization of residual streams from pulp and paper mills: pretreatment and bioconversion of primary sludge to bioethanol. Ind Eng Chem Res 53:19398–19404. https://doi.org/10.1021/ie503021y

Mendes CVT, Cruz CHG, Reis DFN et al (2016) Integrated bioconversion of pulp and paper primary sludge to second generation bioethanol using Saccharomyces cerevisiae ATCC 26602. Bioresour Technol 220:161–167. https://doi.org/10.1016/j.biortech.2016.07.140

Mishra VK, Goswami R (2018) A review of production, properties and advantages of biodiesel. Biofuels 9:273–289. https://doi.org/10.1080/17597269.2017.1336350

Mittelbach M (2015) Fuels from oils and fats: recent developments and perspectives. Eur J Lipid Sci Technol 117:1832–1846. https://doi.org/10.1002/ejlt.201500125

Mohiddin MNB, Tan YH, Seow YX, et al (2021) Evaluation on feedstock, technologies, catalyst and reactor for sustainable biodiesel production: A review. J Ind Eng Chem 98:60–81.https://doi.org/10.1016/j.jiec.2021.03.036

Nigam PS, Singh A (2011) Production of liquid biofuels from renewable resources. Prog Energy Combust Sci 37:52–68. https://doi.org/10.1016/j.pecs.2010.01.003

Padella M, O'Connell A, Prussi M (2019) What is still limiting the deployment of cellulosic ethanol? Analysis of the current status of the sector. Appl Sci 9:1–13. https://doi.org/10.3390/app9214523

Panoutsou C, Germer S, Karka P, et al (2021) Advanced biofuels to decarbonise European transport by 2030: markets, challenges, and policies that impact their successful market uptake. Energy Strateg Rev 34.https://doi.org/10.1016/j.esr.2021.100633

Paris B, Papadakis G, Janssen R, Rutz D (2021) Economic analysis of advanced biofuels, renewable gases, electrofuels and recycled carbon fuels for the Greek transport sector until 2050. Renew Sustain Energy Rev 144:1–10. https://doi.org/10.1016/j.rser.2021.111038

Pereira SR, Ivanuša Š, Evtuguin DV et al (2012) Biological treatment of eucalypt spent sulphite liquors: a way to boost the production of second generation bioethanol. Bioresour Technol 103:131–135. https://doi.org/10.1016/j.biortech.2011.09.095

Pereira SR, Sànchez I, Nogué V, Frazão CJR et al (2015) Adaptation of *Scheffersomyces stipitis* to hardwood spent sulfite liquor by evolutionary engineering. Biotechnol Biofuels 8:1–8.https://doi.org/10.1186/s13068-015-0234-y

Puricelli S, Cardellini G, Casadei S et al (2021) A review on biofuels for light-duty vehicles in Europe. Renew Sustain Energy Rev 137:1–19. https://doi.org/10.1016/j.rser.2020.110398

REN21 (2020) Renewables 2020 global status report. https://www.ren21.net/wp-content/uploads/2019/05/GSR2021_Full_Report.pdf

República Portuguesa, Fundo Ambiental, Agência Portuguesa do Ambiente (2019) Roadmap for carbon neutrality 2050 (RNC2050) long-term strategy for carbon neutrality of the portuguese economy by 2050. In: Rep. R 262/2019. https://unfccc.int/sites/default/files/resource/RNC2050_EN_PTLongTermStrategy.pdf

Romaní A, Michelin M, Domingues L, Teixeira JA (2018) Valorization of wastes from agrofood and pulp and paper industries within the biorefinery concept: Southwestern Europe scenario. In: Bhaskar T, Pandey A, Mohan SV et al (eds) Waste biorefinery: potential and perspectives. Elsevier B.V., Amsterdam, The Netherlands, pp 487–504. http://hdl.handle.net/1822/58954

Romaní A, Larramendi A, Yáñez R et al (2019) Valorization of *Eucalyptus nitens* bark by organosolv pretreatment for the production of advanced biofuels. Ind Crops Prod 132:327–335. https://doi.org/10.1016/j.indcrop.2019.02.040

Rosales-Calderon O, Arantes V (2019) A review on commercial-scale high-value products that can be produced alongside cellulosic ethanol. Biotechnol Biofuels 12:1–58. https://doi.org/10.1186/s13068-019-1529-1

Sandesh K, Ujwal P (2021) Trends and perspectives of liquid biofuel—process and industrial viability. Energy Convers Manag X 10:1–14. https://doi.org/10.1016/j.ecmx.2020.100075

Santos AS, Gilio L, Halmenschlager V et al (2018) Flexible-fuel automobiles and CO_2 emissions in Brazil: parametric and semiparametric analysis using panel data. Habitat Int 71:147–155. https://doi.org/10.1016/j.habitatint.2017.11.014

Sarris D, Papanikolaou S (2016) Biotechnological production of ethanol: biochemistry, processes and technologies. Eng Life Sci 16:307–329. https://doi.org/10.1002/elsc.201400199

Schievano A, Adani F, Buessing L et al (2015) An integrated biorefinery concept for olive mill waste management: supercritical CO_2 extraction and energy recovery. Green Chem 17:2874–2887. https://doi.org/10.1039/c5gc00076a

Searle S, Malins C (2013) Availability of cellulosic residues and wastes in the EU. Int Counc Clean Transp. https://theicct.org/sites/default/files/publications/ICCT_EUcellulosic-waste-residues_20131022.pdf

Senila L, Kovacs E, Scurtu DA et al (2020) Bioethanol production from vineyard waste by autohydrolysis pretreatment and chlorite delignification via simultaneous saccharification and fermentation. Molecules 25:1–22. https://doi.org/10.3390/molecules25112606

Singh D, Sharma D, Soni SL et al (2020) A review on feedstocks, production processes, and yield for different generations of biodiesel. Fuel 262:116553. https://doi.org/10.1016/j.fuel.2019.116553

Statista (2021) Generation of municipal waste in the European Union (EU-28) from 2010 to 2018. https://www.statista.com/statistics/983592/municipal-waste-generation-european-union-eu-28/

Sydney EB, Letti LAJ, Karp SG et al (2019) Current analysis and future perspective of reduction in worldwide greenhouse gases emissions by using first and second generation bioethanol in the transportation sector. Bioresour Technol Reports 7:1–9. https://doi.org/10.1016/j.biteb.2019.100234

Temmes A, Peck P (2020) Do forest biorefineries fit with working principles of a circular bioeconomy? A case of Finnish and Swedish initiatives. For Policy Econ 110:1–12. https://doi.org/10.1016/j.forpol.2019.03.013

Tian SQ, Zhao RY, Chen ZC (2018) Review of the pretreatment and bioconversion of lignocellulosic biomass from wheat straw materials. Renew Sustain Energy Rev 91:483–489. https://doi.org/10.1016/j.rser.2018.03.113

Uslu A, Veum K, Detz R (2019) Monitoring RESfuels D1.4 Monitoring framework and the KPIs for advanced renewable liquid fuels (RESfuels). Amsterdam. http://www.advancefuel.eu/contents/reports/advancefuel-monitoring-framework-and-kpis.pdf

Vásquez MC, Silva EE, Castillo EF (2017) Hydrotreatment of vegetable oils: A review of the technologies and its developments for jet biofuel production. Biomass Bioenerg 105:197–206. https://doi.org/10.1016/j.biombioe.2017.07.008

Williams DL, Schückel J, Vivier MA et al (2019) Grape pomace fermentation and cell wall degradation by *Kluyveromyces marxianus* Y885. Biochem Eng J 150:1–11. https://doi.org/10.1016/j.bej.2019.107282

Xavier AMRB, Correia MF, Pereira SR, Evtuguin DV (2010) Second-generation bioethanol from eucalypt sulphite spent liquor. Bioresour Technol 101:2755–2761. https://doi.org/10.1016/j.biortech.2009.11.092

Zabed H, Sahu JN, Boyce AN, Faruq G (2016) Fuel ethanol production from lignocellulosic biomass: an overview on feedstocks and technological approaches. Renew Sustain Energy Rev 66:751–774. https://doi.org/10.1016/j.rser.2016.08.038

Zabed H, Sahu JN, Suely A et al (2017) Bioethanol production from renewable sources: current perspectives and technological progress. Renew Sustain Energy Rev 71:475–501. https://doi.org/10.1016/j.rser.2016.12.076

Zhao X, Wei L, Cheng S, Julson J (2017) Review of heterogeneous catalysts for catalytically upgrading vegetable oils into hydrocarbon biofuels. Catalysts 7.https://doi.org/10.3390/catal7030083

Chapter 7
Bio-derived and Waste Fats Use for the Production of Drop-In Fuels

Mattia Bartoli, Mauro Giorcelli, Ruggero Vigliaturo, Pravin Jagdale, Massimo Rovere, and Alberto Tagliaferro

7.1 Introduction

Nowadays, the estimation of global primary energy demand is rising year by year (Semieniuk et al. 2021). With the increment of energy consumption, the greenhouse gas emissions marked a new peak records with increment higher than 5% on yearly base (Liu and Raftery 2021). The continuos growth of energy demand and greenhouse gas emissions is drastically affecting the environment and the global energy reserves (Nakicenovic et al. 1998) considering that, even it will far from exhausted, there is a finite supply of crude oil in the world (Welsby et al. 2021). Consequently, the research for renewable resources able to supply alternatives to oil has become very attractive (Dorian et al. 2006). Biomasses represent a very promising source to provide a more sustainable chemicals and fuels for supporting the growing demand for energy and raw materials. Furthermore, biomass use could reduce environmental impact of a oil based society (Kajaste 2014). Nonetheless, the direct use of lignocellulosic biomass as feedstock in the established productive processes is slowed down by their high oxygen content (Cherubini 2010). Accordingly, sustainable fuels industry has actively searched for alternative sustainable feedstocks. Among all the available ones

M. Bartoli (✉) · M. Giorcelli · P. Jagdale
Center for Sustainable Future Technologies @POLITO, Fondazione Istituto Italiano di Tecnologia, Via Livorno 60, 10144 Turin, Italy
e-mail: mattia.bartoli@polito.it

M. Bartoli · M. Giorcelli · P. Jagdale · M. Rovere · A. Tagliaferro
National Consortium for Materials Science and Technology (INSTM), Via G. Giusti 9, 50121 Florence, Italy

R. Vigliaturo
Department of Earth Sciences, University of Turin, Via Valperga Caluso, 35, 10125 Torino, Italy

M. Rovere · A. Tagliaferro
Department of Applied Science and Technology, Politecnico di Torino, C.so Duca degli Abruzzi 24, 10129 Turin, Italy

G. Di Blasio et al. (eds.), *Clean Fuels for Mobility*, Energy, Environment, and Sustainability, https://doi.org/10.1007/978-981-16-8747-1_7

fatty acids and their derivatives are the most attractive due to the low oxygen content and they similarity with linear hydrocarbons (Atabani et al. 2013). First and foremost, the fats conversion processes have been based on the production of biodiesel through transesterification of triglycerides by using short chain alcohols (Meher et al. 2006). In the first decade of twenty-first century, biodiesel commodities have been one of the main driving forces of the green economy (Gasparatos et al. 2017; Mahmudul et al. 2017). Nevertheless, the economical feasibility of biodiesel production is too poor and lead to a crisis of biodiesel producer in the advanced countries (Firoz 2017). An alternative route to exploit the great potential of fats is represented by their thermal cracking through pyrolytic processes (Maher and Bressler 2007). This approach could lead to the formation of high-quality oil that display the properties of drop-in fuels.

Accordingly, in the next sections we overview the main achievement of pyrolytic conversion of fats (triglycerides, fatty acids, waste fats) considering both thermal and thermocatalytic conversion approaches.

7.2 About the Mechanism of Fats Cracking

A natural fatty acid is composed by a carboxylic functionality directly bonded to a long saturated or unsaturated alkyl chain and to a polyalcohol (Das 2006). A comprehensive insight on cracking process of fatty acids esters is reported in Fig. 7.1.

The cracking process of fatty acids esters start with the heterolytic cleavage of O–C bond in the carboxylic functionality occurring for temperatures higher than 280 °C (Zhenyi et al. 2004). Both pathways shown in Fig. 7.1 lead to the production

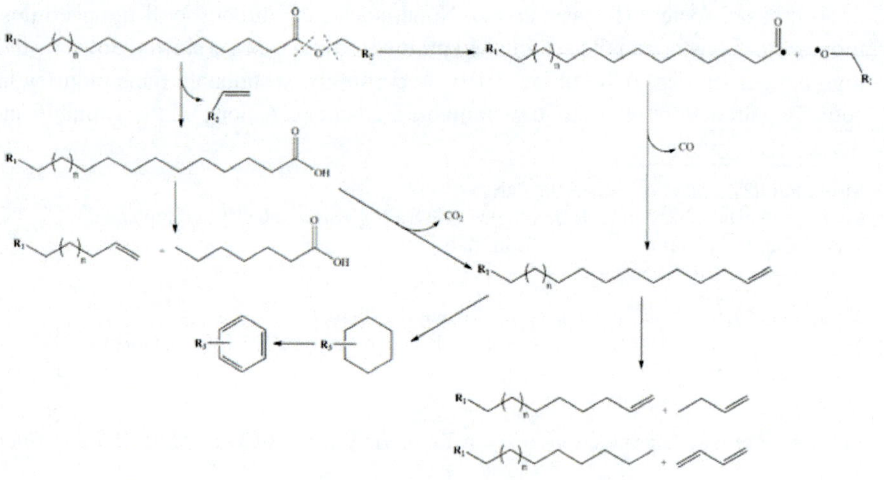

Fig. 7.1 Main pyrolytic degradation pathway of fatty acids esters and main products formed accordingly with chemical pathways reported by Chang and Wan (1947)

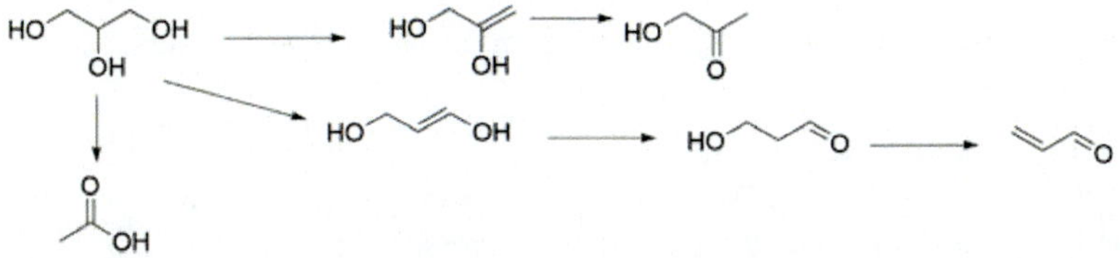

Fig. 7.2 Main pyrolytic degradation pathway of glycerol and main products formed

of terminal alkenes with different intermediates. The first degradative mechanism is based on the formation of free carboxylic acids and unsaturated species originated by the thermal decomposition of polyalcohol. The free fatty acid rapidly undergoes to a complex free-radicals reactions degradation starting with the release of a molecule of carbon dioxide at temperature higher than 400 °C. The other initial degradative route produces carbon monoxide at temperature higher the 300 °C. The radical cleavage of the carboxylic group with the formation of carbon dioxide or carbon monoxide represents a very important step that leads to the formation of linear, cyclic and aromatics hydrocarbons (Gunstone 2009). The formation of terminal alkene open up the way to reactions very close to those observed in the pyrolysis of polyolefins such poly(ethylene) (Aboulkas and El Bouadili 2010). The unsaturated species could be formed cyclic compounds that though further dehydrogenation reactions produce aromatic moieties. Also, advanced cracking of long saturated chains takes place with the formation of small hydrocarbons collected in the produced gaseous phase.

Glycerol represents the most common polyalcohol in the fatty acids esters and its pyrolysis has been deeply studied (Geng et al. 2012) and a summary of the different pathways are reported in Fig. 7.2.

Glycerol cracking produces a great variety of oxygenated compounds that could cross react with intermediates produced from fatty acids cracking with the formation of functionalized cyclic and linear compounds (Kitamura 1971; Maher and Bressler 2007).

IN accordance, the proposed mechanisms fats thermal cracking is a very sound route to produce drop-in fuels. Nonetheless, several issue should be solved prior the commercialization of fat derived fuels due to the presence of unreacted fatty acids and several impurities.

As reported by Gad et al. (2021), the hydrocarbon mixture recovered from fat pyrolysis showed values comparable with those of a traditional diesel but with a considerably lower Flash points as shown in Table 7.1.

Gad and co-workers were able to use the fats derived fuels in model diesel engine achieving the same output reached by using the commercial fuel. Similar results were achieved by Krishania et al. (2020) using a *Jatropha* derived pyrolytic oil but they clearly showed that the liquid fraction from pyrolysis could be at maximum 20 wt% of the fuel. By increasing the amount of fats derived oil the overall outputs of the engine decreased even if the amount used is considerably higher than other bioderived mixtures such long chain alcohols (Kumar et al. 2016) and glycerol ethers (Frusteri et al. 2013).

Table 7.1 Comparison between properties of a commercial diesel and a fat derived oil as reported by Gad et al. (2021)

Property	Diesel standard	Fat derived oil
Density measured at 15 °C (kg/m^3)	832	817
Kinematic viscosity measured at 40 °C (cSt)	Up to 7	2.288
Flashpoint point (°C)	>55	28
Pour point (°C)	Up to15	−27
Cetane number	>46	54.6
Calorific value (MJ/kg)	>44.3	46.6

7.3 Pyrolytic Conversion of Fatty Acids to Drop-In Fuels

7.3.1 A Concise History of the First Studies on Triglycerides Cracking

First studies of triglycerides pyrolysis were carried out in the first half of the twentieth century by Chang and Wan (1947) evaluating the use of cracked tung oil as engine fuels. In the second half of the century, many authors extended this first pioneering study on a plethora of triglycerides model compounds (Kitamura 1971; Nawar 1969; Nichols and Holman 1972) evaluated the effects of hydrocarbon chains structure and chemical degradative pathway.

These first pyrolysis experiments were generally carried out in pressurized batch reactors at temperatures ranging from 300 up to 500 °C and atmospheric pressure. Alencar et al. (1983) studied the pyrolysis products of tropical vegetable oils reporting the formation of mainly alkanes and terminal unsaturated alkenes. Schwab et al. (1988) used a distilling cracking approach to crack a mixture of high-oleic safflower oil and soybean. Authors recovered up to 79 wt% of liquid fraction composed by linear and cyclic hydrocarbon with up to 2% of aromatics and up to 16 of free carboxylic acids from 9.6–16.1%. A similar approach was used by Lima et al. (2004) studying the pyrolytic behaviour of vegetable oils mix in a temperature range from 350 up to 400 °C. The most abundant products recovered were linear alkane and alkenes but also alkadienes and free-carboxylic acids were observed. Surprisingly, no aromatic compounds were observed probably due to the insufficient high temperatures used.

7.3.2 Thermals Cracking of Fatty Acids

Moving from triglycerides to fatty acids, Jaw et al. (2001) studied the pyrolytic behaviour of stearic acid by using a simple thermogravimetric approach evaluating the main kinetic parameters. Alencar et al. (1983) pyrolyzed oleic acid produced only a 11% yield by volume of liquid through cracking distillation achieving a composition rich in cycloalkanes. An improvement in free oleic acid was reported by Asomaning et al. (2014c). In their work, authors used a pressurized reactor for performing the pyrolysis in the range 350–450 °C and processing time from 30 min up to 8 h. Data reported suggested that decarboxylation occurring as very first step of degradation of oleic acid together with the breaking of unsaturated carbon–carbon bonds. Decarboxylation was concurrent with decarbonylation pathway that was favorited for lower temperature. These results suggested that the optimum conditions for recover the highest amount of poor free carboxylic acid were a temperature of up to 410 °C and a residence time of 4 h. Asomaning et al. (2014a) studied also the pyrolysis of linoleic acid in the same condition previously adopted for oleic acid. Authors described the production of gaseous hydrocarbon mixture with up to five carbon atoms together with carbon dioxide and carbon monoxide. The liquid fractions recovered showed a considerable amount of linear hydrocarbons with six and ten carbon atoms suggesting that the allylic bonds were the preferentially breaking point. This also induced the formation of an appreciable amount of cyclic saturated and unsaturated species. Similarly, Maher et al. (2008) convert in the same pressurized reactor stearic acid by using a temperature ramp from 370 up to 450 °C observing up to 36 wt% of unreacted feedstock probably due to the combination of limited process times and temperature value. Nonetheless, the liquid recovered contained an interesting amount of linear and cyclic saturated and unsaturated hydrocarbons together with a limited amount of aromatic species. A more complex mixture of abietic acids and other chemical species known as tall oil was studied by Jenab et al. (2014) in temperature range from 370 up to 450 °C. Authors reported the formation of aromatic compounds with three membered rings at all temperature adopted for the conversion. Nonetheless, the yield of aromatics decreased with the decreasing of process temperature. At lower temperature the liquid composition was dominated by the presence of alkane and alkane with a negligible amount of aromatic species.

Fukushima et al. (1983) investigate the cracking of fatty acids by using vacuum pyrolysis reporting a concentration of free fatty acids of up to 13 mg/g and an alcohol amount of up to 1 mg/g. Simple esters pyrolysis was studied by Mao et al. (2021) by using a tubular reactor. The intrinsically limitations of this approach were overcome by developing an inductively heated reactor coupled with atomization feeding. This system removed any kind of appreciable thermal gradient allowing a very uniform thermal profile. Authors reached a liquid yield of high-quality drop-in fuel of up to 60 wt% at a temperature of 520 °C.

The other class of fatty acids-based compounds studied for the drop-in fuel production is represented by the fatty acid soaps. Fatty acids soaps are salts recovered after the reaction of free fatty acids with alkali such sodium or potassium hydroxides.

In the middle of the twentieth century, Hsu et al. (1950) reported a study on the pyrolysis of calcium derived soaps of a mixture of tung and stearic acids using a cracking distillation system recovering an oil with a 62 wt% of unsaturated hydrocarbons and up to 13 wt% of aromatics. Authors also suggest a mechanism for the formation of olefins though the rearrangement of cyclic diolefins formed though retro Diels–Alder reaction. Demirbaş (2002) compared the methyl ester derivatives with soap derived pyrolytic oil as diesel fuel replacements. The viscosity and the total acid number (TAN) of both were comparable with a diesel fuel type 2. Similarly, Fortes and Baugh (1994) evaluate with a comprehensive set of analytical measurements the oil composition recovered from a calcium soap pyrolysis proving that the composition was totally compatible with a diesel like fuel.

Lappi and Alén (2009) evaluated the pyrolytic behaviour of stearic, oleic and linoleic sodium salts in a temperature range from 450 up to 750 °C with residence time from 20 up to 80s. Sodium stearate pyrolysis led to the formation of a linear alkanes and alkenes while unsaturated soaps led to the formation of cyclic and aromatic compounds in an appreciable amount. Curiously, linoleate pyrolysis produced a great amount of volatile.

7.3.3 From Waste to Value: Use of Waste Streams Source for Drop-In Fuels Production via Fats Pyrolysis

The use of vegetable oils rose up a problem due to the sustainability of the drop-in fuel production. The high cost and the other use of these feedstocks slowed down the spread of thermal lipid conversion technology. Nonetheless, many waste fat source are currently available for the production of more sustainable waste derived fuels. A first case of study was reported by Dandik and Aksoy (1998) by using waste cooked oils using a cracking-distillation process with temperature ranging from 400 up to 420 °C. By using a process time of up to 3 h, authors produced a pyrolytic oil rich in long chain hydrocarbons including paraffins, terminal and internal olefins, cyclic compounds and aromatic ones. Interestingly, the boiling range of pyrolysis oil matched the one of gasoline. Adebanjo et al. (2005) pyrolyzed lard in a continuous microreactor filled with quartz pellets in a temperature range from 600 up to 800 °C. They reported the production of oil up to 61 wt% with a low TAN and aromatic content.

Another interesting waste feedstock for drop-in fuel production is brown grease. Brown grease is defined as the waste oil collected from grease traps installed in commercial, industrial or municipal sewage facilities to separate oily fraction from the watery one.

Brown grease was studied by Sim et al. (2017) by suing model compounds such as palmitic and oleic acids in the ratio commonly found in real brown grease sample and by adding iron salts that is also commonly present in it. The drop-in fuel produced was composed by aliphatic hydrocarbons and some palmitone from the pyrolysis of palmitc acid in the presence of iron with palmitone that rose up to 50 wt% of the oil in the absence of iron salts. Oleic acid pyrolysis led to the formation of short saturated fatty acids, saturated and unsaturated hydrocarbons and oxygen-based compounds (i.e. alcohols and carbonyls).

Strothers et al. (2019) evaluated the effect of different temperature profiles in during the pyrolysis of brown grease producing a homologous series of terminal alkenes with the maximum chain length of seven carbon atoms by using a temperature ramp starting at 100 °C and reaching up to 350 °C over 5 h. Furthermore, the presence of cyclic compounds, both saturated and unsaturated, together with ketones and alcohols and was observed suggesting that the presence of oxygenated compounds was reasonably due to the water removal from the greasy mixture. A reactor preheating led to the degradation of fatty acids to unsaturated fatty acids and low molecular weight hydrocarbons together with high molecular weight ketones. The influence of residual pressure on the product distribution was investigated by Pratt et al. (2021). The pressurized reactors led to the production of high yields of liquid fraction without a massively gasification even at the higher temperature used.

Ratton Coppos et al. (2018) pyrolyzed brown grease soap in a stirred reactor at 500 °C at atmospheric pressure. Furthermore, processing brown grease soap led to high liquid yield close to 60 wt% with very low TAN. The lowest acidity achieved was close to 0.5 mg/g of KOH equivalent.

Nonetheless, the great variability of brown grease together with the presence of fatty acids ester and hydrosoluble species represent an unneglectable issue. Accordingly, Asomaning et al. (2014d) developed a two stage technology for processing brown and other waste derived greases. The first step of the process is represented by a hydrolytic treatment in a pressurized reactor at 280 °C for 1 h. The organic phase recovered was composed by free fatty acids and few small organic molecules. This material was used as feedstock in the second pyrolytic stage carried out in a pressurized microreactor. The oils recovered from all the greases were quite close to each other with a similarity of up to 80% with a high amount of linear non branched alkanes. The great homogeneity was reasonably due to the hydrolytic treatment that removed induced a hydroxylation of double bonds and a removal of contaminant. The results showed that up to 30 wt% of pyrolytic oil was compatible with gasoline uses while up to 50 wt% was indistinguishable from diesel. This approach was very promising but also costly. In order to mitigate the cost of the hydrolytic stage Omidghane et al. (2020) substituted the water with biosolids. Biosolids are a residual fraction from water treatments with a water content of up to 90 wt% (Wang et al. 2008). Omidghane and co-workers investigate the pyrolysis of biosolids hydrolyzed brown grease in a microreactor in temperature range from 350 up to 450 °C using residence time from 1 up to 3 h. Authors proved the complete similarity of product distribution in pyrolytic oils recovered with the ones produced by Asomaning et al. (2014d). Nevertheless, they observed a rising of sulphur content up to hundred ppm suggesting that was due to the initial presence of sulpholipids and sulphonated organic species that were retained into the pyrolytic fractions. Even if fatty acids and triglycerides pyrolytic oils could be very close to commercial fuels they could lack in aromatic content or they have a very high TAN. Accordingly, several strategies have been developed for improved their quality.

Asomaning et al. (2014b) improved the branching degree of hydrocarbon produced by pyrolysis of oleic acid by running the pyrolytic conversion in a pressurized reactor with a residual atmosphere composed by methane, ethane, propane, ethylene or propylene. Asomaning et al. reported the presence of the common product distribution previously observed for pyrolysis of oleic acid under inert atmosphere by using residual atmosphere of saturated hydrocarbons. Oils recovered from these runs were composed by both saturated and unsaturated linear hydrocarbons together with free fatty acids and cyclic compounds. The use of unsaturated hydrocarbon as residual atmosphere inside the reactor induced firstly an increment of liquid yield up to 98% and induced the formation of branched hydrocarbons. These latter species are highly interesting to improve the fuel properties of any kind of drop-in fuels and with this approach were produced without any catalyst addition.

7.3.4 Catalytic Upgrading of Thermal Cracked Fats

Pyrolytic routes are very interesting but an in-situ or ex-situ catalytic upgrading of drop-in fuels produced from fats pyrolysis helps to match the fuel market requirements.

A first report about catalytic conversion of fats was described by Rao (1978). In this patent, the author reported a comprehensive list of catalyst used for the advance cracking of triglycerides. Among the most used materials, zeolites, alumina and silica based materials represent the most performing materials (Twaiq et al. 1999; Tan et al. 2019).

Dandik et al. (1998) pyrolyzed sunflower oil was a 400–420 °C in reactor packed with zeolite HZSM-5 and equipped with a fractionating packed column achieving a feedstock conversion of up to 97 wt% with a production of 33 wt% of liquid hydrocarbon and a variable amount of aromatics based on the fractionating column length.

By using a similar system, Katikaneni et al. (1996) evaluated the influence of HZSM-5 residual acid sites on the pyrolysis of canola oil in a temperature range from 400 up to 500 °C testing also the effect of space velocity observing that the incorporation of potassium into the zeolite structure improved the catalyst toughness preserving the activity of acidic sites during the catalytic pyrolysis process. Furthermore, they proved that the zeolite catalyst induced a strong oligomerization and aromatization of unsaturated species formed during the cracking of fats. Additionally, authors reached the optimum conditions for the production of small chains hydrocarbon with less than five carbon atoms operating at 500 with a space velocity of up to 1.8 h^{-1}.

Prasad et al. (1986a, b) deeply studied the canola oil pyrolysis over HZSM-5 with and without a steam stream. Authors reported a conversion of over up to 96% with an aromatic yield up to 60 wt%. Furthermore, they proved that the spent catalyst could be easily regenerated with a thermal treatment at 600 °C in dry air for 1 h. The steam stream also reduced the formation of both coke and tar with a further improvement of the overall quality of the process. Similarly, Idem et al. (1997) studied the effect of acidic/basic sites on the catalytic pyrolysis of canola in an atmospheric pressure fixed-bed microreactor at temperatures of 400 and 500 °C over several catalysts (i.e. HZSM-5, silica and silica derivatives, γ-alumina, basic oxide) clearly showing that the first steps of canola oil decomposition were independent from the catalyst selection. Nevertheless, the advance cracking processes were boosted by amorphous and disordered materials such as alumina and silica alumina hybrids. Highly ordered materials such as zeolites induced mild advanced cracking reactions with low gas and greater liquids yields. Furthermore, authors observed the inhibition of secondary and advance cracking reactions in the presence of highly basic materials such magnesium and calcium oxides with a greater formation of paraffins. Contrary, acidic sites rich materials promoted the formation of olefins and aromatic compounds with an increment of hydrocarbon chain length with increased of ordering of catalyst surface. Shitao et al. (2020) modified the magnesium oxide.

Wang et al. (2021)modified an aluminium MCM-41 zeolite by doped it with a phosphorous or metals (i.e. lanthanum, nickel, zirconium, cerium) for improved the quality of pyrolytic oil produced from oleaginous biomasses observing an improved ability of the modified MCM-41 to induced the decarboxylation of fatty acids with a greater production of alkane and aromatics. Among all catalysts prepared, nickel based one was the most performing in the in deoxy- and in dehydrogenation while lanthanum based one was the best in preventing the coke formation. Zirconium and cerium based MCM-41 promote and efficient decarbonxylation but the phosphorous doped catalyst was better. Authors suggested that this was due to the higher acidity with a consequently magnification of hydrocarbon yield that reached of up to 82 wt% and TAN close to zero. Furthermore, the oil recovered displayed a very low oxygen content down to 4wt% and a very high calorific value of up to 45 MJ/kg.

Twaiq et al. (2004) compared a HZSM-5/molecular sieve hybrid materials with pure HZSM-5 and MCM-41 zeolites as cracking catalyst for the pyrolysis of palm oil. Authors proved the improvement induced by the addition of the high surface molecular sieves material than allowed to reach a near to completion conversion with a gasoline like fraction recovered up to 47 wt%.

Similarly, Vonghia et al. (1995) studied the catalytic pyrolysis of triglycerides over alumina at 450 °C in a continuous reactor with a hourly space velocity of up to 0.46 h^{-1} by using model compounds accordingly with the composition of vapor phase reported in literature. As emerging from the collected data, authors suggested that triglycerides were split out the carboxylic acids by elimination and formed alkanes though hydrogen transfer mechanism. Furthermore, they observed the formation of ketones as intermediates of fatty acids decarboxylation. These species can undergo to further hydrogen transfer process with the formation or monounsaturated hydrocarbons and methyl ketones and methyl ketones. An unknown reductive mechanism followed by dehydration produces monoalkenes from the methyl ketones. Further speculative reaction pathways were proposed to summarize the evolution of methyketone species but they did not find any support in other published papers.

Snåre et al. (2008) studied the deoxygenation of pyrolytic oil produced from oleic, linoleic and methyl oleate over a carbon supported palladium catalyst in a semi-batch reactor carrying out the catalytic run in a pressure range from 15 up to 27 bar and in a temperature range from 300 up to 360 °C. Authors observed an initial hydrogenation of carbon–carbon double bonds and subsequent deoxygenation process. They also reported the isomerization of the unsaturations.

Morgan et al. (2010) studied the deoxygenation of triglycerides using tristearin, triolein and soybean oil by suing several carbon supported metallic catalysts with a variable loading ranging from 1 up to 20 wt% testing nickel, palladium and platinum in batch conditions. Nickel based material reached a quasi-quantitative conversion of triglyceride in each case either a liquid fraction rich in linear hydrocarbon with chain lengths comprised from five up to seventeen carbon atoms. Palladium and platinum catalyst showed an inferior performances in both cracking and upgrading process. Nevertheless, the consistency of the data was questionable due to the different loading used. Authors tested 20 wt% nickel loaded catalysts and compared it with palladium and platinum loaded of up to 1 and 5 wt%. This drastically decreased the trustworthily of the conclusion reported. Anyhow, it was at least noticeably the activity of the nickel material.

Other exotic catalyst for conversion of brown grease was described by Noshadi et al. (2014) using of ionic liquids as functionalizing agent for tailoring the surface activity of a mesoporous polymeric support. This material led to the near quantitative conversion of brown grease to a real drop-in fuel after a pyrolytic conversion at 600 °C. Bashir et al. (2020) converted brown grease into a drop-in fuel by using a more classical acid catalyst based on iron sulphate producing a biodiesel quality fuel in batch condition using methanol as solvent and reactant.

Zheng et al. (2020) proposed a different approach based on the one stage conversion where the vapor produced from carking of oleic acid directly pass through over a nickel/copper based catalyst on the reactor exit supporting the metallic catalyst over a several support (i.e. activated carbon, zirconia, alumina, zeolites) changing the nickel/copper ratio constant. Authors suggested that Brønsted acid sites of zeolites promoted hydrodeoxygenation while the Lewis acid sites of zirconia and alumina promoted hydrogenation and decarbonylation pathways. Furthermore, Zheng and co-workers supposed that the presence of copper inhibit the coke formation increasing the durability of the catalyst. They reached the maximum hydrocarbon yields, very close by to 100%, under pyrolytic temperature of up to 400 °C and a catalytic bed temperature of up to 500 °C using a catalyst supported onto zeolite MCM-41. In this case, authors reported a selectivity versus linear hydrocarbon with seventeen carbon atoms, mainly olefins, up to 27% and a total deoxygenation calculated on the initial amount of feedstock up to 87%.

Wang et al. (2020) used a similar approach to convert soaps (sodium stearate, palmitate, oleate and linoleate) over HZSM-5 catalyst reporting an appreciable increment of toluene and xylene amount. They also observed an increment of polycyclic aromatic compounds with the increment of soaps unsaturation. They suggested a rearrangement mechanism involving the olefins condensation and dehydrogenation over the catalyst. Another approach related to conversion of fatty acids and fatty acids derivatives is represented by microwave assisted pyrolysis. Microwave heating is a very promising methodology for the conversion of a plethora of different biomasses (Giorcelli et al. 2021) but the direct conversion of fatty acids and fatty acids derivative is quite challenging. This was due to the poor interactions occurring between microwave radiation and fats with a consequently poor heating phenomena (Bartoli et al. 2019). Nonetheless the use of soap instead of neat materials could be usefully to perform MAP. This is due to the improved ability of ionic substances to efficiently interact with microwave radiation and dissipating heat.

Wang et al. (2016) evaluated the soap behavior during MAP of several materials such as sodium stearate, oleate, and linoleate clearly enlighten the massively production of highly unsaturated species and aromatic compounds. Furthermore, the oil recovered showed a very low TAN.

Rapeseed oil was also used as feedstock for MAP conversions as reported by Omar and Robinson (2014) comparing the outputs of MAP with the ones of traditional distillative pyrolysis in a temperature range from 400 up to 500 °C. They reported the production of high quantity of diesel grade fuel by using MAP with a huge presence of aromatics cyclic and linear unsaturated compounds compared with liquids recovered from the classical pyrolysis.

7.4 Conclusions

Fats pyrolysis could represent a real game change event in the field of sustainable fuels. The high yields together with similarity with oil derived fuels have risen the fats derived pyrolytic oils to a strong candidate to lead the mankind transition to a more environmental care society. The possibility of using waste derived fats stream support the he candidacy of fat derived drop-in fuels as main actor in the realm of green fuels overwhelming the biodiesel for all aspects.

Among the main key features of the use of fats feedstock for drop-in fuel production should be highlighted the following points:

- the worldwide availability of the fat feedstock
- the possible replacement of edible and high value fats with fats derived wastestreams
- the very close composition of fats derived hydrocarbon mixtures to an oil derived diesel fraction.

Nevertheless, some issues remain unsolved such as:

- the high total acid number of the fats derived hydrocarbon mixtures
- the utilization of glycerol wastestream produced by the conversion of triglycerides
- the competitiveness of fats derived fuels cost price with oil derived ones.

We believe that further advancement in both reactor technology and catalysts development will guide the fats-based technology to a new shining era. Furthermore, we hoped that the improved environmental accountability will boost the enforcement of regulations abled to promote the spread of fats derived technology.

References

Aboulkas A, El Bouadili A (2010) Thermal degradation behaviors of polyethylene and polypropylene: Part I Pyrolysis kinetics and mechanisms. Energy Convers Manage 51(7):1363–1369

Adebanjo AO, Dalai AK, Bakhshi NN (2005) Production of diesel-like fuel and other value-added chemicals from pyrolysis of animal fat. Energy Fuels 19(4):1735–1741. https://doi.org/10.1021/ef040091b

Alencar J, Alves P, Craveiro A (1983) Pyrolysis of tropical vegetable oils. J Agric Food Chem 31(6):1268–1270

Asomaning J, Mussone P, Bressler DC (2014a) Pyrolysis of polyunsaturated fatty acids. Fuel Process Technol 120:89–95. https://doi.org/10.1016/j.fuproc.2013.12.007

Asomaning J, Mussone P, Bressler DC (2014b) Thermal cracking of free fatty acids in inert and light hydrocarbon gas atmospheres. Fuel 126:250–255. https://doi.org/10.1016/j.fuel.2014.02.069

Asomaning J, Mussone P, Bressler DC (2014c) Thermal deoxygenation and pyrolysis of oleic acid. J Anal Appl Pyrol 105:1–7. https://doi.org/10.1016/j.jaap.2013.09.005

Asomaning J, Mussone P, Bressler DC (2014d) Two-stage thermal conversion of inedible lipid feedstocks to renewable chemicals and fuels. Biores Technol 158:55–62. https://doi.org/10.1016/j.biortech.2014.01.136

Atabani A, Silitonga A, Ong H, Mahlia T, Masjuki H, Badruddin IA, Fayaz H (2013) Non-edible vegetable oils: a critical evaluation of oil extraction, fatty acid compositions, biodiesel production, characteristics, engine performance and emissions production. Renew Sustain Energy Rev 18:211–245

Bartoli M, Frediani F, Briens C, Berruti F, Rosi L (2019) An overview of temperature issues in microwave-assisted pyrolysis. Processes 7(10)

Bashir MJK, Wong LP, Hilaire DS, Kim J, Salako O, Jean MJ, Adeyemi R, James S, Foster T, Pratt LM (2020) Biodiesel fuel production from brown grease produced by wastewater treatment plant: optimization of acid catalyzed reaction conditions. J Environ Chem Eng 8(4):103848. https://doi.org/10.1016/j.jece.2020.103848

Chang C-C, Wan S-W (1947) China's motor fuels from tung oil. Ind Eng Chem 39(12):1543–1548

Cherubini F (2010) The biorefinery concept: using biomass instead of oil for producing energy and chemicals. Energy Convers Manage 51(7):1412–1421

Dandik L, Aksoy HA (1998) Converting used oil to fuel and chemical feedstock through a fractionating pyrolysis reactor. In: Kaseoglu SS, Rhee KC, Wilson RF (eds) Proceedings of the world conference on oilseed and edible oil processing, Istanbul, Turkey

Dandik L, Aksoy HA, Erdem-Senatalar A (1998) Catalytic conversion of used oil to hydrocarbon fuels in a fractionating pyrolysis reactor. Energy Fuels 12(6):1148–1152. https://doi.org/10.1021/ef980012u

Das UN (2006) Essential fatty acids-a review. Curr Pharm Biotechnol 7(6):467–482

Demirbaş A (2002) Diesel fuel from vegetable oil via transesterification and soap pyrolysis. Energy Sources 24(9):835–841

Dorian JP, Franssen HT, Simbeck DR (2006) Global challenges in energy. Energy Policy 34(15):1984–1991. https://doi.org/10.1016/j.enpol.2005.03.010

Firoz S (2017) A review: advantages and disadvantages of biodiesel. Int Res J Eng Technol 4(11):530–533

Fortes I, Baugh P (1994) Study of calcium soap pyrolysates derived from Macauba fruit (Acrocomia sclerocarpa M.). Derivatization and analysis by GC/MS and CI-MS. J Anal Appl Pyrol 29(2):153–167

Frusteri F, Cannilla C, Bonura G, Spadaro L, Mezzapica A, Beatrice C, Di Blasio G, Guido C (2013) Glycerol ethers production and engine performance with diesel/ethers blend. Top Catal 56(1):378–383. https://doi.org/10.1007/s11244-013-9983-7

Fukushima K (1983) Investigation of the possible role of fatty acids and alcohols as precursors of normal hydrocarbons generated during vacuum pyrolysis of a young kerogen. J Anal Appl Pyrol 5(3):245–256. https://doi.org/10.1016/0165-2370(83)80031-X

Gad MS, Abu-Elyazeed OS, Mohamed MA, Hashim AM (2021) Effect of oil blends derived from catalytic pyrolysis of waste cooking oil on diesel engine performance, emissions and combustion characteristics. Energy 223:120019. https://doi.org/10.1016/j.energy.2021.120019

Gasparatos A, Doll CNH, Esteban M, Ahmed A, Olang TA (2017) Renewable energy and biodiversity: Implications for transitioning to a green economy. Renew Sustain Energy Rev 70:161–184. https://doi.org/10.1016/j.rser.2016.08.030

Geng Z, Zhang M, Yu Y (2012) Theoretical investigation on pyrolysis mechanism of glycerol. Fuel 93:92–98. https://doi.org/10.1016/j.fuel.2011.08.021

Giorcelli M, Das O, Sas G, Försth M, Bartoli M (2021) A review of bio-oil production through microwave-assisted pyrolysis. Processes 9(3):561

Gunstone F (2009) The chemistry of oils and fats: sources, composition, properties and uses. John Wiley & Sons, New York, U.S.

Hsu H-L, Osburn J, Grove C (1950) Pyrolysis of the calcium salts of fatty acids. Ind Eng Chem 42(10):2141–2145

Idem RO, Katikaneni SPR, Bakhshi NN (1997) Catalytic conversion of canola oil to fuels and chemicals: roles of catalyst acidity, basicity and shape selectivity on product distribution. Fuel Process Technol 51(1):101–125. https://doi.org/10.1016/S0378-3820(96)01085-5

Jaw K-S, Hsu C-K, Lee J-S (2001) The thermal decomposition behaviors of stearic acid, paraffin wax and polyvinyl butyral. Thermochim Acta 367:165–168

Jenab E, Mussone P, Nam G, Bressler D (2014) Production of renewable hydrocarbons from thermal conversion of abietic acid and tall oil fatty acids. Energy Fuels 28(11):6988–6994. https://doi.org/10.1021/ef501746b

Kajaste R (2014) Chemicals from biomass–managing greenhouse gas emissions in biorefinery production chains–a review. J Clean Prod 75:1–10

Katikaneni SPR, Adjaye JD, Idem RO, Bakhshi NN (1996) Catalytic conversion of canola oil over potassium-impregnated HZSM-5 catalysts: C2–C4 olefin production and model reaction studies. Ind Eng Chem Res 35(10):3332–3346. https://doi.org/10.1021/ie950740u

Kitamura K (1971) Studies of the pyrolysis of triglycerides. Bull Chem Soc Jpn 44(6):1606–1609

Krishania N, Rajak U, Verma TN, Birru AK, Pugazhendhi A (2020) Effect of microalgae, tyre pyrolysis oil and Jatropha biodiesel enriched with diesel fuel on performance and emission characteristics of CI engine. Fuel 278:118252

Kumar BR, Saravanan S, Rana D, Nagendran A (2016) A comparative analysis on combustion and emissions of some next generation higher-alcohol/diesel blends in a direct-injection diesel engine. Energy Convers Manage 119:246–256

Lappi H, Alén R (2009) Production of vegetable oil-based biofuels—thermochemical behavior of fatty acid sodium salts during pyrolysis. J Anal Appl Pyrol 86(2):274–280. https://doi.org/10.1016/j.jaap.2009.07.005

Lima DG, Soares VC, Ribeiro EB, Carvalho DA, Cardoso ÉC, Rassi FC, Mundim KC, Rubim JC, Suarez PA (2004) Diesel-like fuel obtained by pyrolysis of vegetable oils. J Anal Appl Pyrol 71(2):987–996

Liu PR, Raftery AE (2021) Country-based rate of emissions reductions should increase by 80% beyond nationally determined contributions to meet the 2 °C target. Commun Earth Environ 2(1):29. https://doi.org/10.1038/s43247-021-00097-8

Maher K, Bressler D (2007) Pyrolysis of triglyceride materials for the production of renewable fuels and chemicals. Biores Technol 98(12):2351–2368

Maher KD, Kirkwood KM, Gray MR, Bressler DC (2008) Pyrolytic decarboxylation and cracking of stearic acid. Ind Eng Chem Res 47(15):5328–5336. https://doi.org/10.1021/ie0714551

Mahmudul H, Hagos F, Mamat R, Adam AA, Ishak W, Alenezi R (2017) Production, characterization and performance of biodiesel as an alternative fuel in diesel engines–a review. Renew Sustain Energy Rev 72:497–509

Mao X, Xie Q, Yi X, Duan Y, Yu S, Wu Z, Liang X, Nie Y (2021) Fast pyrolysis of methyl ricinoleate in an inductively heated reactor coupled with atomization feeding. Appl Therm Eng 194:117093. https://doi.org/10.1016/j.applthermaleng.2021.117093

Meher L, Sagar DV, Naik S (2006) Technical aspects of biodiesel production by transesterification—a review. Renew Sustain Energy Rev 10(3):248–268

Morgan T, Grubb D, Santillan-Jimenez E, Crocker M (2010) Conversion of triglycerides to hydrocarbons over supported metal catalysts. Top Catal 53(11–12):820–829

Nakicenovic N, Grübler A, McDonald A (1998) Global energy perspectives. Cambridge University Press, Cambridge, U.K.

Nawar WW (1969) Thermal degradation of lipids. J Agric Food Chem 17(1):18–21

Nichols PC, Holman RT (1972) Pyrolysis of saturated triglycerides. Lipids 7(12):773–779

Noshadi I, Kanjilal B, Du S, Bollas GM, Suib SL, Provatas A, Liu F, Parnas RS (2014) Catalyzed production of biodiesel and bio-chemicals from brown grease using ionic liquid functionalized ordered mesoporous polymer. Appl Energy 129:112–122. https://doi.org/10.1016/j.apenergy.2014.04.090

Omar R, Robinson JP (2014) Conventional and microwave-assisted pyrolysis of rapeseed oil for bio-fuel production. J Anal Appl Pyrol 105:131–142. https://doi.org/10.1016/j.jaap.2013.10.012

Omidghane M, Bartoli M, Asomaning J, Xia L, Chae M, Bressler DC (2020) Pyrolysis of fatty acids derived from hydrolysis of brown grease with biosolids. Environ Sci Pollut Res:1–11

Prasad Y, Bakhshi N, Mathews J, Eager R (1986a) Catalytic conversion of canola oil to fuels and chemical feedstocks: Part I Effect of process conditions on the performance of HZSM-5 catalyst. Can J Chem Eng 64(2):278–284

Prasad Y, Bakhshi N, Mathews J, Eager R (1986) Catalytic conversion of canola oil to fuels and chemical feedstocks: Part II effect of co-feeding steam on the performance of HZSM-5 catalyst. Can J Chem Eng 64(2):285–292

Pratt LM, Kim J, Lo H-Y, Xiao D (2021) Brown grease pyrolysis under pressure: extending the range of reaction conditions and hydrocarbon product distributions. Fuel 289:119782. https://doi.org/10.1016/j.fuel.2020.119782

Rao KVC (1978) Production of hydrocarbons by thermolysis of vegetable oils

Ratton Coppos AR, Kahn S, Borges LEP (2018) Biofuels production by thermal cracking of soap from brown grease. Ind Crops Prod 112:561–568. https://doi.org/10.1016/j.indcrop.2017.12.010

Schwab A, Dykstra G, Selke E, Sorenson S, Pryde E (1988) Diesel fuel from thermal decomposition of soybean oil. J Am Oil Chem Soc 65(11):1781–1786

Semieniuk G, Taylor L, Rezai A, Foley DK (2021) Plausible energy demand patterns in a growing global economy with climate policy. Nat Clim Chang 11(4):313–318. https://doi.org/10.1038/s41558-020-00975-7

Shitao Y, Cao X, Wu S, Chen Q, Li L, Li H (2020) Effective pyrolysis of waste cooking oils into hydrocarbon rich biofuel on novel mesoporous catalyst with acid and alkali coexisting. Ind Crops Prod 150:112362. https://doi.org/10.1016/j.indcrop.2020.112362

Sim Y-L, Meyappan N, Yen NS, Swarna Kamala S, Khoo CH, Cheah WL, Hilaire DS, Pinnock T, Bacolod B, Cai ZB, Gurung D, Hasnat R, Strothers J, Remy CT, Gentles PK, Groveman S, Vittadello M, Kim J, Pratt LM (2017) Chemical reactions in the pyrolysis of brown grease. Fuel 207:274–282.https://doi.org/10.1016/j.fuel.2017.06.101

Snåre M, Kubičková I, Mäki-Arvela P, Chichova D, Eränen K, Murzin DY (2008) Catalytic deoxygenation of unsaturated renewable feedstocks for production of diesel fuel hydrocarbons. Fuel 87(6):933–945. https://doi.org/10.1016/j.fuel.2007.06.006

Strothers J, Matthews RB, Toney A, Cobham MR, Cox S, Ford W, Joseph S, Joyette W, Khadka S, Pinnock S, Burns M, Noel M, Tamang MG, Saint Hilaire D, Kim J-H, Pratt LM (2019) Hydrocarbon fuel from brown grease: effects of reaction temperature profile on yields and product distribution. Fuel 239:573–578. https://doi.org/10.1016/j.fuel.2018.11.053

Tan Y, Abdullah A, Hameed B (2019) Product distribution of the thermal and catalytic fast pyrolysis of karanja (Pongamia pinnata) fruit hulls over a reusable silica-alumina catalyst. Fuel 245:89–95

Twaiq FA, Zabidi NA, Bhatia S (1999) Catalytic conversion of palm oil to hydrocarbons: performance of various zeolite catalysts. Ind Eng Chem Res 38(9):3230–3237

Twaiq FAA, Mohamad AR, Bhatia S (2004) Performance of composite catalysts in palm oil cracking for the production of liquid fuels and chemicals. Fuel Process Technol 85(11):1283–1300. https://doi.org/10.1016/j.fuproc.2003.08.003

Vonghia E, Boocock DG, Konar SK, Leung A (1995) Pathways for the deoxygenation of triglycerides to aliphatic hydrocarbons over activated alumina. Energy Fuels 9(6):1090–1096

Wang H, Brown SL, Magesan GN, Slade AH, Quintern M, Clinton PW, Payn TW (2008) Technological options for the management of biosolids. Environ Sci Poll Res Int 15(4):308–317

Wang Y, Dai L, Shan S, Zeng Q, Fan L, Liu Y, Ruan R, Zhao Y, Zhou Y (2016) Effect of unsaturation degree on microwave-assisted pyrolysis of fatty acid salts. J Anal Appl Pyrol 120:247–251. https://doi.org/10.1016/j.jaap.2016.05.012

Wang Y, Ke L, Peng Y, Yang Q, Du Z, Dai L, Zhou N, Liu Y, Fu G, Ruan R, Xia D, Jiang L (2020) Characteristics of the catalytic fast pyrolysis of vegetable oil soapstock for hydrocarbon-rich fuel. Energy Convers Manage 213:112860. https://doi.org/10.1016/j.enconman.2020.112860

Wang F, Yu F, Wei Y, Li A, Xu S, Lu X (2021) Promoting hydrocarbon production from fatty acid pyrolysis using transition metal or phosphorus modified Al-MCM-41 catalyst. J Anal Appl Pyrol 156:105146. https://doi.org/10.1016/j.jaap.2021.105146

Welsby D, Price J, Pye S, Ekins P (2021) Unextractable fossil fuels in a 1.5 °C world. Nature 597(7875):230–234. https://doi.org/10.1038/s41586-021-03821-8

Zheng Y, Wang J, Liu C, Lu Y, Lin X, Li W, Zheng Z (2020) Efficient and stable Ni-Cu catalysts for ex situ catalytic pyrolysis vapor upgrading of oleic acid into hydrocarbon: Effect of catalyst support, process parameters and Ni-to-Cu mixed ratio. Renew Energy 154:797–812. https://doi.org/10.1016/j.renene.2020.03.058

Zhenyi C, Xing JI, Shuyuan LI, Li LI (2004) Thermodynamics calculation of the pyrolysis of vegetable oils. Energy Sources 26(9):849–856. https://doi.org/10.1080/00908310490465902

Chapter 8
Biodiesel as a Clean Fuel for Mobility

Ayat Gharehghani and Amir Hossein Fakhari

8.1 Introduction

According to IEA (2019), biofuel production in 2019 has increased by 6% compared to last year and is expected to grow by 3% annually over the next 5 years. Of course, this is less than 10% of the annual sustainable growth that is needed to align with the Sustainable Development Scenario (SDS) outline by 2030 (Fig. 8.1). Therefore, biofuel consumption will almost triple by 2030 to align with the SDS outline. This value is equal to 9% of the global fuel demand for transportation, compared to the 2018 level. Figure 8.2 compares the amount of biofuel production in 2018 with the amount of biofuel production required for the SDS in some regions of the world (IEA 2019).

According to Fig. 8.3, published research papers that were focused on the development of biofuels and their applications in several fields have increased from 1988 number of studies in 2010 to 4061 studies in 2018 (Ogunkunle and Ahmed 2019). Considering the growth of research interest regarding biofuels, it seems that the use of these fuels can be a suitable solution for reducing the consumption of fossil fuels and the consequent environmental concerns. The economic aspect of conventional biodiesel depends on raw materials, processing, transportation, crude oil prices, etc. (Koh and Ghazi 2011). Several studies (Atabani et al. 2012; Lin et al. 2011) have reported that raw materials are the main cost of biodiesel, which is approximately 75% of the production cost of biodiesel, as shown in Fig. 8.4. Chemicals and catalysts also affect the cost of producing biodiesel.

Biodiesel is an alternative fuel that can be produced from renewable sources such as vegetable and animal fats for compression ignition engines. It is an eco-friendly fuel that is non-toxic and easily degraded in nature. Moreover, this fuel presents a higher ignition temperature, more clean combustion, and higher efficiency than

A. Gharehghani (✉) · A. H. Fakhari
School of Mechanical Engineering, Iran University of Science and Technology, Tehran, Iran
e-mail: ayat_gharehghani@iust.ac.ir

G. Di Blasio et al. (eds.), *Clean Fuels for Mobility*, Energy, Environment, and Sustainability, https://doi.org/10.1007/978-981-16-8747-1_8

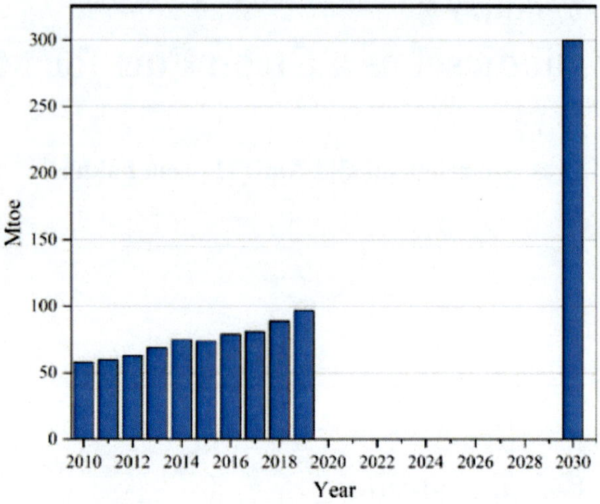

Fig. 8.1 Production of biofuel in 2010–2020 compared to the proposed amount in the SDS outline (IEA 2019)

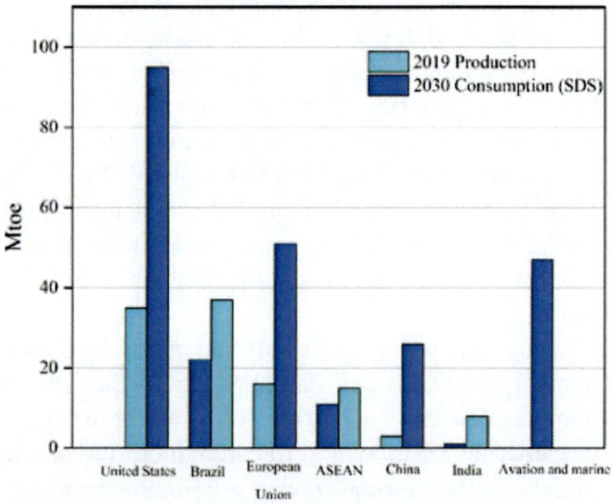

Fig. 8.2 Biofuel production in 2019 compared to consumption in 2030 under the SDS outline in some parts of the world (IEA 2019)

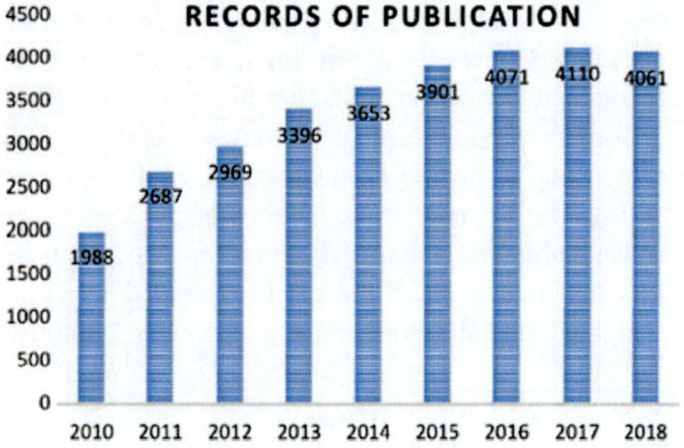

Fig. 8.3 Number of published research studies on biofuels (Ogunkunle and Ahmed 2019)

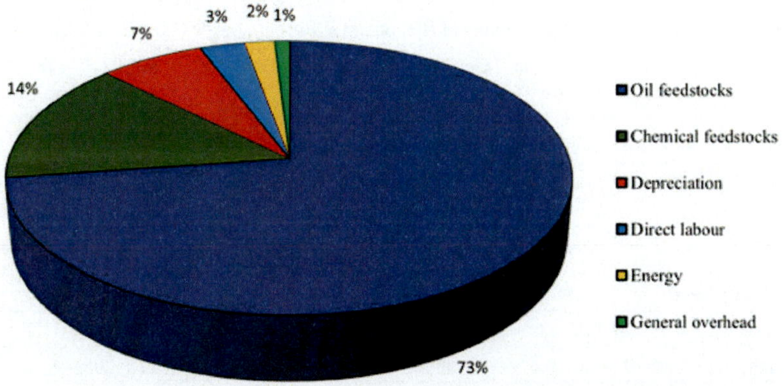

Fig. 8.4 Distribution of the total biodiesel production cost (Atabani et al. 2012; Lin et al. 2011)

diesel fuel (Murray et al. 2019). Biodiesel also has some disadvantages compared to conventional diesel, for example, higher density, viscosity, and NO_x emission production, and a lower amount of lower heating value (LHV) (Rashedul et al. 2014). The use of biodiesel with a certain ratio of diesel fuel in diesel engines has different effects on engine performance and emissions; depending on the operating conditions, engine structure, and biodiesel fuel characteristics (Kumar 2017). In general, the heating value of biodiesel fuel is lower than that of diesel fuel, and the density of biodiesel fuel is slightly higher than that of diesel fuel. In biodiesel and diesel fuel mixtures, the emission level decreases, and the brake-specific fuel consumption rises (Ganesan and Masimalai 2020). Biodiesel is produced from animal fats and vegetables (fatty grains such as rapeseed, sunflower, soybean, and safflower). This fuel can be mixed with certain proportions of diesel fuel or it can be used in the pure form (Temizer et al. 2020). Liquid biofuels are generally categorized into three groups: Vegetable oils, which are obtained directly from the seeds, and using them has many physical problems, one of which is high viscosity; Methyl ester fatty acids, which are derived from plant seeds and have properties close to diesel; and alcohols, such as ethanol and butanol, which are obtained from the fermentation of sugar corn, and agricultural wastes. Methyl ester fatty acid or biodiesel is a fuel with combustion properties similar to diesel fuel, and it is produced from vegetable seed oils, animal fats, and waste oils (Ogunkunle and Ahmed 2019). To discuss how biodiesel fuel is produced, a brief look is taken at the biodiesel production process. At first, the grains must be prepared. The plant is dried and decomposed with the help of the sun, subsequently, it is roasted for about 10 min and then the grains are mechanically extracted with solvents, etc. The next stage is oil refining, which is done by depositing in boiling water or filtering. Then the oil processing stage begins. The use of this oil is generally divided into four methods of pyrolysis, micro-emulsification, dilution, and transesterification (Murray et al. 2019). The thermophysical properties of biodiesel influence the performance and emission of an engine. Cetane number, heating value, ignition temperature, density, pour point, and viscosity can be considered as the most important properties of biodiesel (Ashraful et al. 2014). Table 8.1 shows the

Table 8.1 Properties of various biodiesel fuels from different sources (Datta and Kumar 2016)

Fuel	Density (kg/m^3)	Kinematic viscosity at 40 °C (mm^2/s)	Cetane No.	Heating value (MJ/kg)	Cloud point (°C)	Flash point (°C)	Fire point (°C)	Pour point (°C)
Diesel	850	2.44–2.60	47–50	42–44.3	–	68–75	80	−20
Camelina	918	24	50.4	38	3	>220	–	−7
Coconut	877	3.18	60	36.98	1	136.5	–	−4
Safflower	920	26.64	51.1	–	−4	174	–	−7
Canola	872	4.22	53.7	39.28	−4	153	–	−6
Karanja	880–890	4.37–9.6	48–58	36.12–42.13	−2–14.6	170–205	–	−6–5.1
Mahua	880–916	3.98–5.72	–	37–39.4	–	129–208	141	6
Cottonseed	850–885	6–9.6	52	37.5–41.68	−2	–	–	−4
Palm	870–878	4.5–5.11	50–62	37.2–39.9	14	173	182	8
Jatropa	873	4.23	–	42.67	10.2	148	–	4.2
Polanga	869	3.99	–	41.39	13.2	140	–	4.3
Soybean	885–914	4.075–39.5	37–51.3	37.3–39.66	–	69–163	–	–
Sunflower	880–885	4.38–4.4	50–51.6	37.5–39.9	–	183	–	–
Rapeseed	872–885	4.58–11	37.6–54.5	37.3–39.9	–	177–275	–	–
Honge	890	5.6	45	36.01	–	163	–	–
Peanut	886.4	5.25	54	39.7	–	193	–	–
Corn	885.8	4.36	55.4	39.87	–	167	–	–
Palm Kernel	876.6	3.24	62.1	38.5	–	131	–	–
Tallow	832	4.89	58.9	37.2	13	124	–	10
Waste Fried	884.2	4.86	55	39.68	–	167	–	–
Jojoba	866	19.2	63.5	43.38	–	61	–	–
Neem	820	8.8	51	40.1	–	–	–	–
Chicken fat	869	2.8	48	–	−7	74	–	–
Mutton fat	856	8.15	59	–	−4	–	–	−5

different properties of biodiesel fuels produced from different sources (Datta and Kumar 2016).

Figure 8.5 illustrates the power output of an engine that works with four types of biodiesel under various load states. In all tests, the sunflower biodiesel resulted in the best performance (Jiaqiang et al. 2016a).

In terms of fuel storage and transportation, low oxidation stability and high cloud point (cold flow characteristics) are the most important challenges that should be faced during the general use of biodiesel. These problems can reduce the fuel quality, and have negative impacts on the engine performance and the durability of its components. Oxidation stability is a problem in biodiesel fuel due to its high percentage of

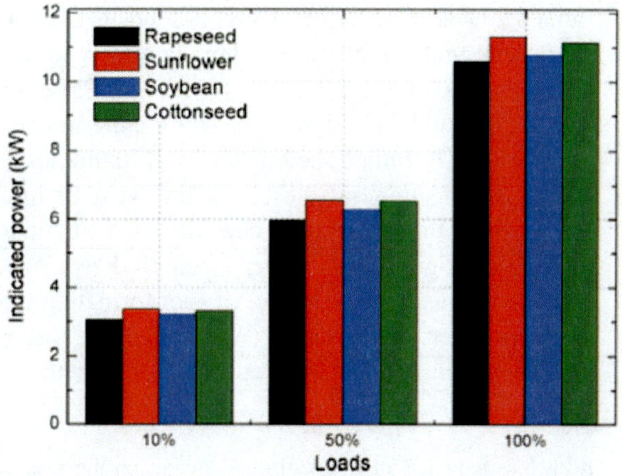

Fig. 8.5 The indicated power in a diesel engine with four different types of biodiesel (Jiaqiang et al. 2016a)

unsaturated fatty acids. This is one of the most important obstacles in the generalization of this fuel. When the fuel is exposed to the air, during storage or consumption, the quality of the fuel is gradually reduced and its combustion characteristics change (Kumar 2017). The most important factors in the oxidation of this fuel are the structure of fatty acids, environmental conditions, and materials in the storage tanks. The structure of fatty acids and the amount of unsaturated fatty acids in different variants of this fuel determine how serious this problem can be. Adding antioxidant additives in the fuel is one way to cope with this problem (Ganesan and Masimalai 2020). There are several ways to overcome the challenge of cold flow characteristics. In terms of fuel structure modification, isomerization processes for branching the structure, and processes regarding reducing and increasing the double bond are some of these ways that can be used (Verma and Dwivedi 2016). Improving environmental conditions in the production process of oilseed is another usable solution. The most common way, which has a positive effect on fuel stability besides improving the cold flow characteristics, is to combine biodiesel with diesel. This can improve the cold flow characteristics to an acceptable level (Reaume and Ellis 2013; Gómez et al. 2002).

Later in this chapter, the sustainability, power output, and emissions for combustion of biodiesel in CI engines, as well as the application of biodiesel in LTC engines will be investigated.

8.2 Sustainability of Biodiesel Engines

Most of the research studies in the field of biodiesel engines have been conducted on performance and emission, so there are not many studies that focused on biodiesel engine sustainability, since performing research on this subject is time-consuming and costly. For this purpose, a brief review has been performed on the research studies regarding the sustainability of biodiesel engines, and the findings are tabulated in

Table 8.2 (Xue et al. 2011). Carbon deposits are connected to the formation of soot throughout fuel combustion and oxidation. When biodiesel is used, a lower amount of soot is formed, therefore a lower level of PM emission is observed with biodiesel fuel (Xue et al. 2011). During a 100-h experiment, Sinha and Agarwal (Sinha and Agarwal 2010) studied the impacts of B20 biodiesel, a mixture of 20% methyl ester of rice oil and mineral diesel, on the wear of in-cylinder engine components. It was reported that the carbon deposits on the cylinder head, injector tip, and piston crown were significantly lower when biodiesel was used instead of mineral diesel. The reason was the lower level of soot formation throughout the combustion of this fuel. Moreover, some studies (Agarwal 2005; Agarwal and Das 2003) reported that biodiesel minimizes carbon deposits in the combustion chamber.

Table 8.2 A review of the performed studies on the sustainability of biodiesel in engines (Xue et al. 2011)

Content and feedstock	Ref. diesel	Engine tested	Operation condition	Duration	Test results
20% Rice bran oil	Conventional	4-cylinder, NA, WC, DI	Ten nonstop running cycles 1500 rpm	100 h	CD: significantly lower Wear: Lower
20% Linseed oil	Agricultural	1-Cyliner, WC, portable	1500 rpm	512 h	IJ: no coking, no filter Plugging: Wear: lower
20% Linseed oil 100%, 15%, 7.5% palm oil	Agricultural conventional	1-Cylinder,WC, portable 4-Cylinder,NA, WC, IDI	1500 rpm 2000 rpm	512 h 100 h	Wear: lower The reduction of wear with the increased content of biodiesel
100%, 50% soybean oil	No. 2 (EN 590)	TC, DI, 1.9 L	NEDC driving cycle	1350 km	Wear: higher except piston
100% waste olive oil	No. 2 (EN 590)	3-Cylinder, WC, DI, 2.5 L	8–15 Kw and 1800–2100 rpm	50 h	CD: no visual Difference: Wear: no visual difference
100% rapeseed oil	No. 2 (EN 590)	6-Cylinder, WC, DI, 11 L	–	110 h	CD: similar: IJ: cleaner than that of D2
100% Mahua, Karanja oil	High speed diesel	–	Static immersion test at ambient temperature	300 D	No corrosion on piston metal and piston liner

IJ is the injector, CD is the carbon deposit

When lower levels of biodiesel are used in the mixture with diesel fuel, it is effective in reducing friction (Xue et al. 2011). For instance, based on tribological research on lubricating oil that was conducted by Agarwal (2005), it was found that the level of potential contaminants such as abrasion residues, soot, resinous compounds, oxidation products, and moisture content is lower in a biodiesel engine compared to those of a diesel engine. The enhanced operation of the biofuel fuel system can also be related to the inherent lubrication of biodiesel, which leads to a lower level of wear in the major moving components. Research studies (Agarwal 2005; Agarwal and Das 2003) have shown that high concentrations of biodiesel dissolve lubricants to some extent, therefore it leads to an increased friction factor for the moving engine parts. Some acidic compounds may be created throughout the combustion process which may be dissolved in the lubricant. The majority of articles report that if biodiesel or its mixtures are used in an engine, carbon sedimentation and engine wear can be reduced. Since biodiesel engines are durable, they can cope with the durability problems of vegetable oils such as plugins, fuel filters, and injector coking. In general, although there are not suitable reports regarding wear, it is expected that the use of biodiesel will enhance the engine durability compared to diesel fuel because of the low soot formation and inherent lubrication (Xue et al. 2011). However, to clarify the cause and mechanism of wear, more studies are needed to be done on the durability tests of biodiesel engines, since the research studies on these aspects are not yet sufficient.

8.3 The Power Output in the CI Combustion of Biodiesel

Combustion challenges for the wider use of biodiesel can be examined in two categories of power output and emission. The heating value of biodiesel is lower than that of diesel, but biodiesel has oxygen in its structure and a higher Cetane number than diesel fuel. In 2001, Altin et al. (2001) investigated different vegetable oils as well as their esters in the combustion of a compression ignition engine. In comparison to when the diesel fuel was used, the lowest power reduction and the lowest raise in the brake-specific fuel consumption (BSFC) were 3% and 14.5%, respectively. In 2006, in a study of engine performance with sunflower oil biodiesel, Kaplan et al. (2006) achieved a 5% reduction in power compared to diesel engine performance at low engine speeds and full load. In 2006, Lin et al. (2006) worked on palm oil biodiesel. As the highest power output, they observed a 3.5% reduction compared to the diesel engine performance. When B20 fuel was used the power reduction was about 1%. In 2011, Zhu et al. (2011) performed a study in which they examined the impacts of combining biodiesel from waste oil with ethanol. They added ethanol to the fuel to enhance the viscosity and density of biodiesel and to improve its atomization process. The study results suggested that compared to diesel, the mixture of biodiesel and ethanol (95–5%) raises the BSFC, however with higher brake mean effective pressure (BMEP), the difference becomes smaller. It was also indicated that the thermal efficiency increases compared to diesel and by increasing the BMEP, the difference between the obtained results from these two fuels increases as well. In 2018, in a study on biodiesel made from rapeseed oil, Kurczyński and Łagowski (2019) found that at 1750 rpm, the BSFC of B20 is slightly higher than that for diesel fuel, but the B30 fuel has a more considerable difference with them. They also observed that at 4000 rpm, the difference of the results with B30 fuel was drastically reduced, indicating good biodiesel performance at high speeds. In 2019, in a study on soybean oil biodiesel, Seraç et al. (2020) found that the brake thermal efficiency at the power of 3.6 kW is lower for B20 than that of the diesel fuel, but at the power of 10 kW, this efficiency for the B20 improves for about 3.1% compared to that of the diesel fuel. Lapuerta et al. (2008) presented a review paper on the use of biodiesel as a fuel. In most reports, a 7% reduction in power was observed when pure biodiesel was used and in special mixtures such as B20, 1–3% of reduction or even sometimes a rise in power was reported. Regarding the brake power, compensation methods for the reduction in heating value in biodiesel combustion were examined. One of the most important compensations was the mass and volume of injected fuel in all load conditions, during the operation of the engine with biodiesel and diesel fuels. They stated that the higher viscosity of biodiesel reduces the reversed flow between the piston and the pump, resulting in the injection of a higher amount of this fuel at the same injection position. It was stated that another important reason in power compensation is that the better lubrication with the biodiesel fuel reduces friction. Raising the Cetane number starts the combustion slightly earlier. Another important factor in biodiesel combustion in the engine is that the torque peak is pulled towards higher speeds since the high flame speed of the biodiesel at these speeds can result in more complete combustion and better results. This paper also revealed that the oil that is used for preparing biodiesel does not have much effect on the output. In some studies, (Carraretto et al. 2004; Luján et al. 2009) the rise of brake-specific fuel consumption was considerably higher than the proportional amount of the heating value reduction, however, the amount of ester in their biodiesel was unusually low. Overall, research studies (Xue et al. 2011) have suggested that the use of biodiesel reduces engine power and increases fuel consumption, which is normally acceptable. It is also worth noting that over time, new engine designs are likely to become more flexible towards the use of biodiesel. It has been completely confirmed that the main reason for the power loss is that biodiesel presents a lower heating level as compared to diesel. High viscosity and high lubrication of biodiesel also have certain effects on engine power, but the obtained results on these subjects are not compatible. Moreover, if there is no change in the engine, the biodiesel injection characteristics also affect the engine power.

8.4 The Emission Level in the CI Combustion of Biodiesel

Studies (Robbins et al. 2011) have shown that the use of biodiesel (and biodiesel mixtures) has a significant effect on hydrocarbons (HC), carbon monoxide (CO), and particulate matter (PM) emissions. The influence of biodiesel on NO_x emissions is variable and less considerable. Figure 8.6 shows the effects of the biodiesel mixture concentration on these four emissions when this fuel is used in a heavy-duty (HD) engine (Hoekman and Robbins 2012). This evaluation was performed by the EPA nearly two decades ago, and therefore includes data on some of the engine types that are now obsolete. Figure 8.7 shows the results of NO_x emission from the evaluation of a 4-stroke heavy-duty (HD) engine with both biodiesel and petroleum-based diesel fuels (Hoekman and Robbins 2012).

Saddam and Al-lwayzy (2017) studied the use of microalgae biodiesel from chlorella protothecoides as an alternative fuel in CI engines and examined the emissions in this condition. The levels of CO and CO_2 emissions at different engine speeds for the petroleum-based diesel (PD), and various microalgae chlorella protothecoides mixtures including MCP-B100, MCP-B50, and MCP-B20 have been illustrated in Fig. 8.8. As shown in Fig. 8.8, for all fuels, the level of CO and CO_2 emissions are relatively constant for the engine speeds below 2320 rpm. After this point, as the

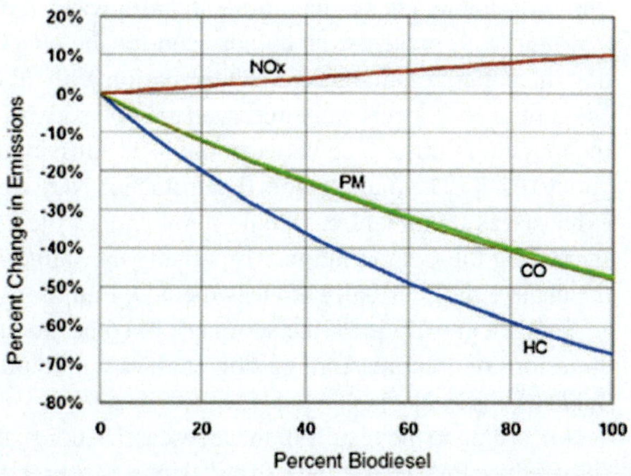

Fig. 8.6 Average effects of using different concentrations of biodiesel in HD engines on the emissions (Hoekman and Robbins 2012)

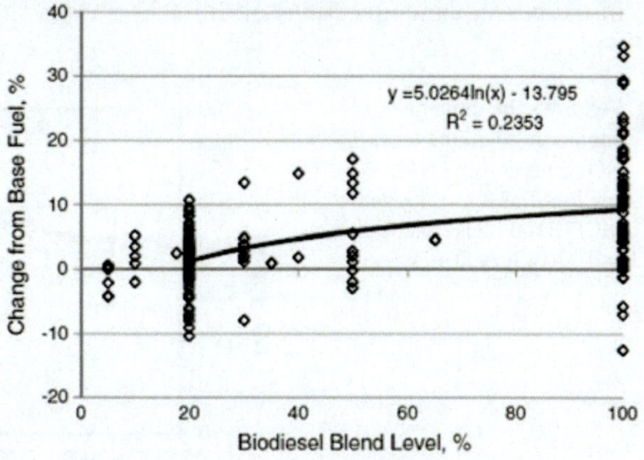

Fig. 8.7 The comparison of NO_x emission with different concentrations of biodiesel fuel compared to petroleum-based diesel fuel in some 4-cylinder HD diesel engines (Hoekman and Robbins 2012)

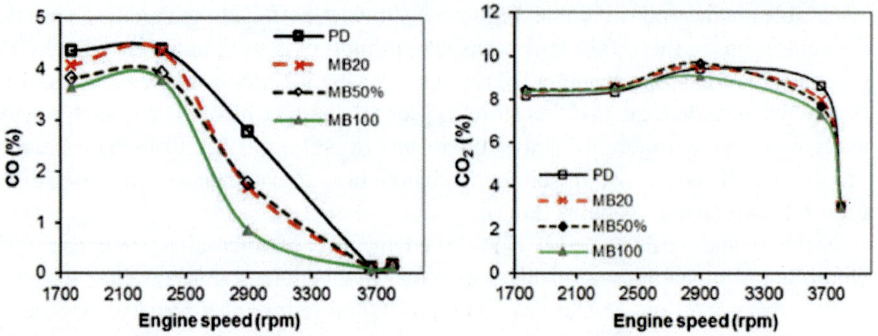

Fig. 8.8 The influence of engine speed on the level of CO_2 and CO emissions with PD, MCP-B20, MCP-B50, and MCP-B100 fuels (Saddam and Al-lwayzy 2017)

engine speed increases, the CO emissions decrease, while CO_2 rises and at 2900 rpm this emission reaches a maximum and afterward, it decreases.

Figure 8.9, presents the comparison for the NO_x emissions. This Figure reveals that the pattern of changes in NO_x emission with all of these fuels is almost similar. Maximum NO_x levels were observed at 2900 and 3670 rpm. According to the figure, the maximum difference between the NO_x emission levels of PD and MCP-B100 is about 16.4%. The main factors that influence NO_x emissions are the oxygen content, exhaust gas temperature, Cetane number, and the chemical structure of biodiesel. Increasing the Cetane number by raising the saturation of the fatty ester chain and the chain length virtually reduces the NO_x emissions.

To get a general idea, the summary of some studies on the production of various emissions of compression ignition engines with biodiesel fuel has been collected in Table 8.3.

According to the results of these research studies, the use of biodiesel can successfully reduce PM, UHC, and CO emissions. Most reports reveal that by using this fuel NO_x emission increases. Besides the working conditions of the engine, the presence of oxygen in the connections of this fuel is one of the most important reasons for

Fig. 8.9 The influence of engine speed on the level of NO_x emissions with PD, MCP-B20, MCP-B50, and MCP-B100 fuels (Saddam and Al-lwayzy 2017)

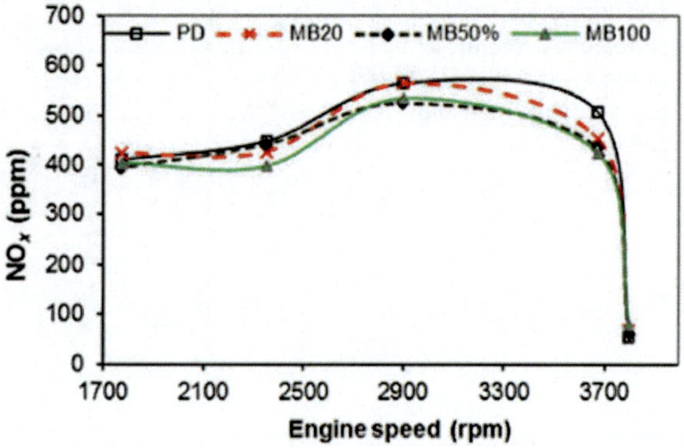

Table 8.3 Research studies on the engine emissions using the biodiesel fuel

Researcher	Publishing year	Oil basis	The change of emissions by using biodiesel instead of diesel fuel (%)				Notes
			CO	UHC	NO_x	CO_2	
Nantha Gopal et al. (2014)	2014	Waste oil	−31	−57	+18.33	–	The values are for B100. For the NO_x emission, there were fluctuations in different loads and different mixture ratios. In some cases and with some ratios of biodiesel the emissions production was reported less than when diesel fuel was used
Can (2014)	2014	Waste oil	−11.8 to −51	−2 to −29	+6.4 to +7.8	+3.3 to +5	B5 and B10 mixtures were compared at different loads
Qi et al. (2009)	2009	Soy	−27	−27	−5	–	The presented values are the average reduction of emission. The test was performed in four different mixture concentrations from 10 to 40%
Hajbabaei et al. (2012)	2012	Soy and animal fats	–	−73	+14.1 to +47.1	–	The values are for B100. NO_x emission for soybean biodiesel is higher than the biodiesel from animal fats. Using biodiesel fuel, the amount of PM emission was reduced by about 69%
Raman et al. (2019)	2019	Canola	−35.4	−42.1	+32.9	–	The values are for B100 and the maximum power
Gad et al. (2020)	2020	Waste oil	−25	−20	+40	–	The presented values are the average of the changes and are for B100 fuel

more complete combustion and, consequently, the reduction of PM, UHC, and CO emissions and sometimes the increase in the NO_x emission level. Since NO_x emission is already an important problem in the combustion of diesel fuel, the rise of NO_x emission in the combustion of biodiesel fuel should be considered a major challenge. Sivalakshmi and Balusamy (2013) investigated that the impacts of biodiesel fuel on NO_x emission. Their study revealed that the use of biodiesel decreases CO, HC, and soot emissions, but the level of NO_x emission increases. It has also been stated that the level of NO_x is mostly influenced by the rapid mechanism of biodiesel combustion (Jiaqiang et al. 2016b; Mueller et al. 2009) (see Fig. 8.10).

In a review paper that was published in 2019, Dabhi et al. (2019) investigated the effect of biodiesel/diesel mixture on the emissions and performance of a single-cylinder compression-ignition engine with a constant compression ratio. They observed that the NO_x emission increased in higher loads. The mixture of biodiesel and diesel has a higher NO_x emission than pure diesel since it has a higher oxygen content and a slightly higher Cetane number. The temperature, oxygen level, ignition delay, fuel nitrogen percentage, and residence time are the most important factors that mainly affect NO_x emission. NO_x emission also increases as the heat release rate (HRR) rises. Due to the higher Cetane number of biodiesel fuel, its ignition delay decreases. The mixture of biodiesel and diesel produces less HC than diesel fuel. As the load and engine speed rise, the amount of HC decreases. Combustion efficiency and temperature are the most important factors affecting HC production. Also, CO production increases with higher engine loads. The temperature, fuel-to-air ratio, and unburned mixture are the most important factors affecting CO emission. It should be noted that as stated before, biodiesel has more oxygen than diesel, which leads to a lower level of CO emission. In 2017, Ghareghani et al. (2017) experimentally studied the combustion characteristics, performance, and emission of a single-cylinder diesel engine using conventional diesel and biodiesel produced from waste fish oil (WFO) and its mixtures with diesel fuel (B25, B50, B75). The results showed that on average biodiesel presents 2.92% less gross thermal efficiency and about 1.1% lower combustion loss compared to diesel fuel. Also, they stated there is a gentle reduction (5–27%) in the CO emission concentrations with biodiesel and its mixtures while a significant decrease (11–70%) occurred for the UHC. On the other hand, higher NO_x emission (average of 12.8% for the entire engine loads)

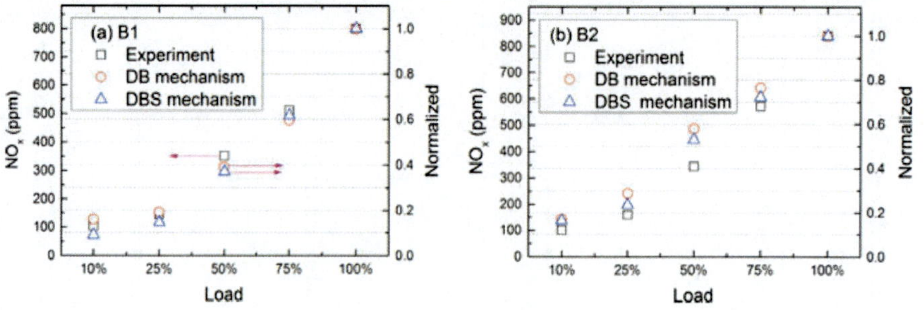

Fig. 8.10 NO_x emission prediction test (Jiaqiang et al. 2016b; Mueller et al. 2009)

and also higher CO_2 emission (due to complete combustion) were observed. Oxygen content is a major factor that can affect the pollution in the CI combustion of biodiesel. Biodiesel has oxygen content levels in the range of 10–12 wt%. This variety is mostly because of the different oxygenation levels and chemical structure of the employed feedstock. When biodiesel is produced for highly saturated oils, the oxygenation level of biodiesel is higher; therefore, its combustion is cleaner and more stable. On the other hand, its heating value is lower. Due to higher levels of oxygen content in biodiesel, the CO, HC, and soot emissions from the combustion of this fuel are lower than the diesel fuel; however, it results in higher levels of NO_x emissions (Jena and GIET 2017). The summary of the findings regarding the emissions from the CI combustion of biodiesel has been presented as follows.

8.4.1 PM Emission

Overall, studies (Ozsezen et al. 2009; Utlu and Süreyya 2008) have shown that with biodiesel, PM emissions are significantly reduced compared to when diesel fuel is used. This reduction becomes smaller as the ratio of biodiesel in the mixed fuel is reduced, and there may be an unusual change in certain contents of biodiesel. The decreasing trend of PM emission with the use of biodiesel in the engine is due to the reduction of aromatic and sulfur compounds and the Cetane number, however, the most important factor is the higher oxygen content. With higher engine loads, the PM emission of biodiesel increases. In other words, as the engine speed increases, the PM emission gets lower. In an optimal condition, advancing the injection of biodiesel is not suitable for a diesel engine. The PM emission with biodiesel is quite lower than it would be with diesel fuel, but even this low amount of PM emission can be reduced using exhaust gas recirculation (EGR). It should be noted that while using biodiesel fuel at low temperatures, the PM emission rises abnormally as compared to that of diesel fuel (Xue et al. 2011).

8.4.2 NO_x Emission

Most researchers reported (Lin et al. 2009; Keskin et al. 2008) that NO_x emission increases when biodiesel fuel is used instead of diesel fuel. This growth is mostly because of the higher oxygen content in biodiesel. Additionally, the Cetane number and different injection characteristics affect NO_x emission from biodiesel combustion. The unsaturated compound content in biodiesel can have a significant influence on NO_x emission. A higher amount of unsaturated compound content leads to a greater reduction in NO_x emission. While using biodiesel fuel, at higher engine loads, the NO_x emission level increases. This is consistent with the mechanism of NO_x formation. By using EGR, NO_x emissions from biodiesel combustion can be reduced (Song and Zhang 2008).

8.4.3 CO Emission

Studies (Kim and Choi 2010; Usta 2005) have shown that when biodiesel is used, CO emissions are reduced because of its higher oxygen content and lower carbon to hydrogen ratio in the structure of biodiesel compared to those of diesel fuel. With higher concentrations of pure biodiesel in the biodiesel-diesel mixture, the CO emission decreases. CO emission from biodiesel combustion is influenced by the raw materials and biodiesel properties such as the Cetane number and advancement in the combustion. Engine load has also been shown to have a considerable impact on CO emission. In other words, while using biodiesel fuel, CO emission rises with higher engine speeds (Gumus and Kasifoglu 2010).

8.4.4 HC Emission

Research (Mahanta et al. 2006; Sahoo et al. 2007) has shown that when pure biodiesel is burned instead of diesel fuel, HC emission is decreased. It was also suggested that with higher biodiesel content, the HC emission of the fuel mixture is reduced. Biodiesel feedstock and its properties affect the HC emission, especially with different chain lengths or saturation levels in the biodiesel fuel. Researchers have shown that advancing biodiesel injection and combustion reduces HC emission. There are various results regarding the impacts of engine load on the HC emission when biodiesel is used as the fuel, and these results are not quite compatible. The HC emission can be reduced by using an oxidative catalytic convertor; however, this reduction technique is old and not much efficient (Baiju et al. 2009).

8.5 Biodiesel and Low-Temperature Combustion Engines

Low-temperature combustion (LTC) internal combustion engines are one of the most important alternatives to conventional compression ignition engines. In the future, the possibility of using the LTC engines in the industry is high due to the increase of biodiesel usage according to the SDS. This indicates that currently more research studies are required on the use of biodiesel in LTC engines. The use of LTC engines may solve some combustion challenges that should be faced while using biodiesel fuel (such as the NO_x emission). In the last two decades, more efforts are taken to improve engine emission levels. Compression ignition engines have high combustion efficiencies, but they also produce large levels of two important pollutants, including NO_x and PM (Yousefi et al. 2015a). As a result, reducing emissions without compromising efficiency, has gained lots of research attention in recent years (Dec 2009). The most important reason for the production of PM and soot is the richness of some regions inside the combustion chamber. As the result of the combustion in these regions, the in-cylinder temperature significantly rises and the required conditions for

NO_x production are created (Fakhari et al. 2020a). Therefore, preventing the formation of rich zones inside the combustion chamber has been considered an important approach for the advancement of compression ignition engines and reducing their emissions. LTC combustion strategy has been introduced as a type of combustion in which fuel and air are pre-mixed to homogenize the mixture in the combustion chamber and reduce the combustion temperature by decreasing the local equivalence ratio. This consequently leads to lower NO_x emission levels (Motallebi Hasankola et al. 2020). This homogenization also has a great effect on reducing PM and soot formation. However, decreasing the combustion temperature increases the UHC and CO emissions to some extent in the engines (Dempsey et al. 2016). Even though the evolution of LTC engines began before 2000, it has grown rapidly during the last two decades. Homogeneous Charge Compression Ignition (HCCI) engines were first introduced as a concept in 1979, then the research began on these engines. The operation in these engines is based on the combustion of a homogeneous mixture, using compression and self-ignition temperature. This strategy has a very fast combustion phase and the entire energy is released in a short crankshaft angle period (Jahanian and Jazayeri 2012). In general, the advantages of HCCI combustion are high thermal efficiency and being economic regarding fuel consumption, reduced PM and NO_x emissions, and high flexibility towards the fuels, which can facilitate the use of renewable fuels (Yousefi et al. 2015b). On the other hand, the main problems of HCCI are low controllability on the combustion phase, limited operating range (which includes a significant pressure rise rate at high load and combustion loss in low loads), high noise level, production of CO and UHC emissions, and cold engine start-up (Neshat and Khoshbakhti 2014). Of course, many solutions have been proposed, such as the use of EGR, changing the air inlet temperature, changing the equivalence ratio, and using different fuels, but combustion controllability and the limited operating range of these engines are still challenging (Machrafi and Cavadiasa 2008; Ganesh and Nagarajan 2010). In the development of LTC engines, other types of combustion strategies in engines have been introduced, such as compression ignition engines with thermal stratification, which focus on creating an environment with step-by-step temperature stratification in the combustion chambers. This can affect the engine noise and rapid changes in the rate of heat release (Lawler et al. 2017; Hardy and Rolf 2006). These engines are a subset of stratified charge compression ignition (SCCI) engines, but the researchers have mostly focused on another variant of an engine in this group, which aims to stratify the fuel in the combustion chamber. This type of engine which is known as the premixed charge compression ignition (PCCI) has fewer technical problems compared to thermally-stratified engines. Several techniques can be used for obtaining this type of combustion. The point that is considered in all of them is the end of fuel injection before the start of combustion (SOC). In PCCI injection, the fuel is injected from the inlet manifold or it is injected directly but the injection is done early in the compression phase, so that it has enough time to mix with the air, and unlike conventional compression ignition engines, it gets homogeneous enough. However, in this case, considering the low pressure and temperature in the combustion chamber when the fuel is injected, the collision of the fuel with the cylinder wall increases (Esfahanian et al. 2017). However, many strategies such as using injectors with low injection angles or changing the geometry of the piston bowl have been presented to solve these problems. Positive results were obtained for this type of combustion by two-stage direct injection or manifold injection along with direct injection (creating fuel stratification), and this combustion strategy was developed (Kaplan et al. 2006). In a combustion strategy, known as Partially Premixed Compression Ignition (PPCI), the fuel is directly injected late and near the top dead center. Overall, even though due to the two-stage injection technology, the PCCI strategy is more successful than HCCI regarding the engine control, they still face challenges in creating a homogeneous mixture, the collision of fuel with the walls, and even combustion control (Boot et al. 2010; Pandey et al. 2019). The fuel stratification works based on creating areas with different concentrated fuel levels in the combustion chamber, which causes successive self-ignition. This reduces the peak heat release rate and increases the combustion duration (Li et al. 2013). As the fuels are more sensitive towards the equivalence ratio, the fuel stratification is more improved. In higher engine loads and under the naturally aspirated state, the fuels that are more similar to gasoline are more sensitive compared to the diesel fuel, so using them instead of diesel in engines with strategies such as PPCI can reduce performance limitations at high loads. Therefore, low-temperature combustion strategies are divided into two groups: diesel and gasoline-based (Sjöberg and Dec 2006). Gasoline fuel performs well in the engines with the PPCI strategy at higher loads but does not perform well at low loads due to its high ignition temperature. In contrast, diesel performs well at low loads due to its high reactivity, and at high loads, it faces the problem regarding fuel stratification and controlling the combustion phase. As a result, a new type of LTC combustion was introduced as Reactivity-controlled compression ignition (RCCI). In this combustion, the ratio of high- and low-reactivity fuel changes according to the operating conditions of the engine to reduce the operating limit of the engine. Low-reactivity fuel is injected through the manifold and high-reactivity fuel is directly injected. Apart from the combustion initiation, the use of two fuels with different reactivities affects the stratification of the equivalence ratio within the chamber (Tanov 2014). Similar to the other LTC strategies, the RCCI strategy has a volumetric ignition that is dependent on self-ignition. As a result, they are much more dependent on the structure and size of the fuel molecules compared to the conventional engines that have flame propagation. It should be noted that the structure and size of the fuel molecules are strongly affected by the fuel injection parameters. Among these parameters, the start of direct injection timing is the basic parameter for evaluating the performance of engines with RCCI strategy (Li et al. 2017) due to the impacts it has on the stratification of the equivalence ratio, as well as starting and controlling the combustion phase ratio (Harari et al. 2020). Figure 8.11 properly illustrates the concepts of advanced combustion strategies as well as conventional diesel combustion in the equivalence ratio-temperature diagram. Also, the concepts of these combustion technologies based on their fuel source and EGR strategy can be observed in Fig. 8.12. Furthermore, Fig. 8.13 schematically shows the operating region for the five combustion concepts based on the diesel injection timing and the NG substitution ratio. The CDC can control the combustion stage by setting the diesel injection timing near the TDC. The maximum combustion temperature in CDC mode is very high due to the high heat release level over a short combustion duration. The HCCI and PCCI strategies make use of the homogeneity of the charge for the premixed combustion. For example, in the HCCI combustion strategy, the diesel fuel must be injected during the intake process from 160 to 220°CA BTDC to create a homogeneous charge mixture.

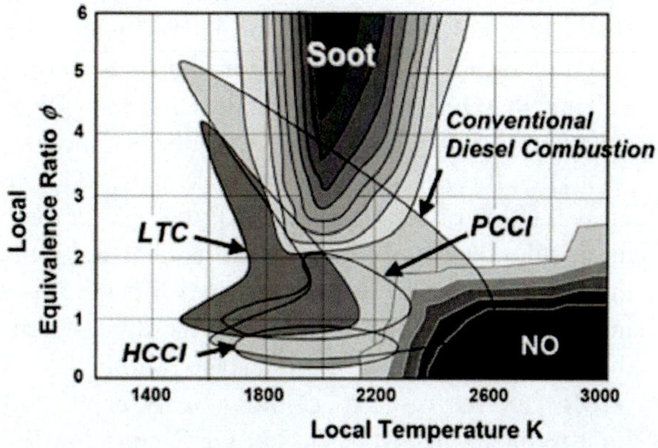

Fig. 8.11 The diagram of soot and NO_x emissions based on Φ-T in conventional and LTC combustion modes (Shim et al. 2020)

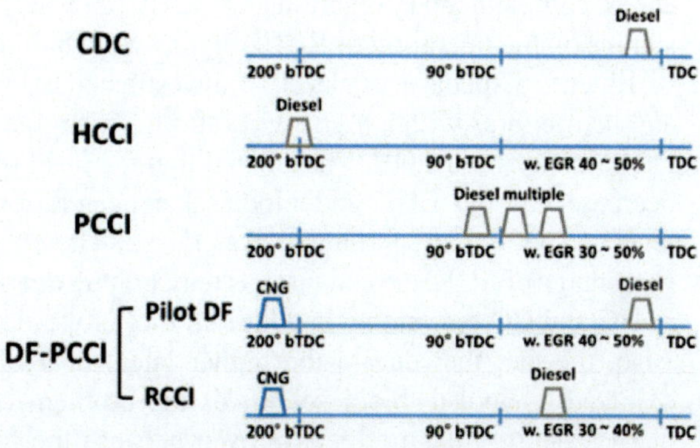

Fig. 8.12 Concepts of diesel combustion and various types of LTC based on the fuel source and EGR strategy (Shim et al. 2020)

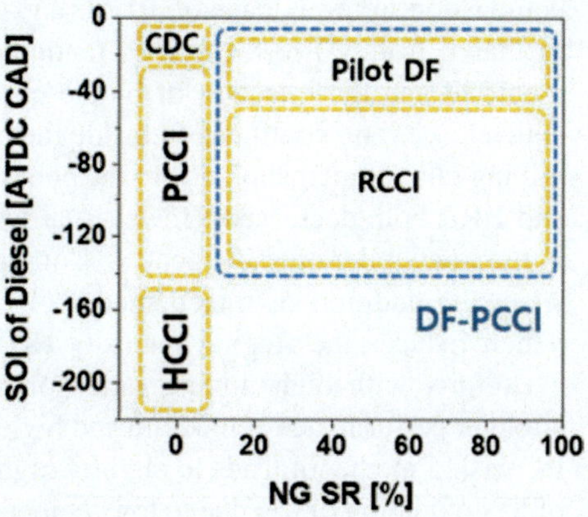

Fig. 8.13 Operating zone of various combustion concepts based on diesel injection timing and NG replacement ratio (Shim et al. 2020)

In the PCCI strategy, diesel fuel must be injected into the combustion chamber from 30 to 160°CA BTDC during the compression stage to mix into the charge. Besides, to make the mixing duration of the fuel–air mixture longer and enhance the combustion stage, the EGR technique and multi-fuel injection systems were performed. The dual-fuel mode and RCCI strategy make use of two different fuels including CNG and diesel as low-reactivity fuel (LRF) and high-reactivity fuel (HRF), respectively (Fakhari et al. 2020b). Considering the method for the control of the combustion stage, dual-fuel PCCI combustion can be categorized into two groups: pilot dual-fuel and RCCI. For the RCCI mode, HRF and LRF are both initially injected into the chamber. The SOC is assessed using the chemical kinetics of the stratified fuel reaction. Therefore, the combustion phase in the RCCI strategy is mainly assessed by the mass fraction of the HRF and LRF throughout the combustion phase. On contrary, in the pilot dual-fuel strategy, LRF is injected from the intake manifold throughout the intake stage and, subsequently, HRF is injected into the chamber through direct injection near the TDC. Therefore, the injection timing of HRF is the main factor that determines the combustion stage (Shim et al. 2020).

Most of the research studies (Neely et al. 2005) on biodiesel have reported improvements in PM, CO, and UHC emission levels, mainly due to the oxygen content in biodiesel fuel. In general, biodiesel combustion and RCCI combustion can be complementary regarding CO, UHC, and NO_x emissions. Hence, the research studies on the operation of RCCI engines with biodiesel fuel are reviewed.

Jiménez-Espadafor et al. (2012) investigated a diesel engine with colza biodiesel and its mixtures with a high swirl ratio in an engine with HCCI combustion mode, late injection, and EGR. They stated that the level of PM and NO_x pollutants were decreased as the EGR and biodiesel concentrations were raised. Although, the level of CO and HC pollutants was raised. They explained that by using a higher percentage of EGR, the exhaust gas temperature decreases, which leads to a decrease in HC and CO emissions. In addition, fuel properties can influence the LTC mode. Also, the fact that due to the higher latent heat of vaporization, the addition of oxidized ethanol to biodiesel solves the problem of high CO and HC emissions in LTC engines has not been proven because the higher latent heat decreases the combustion rate. In 2015, Zhou et al. (2015) numerically investigated the RCCI combustion in three loads of 10, 50 and 100% and with various mass fractions of methanol from 20 to 80%, with methanol fuel entering the cylinder through the manifold and the injection of biodiesel through direct injection for changing the equivalence ratio stratification inside the cylinder. At the load of 10%, with a higher mixing ratio of methanol, due to the poor combustion, the in-cylinder pressure peak, and HRR both decreased. However, at loads of 50 and 100%, an obvious increase in the in-cylinder pressure peak was observed with a higher methanol mixing ratio. Also, the addition of mixed methanol can create a fuel reactivity stratification, which reduces the ringing intensity (RI) and effectively prevents engine knock. Therefore, with high mixing ratios of methanol, RCCI combustion can have a stronger performance at medium and high loads. At the engine loads of 10 and 50%, increasing methanol leads to slightly higher CO emissions. This is mainly because of less oxidation of CO due to low temperature. However, at full load, CO emissions are reduced due to the higher oxygen content of the fuel mixture. At the load of 10%, NO_x emission was dramatically decreased by increasing methanol, while at loads of 50 and 100%, NO_x emission was changed insignificantly. In 2015, Li et al. (2015) numerically investigated the effect of gasoline ratio and the start of injection timing (SOI) on the performance of an RCCI engine with gasoline/biodiesel fuel. In their study gasoline ratio changed in the range of 0.0–0.8 and two SOI timings, including common SOI and advanced SOI were tested. To simulate the combustion process, KIVA4-CHEMKIN was used. By comparing the combustion characteristics with the two SOI timing and with different gasoline ratios, they observed that the A-SOI (advanced SOI) can provide more control at the start of combustion when the gasoline ratio is changed. Regarding the pollutants, increasing gasoline can reduce NO_x and soot emissions by achieving more homogeneous combustion. However, when the A-SOI timing was used with same the injection angle (87°) that was designed for the C-SOI (common SOI), soot formation was increased, especially when the fuel with a higher amount of biodiesel was injected. In 2015, Ghareghani et al. (2015) investigated the RCCI combustion with natural gas (CNG)/diesel fuel and natural gas/biodiesel fuel (waste fish oil biodiesel) with compression ignition of diesel and biodiesel at medium and high engine loads. The results showed that using biodiesel as the high-reactivity fuel in the dual fuel mode leads to a higher in-cylinder pressure and a shorter heat release rate compared to conventional combustion. Also, on average, the CNG/biodiesel dual-fuel mode has about 1.6% less thermal efficiency and 2% less combustion loss as compared to those of the CNG/diesel dual-fuel mode in the entire examined engine loads. Furthermore, at high engine loads, the CO emission for the CNG/biodiesel mode reached the same level of conventional combustion and on average, its UHC emission was virtually 32.5% reduced as compared to that with CNG/diesel mode in the entire engine loads. In 2016, Li et al. (2016) numerically investigated the effects of three types of piston bowl geometries on the performance of an RCCI engine with gasoline/biodiesel fuel at high engine speeds in two concentrations of biodiesel (80% and 60%) and three SOI modes of 11°, 35° and 60° ATDC. By analyzing the comparison of the results regarding the engine performance, combustion characteristics, and emissions with different piston bowl geometries, it was found that as compared to the omega combustion chamber (OCC) and hemispherical combustion chamber (HCC) geometries, the shallow-depth combustion chamber (SCC) geometry is more suitable. Shallow geometry also exhibited better combustion and performance, while it produced lower amounts of CO, NO_x, and soot emissions. Figure 8.14 shows the iso-surfaces of temperature at 1300, 1500, 2000 K and equivalence ratios of 0.14, 0.5, and 1.0 for the piston bowl geometries with the biodiesel concentration of 80% and the SOI timing of −60 at CA50. CA50 is the crank angle at which 50% of the combustion heat is released. It can be observed that due to early injection, a high equivalence ratio is observed near the cylinder liner. The piston bowl of the shallow-depth combustion chamber (SCC) has a shallow depth and a larger diameter so that the distant squish flow that is created in the bowl can be more properly pointed towards the near-wall area (higher equivalence ratio in Fig. 8.6b). This results in better fuel mixing in this area and consequently improves the combustion (Fig. 8.6a).

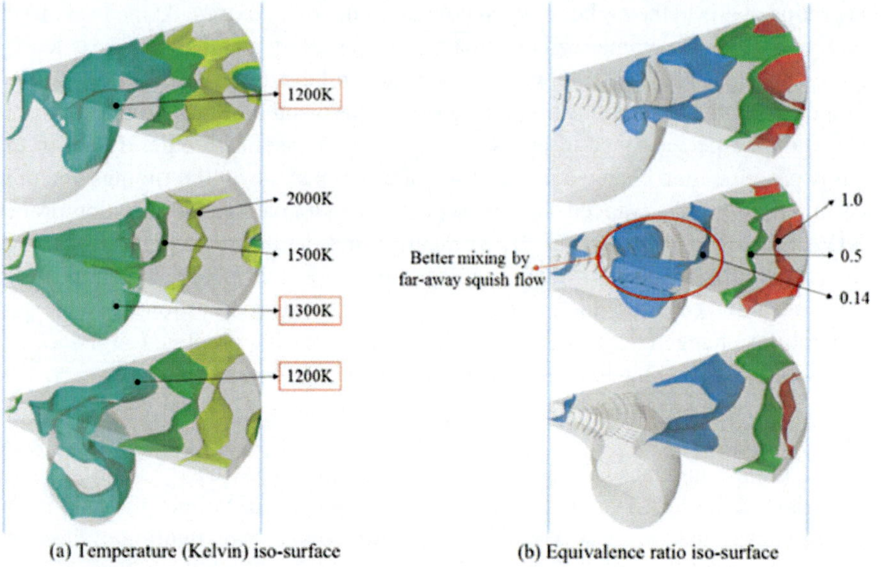

Fig. 8.14 Iso-surfaces of **a** temperature and **b** equivalence ratio, for various piston bowl geometries with B80 fuel and the SOI timing of −60° at CA50 (Li et al. 2016)

According to Fig. 8.14a, SCC geometry presents a higher heat release rate compared to HCC and OCC. Furthermore, one reason for the maximum peak pressure for the SCC geometry can be the decreased heat loss due to the smaller surface area.

In 2019, Charitha et al. (2019) experimentally investigated the performance and combustion characteristics of RCCI combustion with biodiesel fuel as a LRF and diesel as a HRF, with a variable percentage of biodiesel (from cottonseed) of 10–30% at different engine loads. They found that by using this fuel, the NO_x and soot emissions were simultaneously reduced. In low percentages of cottonseed biodiesel, CO_2, and UHC emissions were decreased but with a higher percentage of cottonseed biodiesel, they were increased. According to Fig. 8.15, an increase in the brake thermal efficiency (BTE) of the engine is observed at all loads. Also, the exhaust gas temperature (EGT) and brake-specific fuel consumption (BSFC) were lower in RCCI mode compared to conventional diesel combustion.

In 2020, Harari et al. (2020) investigated the influence of raising the gasoline mass fraction in a Dual-fuel engine at a constant speed of 1500 rpm, using gasoline as the main fuel and diesel, B20, B100 fuels as the direct injection fuels. As can be observed in Fig. 8.16, injection of gasoline considerably reduces the NO_x emission compared to conventional diesel engines.

Soot formation is also significantly reduced by adding gasoline because it leads to the formation of a homogeneous mixture. On the other hand, the hydrocarbons and CO emissions were increased, but these emissions can be eliminated by controlling the injection time of gasoline and keeping the combustion homogeneous. According to Figs. 8.17 and 8.18, gasoline injection increases the specific fuel consumption and decreases thermal efficiency compared to a conventional diesel engine.

In 2020, Thomas et al. (2020) investigated the effect of using hexanol as a low-reactivity fuel with direct injection of vegetable waste oil biodiesel in an RCCI single-cylinder engine. It is noteworthy that to obtain the optimal variable point the hexanol to biodiesel ratio was considered. Experiments were performed at medium and rated loads in three injection pressures of 400 bars, 500 bars, and 600 bars. The results showed that the maximum thermal efficiency was increased by 1.5% compared to when diesel fuel was used. It was found that by using biodiesel-hexanol mixtures instead of diesel fuel the NOx and soot emissions were reduced. With higher hexanol ratios, the peak pressure rise was increased. Based on Fig. 8.19, by increasing the injection pressure, the ignition delay decreases. The injection pressure of 500 bars and hexanol mixture ratio of 30% in medium load mode and 60% in rated load were the most optimal points for reducing emission. This study suggested that the RCCI combustion of biodiesel/hexanol can be an effective alternative to conventional diesel combustion.

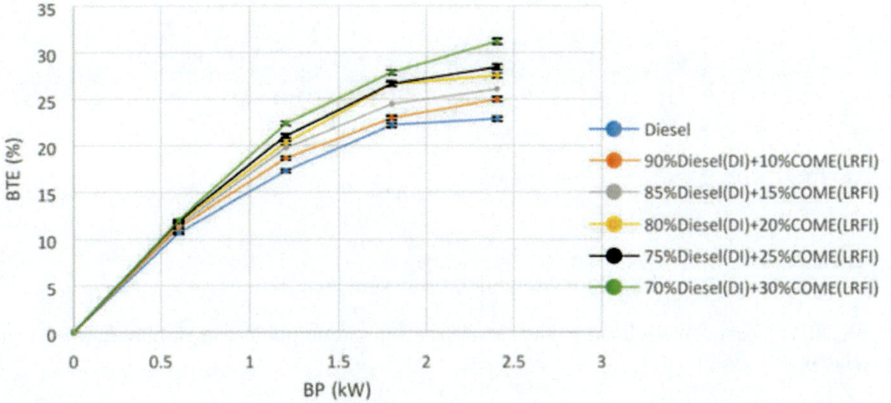

Fig. 8.15 Changes in brake thermal efficiency at different brake powers (Charitha et al. 2019)

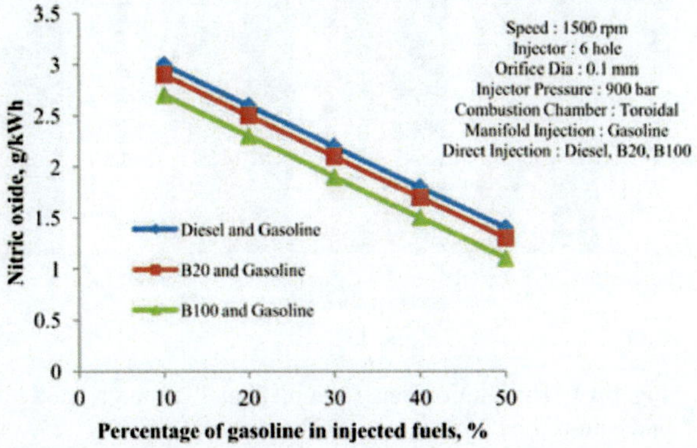

Fig. 8.16 The change of NO_x emission with different gasoline percentages in RCCI mode (Harari et al. 2020)

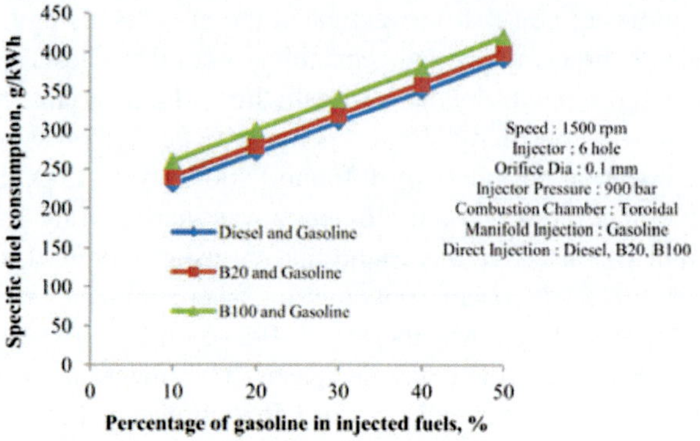

Fig. 8.17 Changes in specific fuel consumption with different percentages of gasoline in RCCI mode (Harari et al. 2020)

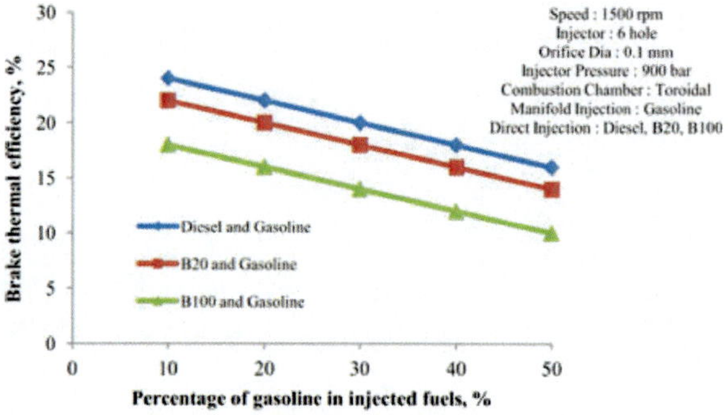

Fig. 8.18 Changes in the brake thermal efficiency with various portions of gasoline in RCCI mode (Harari et al. 2020)

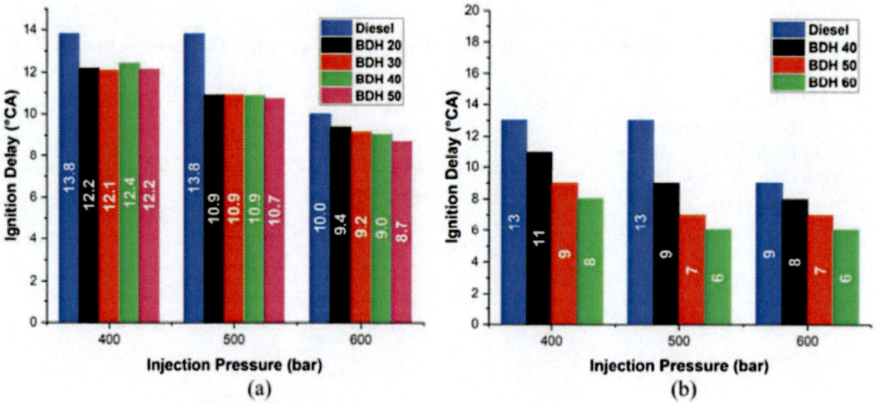

Fig. 8.19 The ignition delay at **a** medium load and **b** rated load, with different injection pressures and various fuel mixture ratios (Thomas et al. 2020)

8.6 Conclusions

In this chapter, the effects of using biodiesel fuels in engines were categorized in different sections based on their ability to achieve sustainable mobility. Combustion phasing, emission characteristics, as well as power output, were the main parameters that were discussed in investigating the role of biodiesel in sustainable mobility. Waste oil biodiesel is known as renewable fuel with a very bright future because of its global availability, and low production costs. This fuel can have a positive economic and environmental impact.

- By examining the power output of biodiesel combustion in compression ignition engines, it was found that the rise of fuel consumption or reduction of engine power is not significant to be a major challenge for the widespread use of biodiesel, especially when the combination of biodiesel and diesel fuel is used.
- In the studies regarding the emissions of biodiesel combustion in compression ignition engines, it was observed that because of the oxygen content in the biodiesel fuel, the use of this fuel reduces the PM, CO, and UHC emissions. In contrast, most reports have shown an increase in NO_x emissions from the combustion of this fuel.
- Using biodiesel with an RCCI combustion strategy can be useful for improving the performance of the RCCI engine in the reduction of CO and UHC, as well as solving the challenge of NO_x emission with biodiesel fuel.
- The use of biodiesel along with the RCCI combustion strategy leads to improvements in the RCCI engine performance and overcoming the NO_x challenge in biodiesel combustion.
- The use of biodiesel reduces carbon sediments and wears in the main engine components. Also, compared to diesel, it results in enhancement of engine durability. Thus, the biodiesel blend can control air pollution and lift the pressure from the rare resources to some extent, without a significant power loss in the engines.

The use of biodiesel in RCCI engines, besides the main goal of substituting fossil fuels with renewable fuels, can reveal some positive combustion characteristics. Therefore, the use of these technologies together seems to have a bright future.

References

Agarwal AK (2005) Experimental investigations of the effect of biodiesel utilization on lubricating oil tribology in diesel engines. Proc Inst Mech Eng, Part D: J Autom Eng 219:703–713. https://doi.org/10.1243/095440705X11239

Agarwal AKB, Das LM (2003) Wear assessment in a biodiesel fueled compression ignition engine. J Eng Gas Turbines Power 125:820–826. https://doi.org/10.1115/1.1501079

Altın R, Çetinkaya S, Yücesu HS (2001) The potential of using vegetable oil fuels as fuel for diesel engines. Energy Convers Manage 42:529–538. https://doi.org/10.1016/S0196-8904(00)00080-7

Ashraful AMM, Kalam MA, Rizwanul Fattah IM, Imtenan S, Shahir SA, Mobarak HM (2014) Production and comparison of fuel properties, engine performance, and emission characteristics of biodiesel from various non-edible vegetable oils: a review. Energy Convers Manage 80:202–228. https://doi.org/10.1016/j.enconman.2014.01.037

Atabani AE, Silitonga AS, Badruddin IA, Mahlia TM, Masjuki H, Mekhilef S (2012) A comprehensive review on biodiesel as an alternative energy resource and its characteristics. Renew Sustain Energy Rev 16:2070–2093. https://doi.org/10.1016/j.rser.2012.01.003

Baiju B, Naik MK, Das LM (2009) A comparative evaluation of compression ignition engine characteristics using methyl and ethyl esters of Karanja oil. Renew Energy 34:1616–1621. https://doi.org/10.1016/j.renene.2008.11.020

Boot M, Rijk E, Luijten C, Somers B (2010) Spray impingement in the early direct injection premixed charge compression ignition regime. SAE Technical Paper. https://doi.org/10.4271/2010-01-1501

Can Ö (2014) Combustion characteristics, performance and exhaust emissions of a diesel engine fueled with a waste cooking oil biodiesel mixture. Energy Convers Manage 87:676–686. https://doi.org/10.1016/j.enconman.2014.07.066

Carraretto C, Macor A, Mirandola A, Stoppato A, Tonon S (2004) Biodiesel as alternative fuel: experimental analysis and energetic evaluations. Energy 29:2195–2211

Charitha VT, Prasad S, Srihari M (2019) Investigation on performance and emissions of RCCI dual fuel combustion on diesel—bio diesel in a light duty engine. Renew Energy 134:1081–1088. https://doi.org/10.1016/j.renene.2018.09.048

Dabhi Y, Brahmbhatt H, Kapadia H, Zalavadiya S (2019) Emission characteristic of diesel and bio-diesel fuelled C.I engine. Int Res J Eng Technol 6:1051–1056

Datta AM, Kumar B (2016) A comprehensive review of biodiesel as an alternative fuel for compression ignition engine. Renew Sustain Energy Rev 57:799–821. https://doi.org/10.1016/j.rser.2015.12.170

Dec JE (2009) Advanced compression-ignition engines understanding the in-cylinder processes. Proc Combust Inst 32:2727–2742. https://doi.org/10.1016/j.proci.2008.08.008

Dempsey AB, Curran SJ, Wagner RM (2016) A perspective on the range of gasoline compression ignition combustion strategies for high engine efficiency and low NO_x and soot emissions: effects of in-cylinder fuel stratification. Int J Engine Res 17:897–917. https://doi.org/10.1177/1468087415621805

Esfahanian V, Salahi MM, Gharehghani A, Mirsalim M (2017) Extending the lean operating range of a premixed charged compression ignition natural gas engine using a pre-chamber. Energy 119:1181–1194. https://doi.org/10.1016/j.energy.2016.11.071

Fakhari AH, Shafaghat R, Jahanian O (2020a) Numerical simulation of natural gas/diesel dual-fuel engine for investigation of performance and emission. J Therm Anal Calorim 139:2455–2464. https://doi.org/10.1007/s10973-019-08560-7

Fakhari AH, Shafaghat R, Jahanian O (2020b) Numerical simulation of a naturally aspirated natural gas/diesel RCCI engine for investigating the effects of injection timing on the combustion and emissions. ASME J Energy Resour Technol 142:072301. https://doi.org/10.1115/1.4046470

Gad MS, El-Seesy AI, Radwan A, He Z (2020) Enhancing the combustion and emission parameters of a diesel engine fueled by waste cooking oil biodiesel and gasoline additives. Fuel 269:117466. https://doi.org/10.1016/j.fuel.2020.117466

Ganesan N, Masimalai S (2020) Experimental investigation on a performance and emission characteristics of single cylinder diesel engine powered by waste orange peel oil biodiesel blended with antioxidant additive. Energy Sources, Part A: Recov Util Environ Effects 42:1412–1423. https://doi.org/10.1080/15567036.2019.1604856

Ganesh D, Nagarajan G (2010) Homogeneous charge compression ignition (HCCI) combustion of diesel fuel with external mixture formation. Energy 35:148–157. https://doi.org/10.1016/j.energy.2009.09.005

Gharehghani A, Hosseini R, Mirsalim M, Jazayeri SA, Yusaf T (2015) An experimental study on reactivity controlled compression ignition engine fueled with biodiesel/natural gas. Energy 89:558–567. https://doi.org/10.1016/j.energy.2015.06.014

Gharehghani A, Mirsalim M, Hosseini R (2017) Effects of waste fish oil biodiesel on diesel engine combustion characteristics and emission. Renew Energy 101:930–936. https://doi.org/10.1016/j.renene.2016.09.045

Gómez MG, Howard-Hildige R, Leahy JJ, Rice B (2002) Winterisation of waste cooking oil methyl ester to improve cold temperature fuel properties. Fuel 81:33–39. https://doi.org/10.1016/S0016-2361(01)00117-X

Gumus M, Kasifoglu S (2010) Performance and emission evaluation of a compression ignition engine using a biodiesel (apricot seed kernel oil methyl ester) and its blends with diesel fuel. Biomass Bioenergy 34:134–139. https://doi.org/10.1016/j.biombioe.2009.10.010

Hajbabaei M, Johnson KC, Okamoto RA, Mitchell A, Pullman M, Durbin TD (2012) Evaluation of the impacts of biodiesel and second generation biofuels on NO_x emissions for CARB diesel fuels. Environ Sci Technol 46:9163–9173. https://doi.org/10.1021/es300739r

Harari PAB, Yaliwal VS, Khan TM, Soudagar Y, Manzoore EM, Sajjan AM (2020) Experimental studies on performance and emission characteristics of reactivity controlled compression ignition (RCCI) engine operated with gasoline and Thevetia Peruviana biodiesel. Renew Energy 160:865–875. https://doi.org/10.1016/j.renene.2020.07.009

Hardy WLR, Rolf D (2006) A study of the effects of high EGR, high equivalence ratio, and mixing time on emissions levels in a heavy-duty diesel engine for PCCI combustion. SAE International. https://doi.org/10.4271/2006-01-0026

Hoekman SK, Robbins C (2012) Review of the effects of biodiesel on NO_x emissions. Fuel Process Technol 96:237–249. https://doi.org/10.1016/j.fuproc.2011.12.036

IEA (2019) Renewables 2019, Market analysis and forecast from 2019 to 2024. Available from https://www.iea.org/reports/renewables-2019/transport

Jahanian O, Jazayeri SA (2012) A comprehensive numerical study on effects of natural gas composition on the operation of an HCCI engine. 67:503–515. https://doi.org/10.2516/ogst/2011133

Jena J, GIET B (2017) Effect of oxygen content in biodiesel blends on the performance of a diesel engine. Int J Eng Manage Res:19–27

Jiaqiang E, Liu T, Yang WM, Li J, Gong J, Deng Y (2016a) Effects of fatty acid methyl esters proportion on combustion and emission characteristics of a biodiesel fueled diesel engine. Energy Convers Manage 117:410–419. https://doi.org/10.1016/j.enconman.2016.03.021

Jiaqiang E, Liu T, Yang W, Deng Y, Gong J (2016b) A skeletal mechanism modeling on soot emission characteristics for biodiesel surrogates with varying fatty acid methyl esters proportion. Appl Energy 181:322–331. https://doi.org/10.1016/j.apenergy.2016.08.090

Jiménez-Espadafor FJT, Miguel V, José AC, Becerra E, José A (2012) Experimental analysis of low temperature combustion mode with diesel and biodiesel fuels: a method for reducing NO_x and soot emissions. Fuel Process Technol 103:57–63. https://doi.org/10.1016/j.fuproc.2011.11.014

Kaplan C, Arslan R, Sürmen A (2006) Performance characteristics of sunflower methyl esters as biodiesel. Energy Sources, Part A: Recov Util Environ Effects 28:751–755. https://doi.org/10.1080/009083190523415

Keskin A, Gürü M, Altıparmak D (2008) Influence of tall oil biodiesel with Mg and Mo based fuel additives on diesel engine performance and emission. Bioresour Technol 99:6434–6438https://doi.org/10.1016/j.biortech.2007.11.051

Kim H, Choi B (2010) The effect of biodiesel and bioethanol blended diesel fuel on nanoparticles and exhaust emissions from CRDI diesel engine. Renew Energy 35:157–163. https://doi.org/10.1016/j.renene.2009.04.008

Koh MY, Ghazi TI (2011) A review of biodiesel production from Jatropha curcas L. oil. Renew Sustain Energy Rev 15:2240–2251. https://doi.org/10.1016/j.rser.2011.02.013

Kumar N (2017) Oxidative stability of biodiesel: causes, effects and prevention. Fuel 190:328–350. https://doi.org/10.1016/j.fuel.2016.11.001

Kurczyński D, Łagowski P (2019) Performance indices of a common rail-system CI engine powered by diesel oil and biofuel blends. J Energy Inst 92:1897–1913. https://doi.org/10.1016/j.joei.2018.11.004

Lapuerta M, Armas O, Rodriguez-Fernandez J (2008) Effect of biodiesel fuels on diesel engine emissions. Progress Energy Combus Sci 34:198–223. https://doi.org/10.1016/j.pecs.2007.07.001

Lawler B, Splitter D, Szybist J, Kaul B (2017) Thermally stratified compression ignition: a new advanced low temperature combustion mode with load flexibility. Appl Energy 189:122–132. https://doi.org/10.1016/j.apenergy.2016.11.034

Li Y, Jia M, Liu Y, Xie M (2013) Numerical study on the combustion and emission characteristics of a methanol/diesel reactivity controlled compression ignition (RCCI) engine. https://doi.org/10.1016/j.apenergy.2013.01.058

Li J, Yang WM, An H, Zhao D (2015) Effects of fuel ratio and injection timing on gasoline/biodiesel fueled RCCI engine: a modeling study. Appl Energy 155:59–67. https://doi.org/10.1016/j.apenergy.2015.05.114

Li J, Yang WM, Zhou DZ (2016) Modeling study on the effect of piston bowl geometries in a gasoline/biodiesel fueled RCCI engine at high speed. Energy Convers Manage 112:359–368. https://doi.org/10.1016/j.enconman.2016.01.041

Li J, Yang W, Zhou D (2017) Review on the management of RCCI engines. Renew Sustain Energy Rev 69:65–79. https://doi.org/10.1016/j.rser.2016.11.159

Lin YC, Lee WJ, Wu TS, Wang CT (2006) Comparison of PAH and regulated harmful matter emissions from biodiesel blends and paraffinic fuel blends on engine accumulated mileage test. Fuel 85:2516–2523. https://doi.org/10.1016/j.fuel.2006.04.023

Lin BF, Huang JH, Huang DY (2009) Experimental study of the effects of vegetable oil methyl ester on DI diesel engine performance characteristics and pollutant emissions. Fuel 88:1779–1785. https://doi.org/10.1016/j.fuel.2009.04.006

Lin L, Cunshan Z, Vittayapadung S, Xiangqian S, Mingdong D (2011) Opportunities and challenges for biodiesel fuel. Appl Energy 88:1020–1031

Luján JM, Bermúdez V, Tormos B, Pla B (2009) Comparative analysis of a DI diesel engine fuelled with biodiesel blends during the European MVEG-A cycle: performance and emissions (II). Biomass Bioenerg 33:948–956

Machrafi H, Cavadiasa S (2008) An experimental and numerical analysis of the influence of the inlet temperature, equivalence ratio and compression ratio on the HCCI auto-ignition process of primary reference fuels in an engine. Fuel Process Technol 89:1218–1226. https://doi.org/10.1016/j.fuproc.2008.05.019

Mahanta P, Mishra SC, Kushwah YS (2006) An experimental study of Pongamia pinnata L. oil as a diesel substitute. Proc Inst Mech Eng, Part A: J Power Energy 220:803–808. https://doi.org/10.1243/09576509JPE172

Motallebi Hasankola S, Shafaghat R, Jahanian O, Nikzadfar K (2020) An experimental investigation of the injection timing effect on the combustion phasing and emissions in reactivity-controlled compression ignition (RCCI) engine. J Therm Anal Calorim 139:2509–2516. https://doi.org/10.1007/s10973-019-08761-0

Mueller CJ, Boehman AL, Martin GC (2009) An experimental investigation of the origin of increased NO_x emissions when fueling a heavy-duty compression-ignition engine with soy biodiesel. SAE Int J Fuels Lub 2:789–816. https://www.jstor.org/stable/26273427

Murray R, King G, Wyse-Mason R (2019) Micro-emulsification versus transesterification: an investigation of the efficacy of methanol use in improving vegetable oil engine performance. Biofuels:1–10. https://doi.org/10.1080/17597269.2019.1598316

Nantha Gopal KP, Sharma A, Samanchi S, Sathyanarayanan C, Elango T (2014) Investigation of emissions and combustion characteristics of a CI engine fueled with waste cooking oil methyl ester and diesel blends. Alex Eng J 53:281–287. https://doi.org/10.1016/j.aej.2014.02.003

Neely D, Sasaki D, Huang Y, Lee JA, Stewart DW (2005) New diesel emission control strategy to meet US Tier 2 emissions regulations reprinted from: diesel exhaust emission control. SAE Paper, 724. https://www.jstor.org/stable/44720976

Neshat ES, Khoshbakhti R (2014) Development of a new multi zone model for prediction of HCCI (homogenous charge compression ignition) engine combustion, performance and emission characteristics. Energy 73:325–339. https://doi.org/10.1016/j.energy.2014.06.025

Ogunkunle O, Ahmed NA (2019) A review of global current scenario of biodiesel adoption and combustion in vehicular diesel engines. Energy Rep 5:1560–1579. https://doi.org/10.1016/j.egyr.2019.10.028

Ozsezen AN, Canakci M, Turkcan A, Sayin C (2009) Performance and combustion characteristics of a DI diesel engine fueled with waste palm oil and canola oil methyl esters. Fuel 88:629–636. https://doi.org/10.1016/j.fuel.2008.09.023

Pandey S, Bhurat S, Chintala V (2019) Combustion and emissions behaviour assessment of a partially premixed charge compression ignition (PCCI) engine with diesel and fumigated ethanol. Energy Procedia 160:590–596. https://doi.org/10.1016/j.egypro.2019.02.210

Qi DHG, Chen H, Bian Y, Liu J, Ren X (2009) Combustion and performance evaluation of a diesel engine fueled with biodiesel produced from soybean crude oil. Renew Energy 34:2706–2713. https://doi.org/10.1016/j.renene.2009.05.004

Raman LAD, Rajakumar S, Sivasubramanian V (2019) Experimental investigation on performance, combustion and emission analysis of a direct injection diesel engine fuelled with rapeseed oil biodiesel. Fuel 246:69–74. https://doi.org/10.1016/j.fuel.2019.02.106

Rashedul HKM, Kalam A, Rahman A, Shahir SA (2014) The effect of additives on properties, performance and emission of biodiesel fuelled compression ignition engine. Energy Convers Manage 88:348–364. https://doi.org/10.1016/j.enconman.2014.08.034

Reaume SJ, Ellis N (2013) Use of hydroisomerization to reduce the cloud point of saturated fatty acids and methyl esters used in biodiesel production. Biomass Bioenergy 49:188–196. https://doi.org/10.1016/j.biombioe.2012.12.008

Robbins C, Hoekman SK, Ceniceros E, Natarajan M (2011) Effects of biodiesel fuels upon criteria emissions (No. 2011-01-1943). SAE Technical Paper

Saddam H, Al-lwayzy TY (2017) Diesel engine performance and exhaust gas emissions using Microalgae Chlorella protothecoides biodiesel. Renew Energy 101:690–701. https://doi.org/10.1016/j.renene.2016.09.035

Sahoo PKD, Babu MKG, Naik SN (2007) Biodiesel development from high acid value polanga seed oil and performance evaluation in a CI engine. Fuel 86:448–454. https://doi.org/10.1016/j.fuel.2006.07.025

Seraç MR, Aydın S, Yılmaz A, Şevik S (2020) Evaluation of comparative combustion, performance, and emission of soybean-based alternative biodiesel fuel blends in a CI engine. Renew Energy 148:1065–1073. https://doi.org/10.1016/j.renene.2019.10.090

Shim E, Park H, Bae C (2020) Comparisons of advanced combustion technologies (HCCI, PCCI, and dual-fuel PCCI) on engine performance and emission characteristics in a heavy-duty diesel engine. Fuel 262:116436. https://doi.org/10.1016/j.fuel.2019.116436

Sinha S, Agarwal AK (2010) Experimental investigation of the effect of biodiesel utilization on lubricating oil degradation and wear of a transportation CIDI engine. ASME J Eng Gas Turbines Power 132. https://doi.org/10.1115/1.3077659

Sivalakshmi S, Balusamy T (2013) Effect of biodiesel and its blends with diethyl ether on the combustion, performance and emissions from a diesel engine. Fuel 106:106–110. https://doi.org/10.1016/j.fuel.2012.12.033

Sjöberg M, Dec JE (2006) Smoothing HCCI heat-release rates using partial fuel stratification with two stage ignition fuels. SAE Trans:318–334. https://www.jstor.org/stable/44687308

Song JT, Zhang CH (2008) An experimental study on the performance and exhaust emissions of a diesel engine fuelled with soybean oil methyl ester. Proc Inst Mech Eng Part D: J Autom Eng 222:2487–2496. https://doi.org/10.1243/09544070JAUTO932

Tanov S (2014) Combustion stratification with partially premixed combustion, PPC, using NVO and split injection in a LD-diesel engine. SAE Int J Engines 7:1911–1919. https://www.jstor.org/stable/26277898

Temizer İ, Cihan Ö, Eskici B (2020) Numerical and experimental investigation of the effect of biodiesel/diesel fuel on combustion characteristics in CI engine. Fuel 270:117523. https://doi.org/10.1016/j.fuel.2020.117523

Thomas JJS, Nagarajan VR, Kumar G, Suraj Basrin G (2020) Influence of waste vegetable oil biodiesel and hexanol on a reactivity controlled compression ignition engine combustion and emissions. Energy 206:118199. https://doi.org/10.1016/j.energy.2020.118199

Usta N (2005) Use of tobacco seed oil methyl ester in a turbocharged indirect injection diesel engine. Biomass Bioenerg 28:77–86. https://doi.org/10.1016/j.biombioe.2004.06.004

Utlu ZK, Süreyya M (2008) The effect of biodiesel fuel obtained from waste frying oil on direct injection diesel engine performance and exhaust emissions. Renew Energy 33:1936–1941. https://doi.org/10.1016/j.renene.2007.10.006

Verma PS, Dwivedi G (2016) Evaluation and enhancement of cold flow properties of palm oil and its biodiesel. Energy Rep 2:8–13. https://doi.org/10.1016/j.egyr.2015.12.001

Xue J, Grift TE, Hansen AC (2011) Effect of biodiesel on engine performances and emissions. Renew Sustain Energy Rev 15:1098–1116. https://doi.org/10.1016/j.rser.2010.11.016

Yousefi A, Birouk M, Lawler B, Gharehghani A (2015a) Performance and emissions of a dual-fuel pilot diesel ignition engine operating on various premixed fuels. Energy Convers Manage 106:322–336. https://doi.org/10.1016/j.enconman.2015.09.056

Yousefi A, Gharehghani A, Birouk M (2015b) Comparison study on combustion characteristics and emissions of a homogeneous charge compression ignition (HCCI) engine with and without pre-combustion chamber. Energy Convers Manage 100:232–241. https://doi.org/10.1016/j.enconman.2015.05.024

Zhou DZ, Yang WM, An H, Li J, Shu C (2015) A numerical study on RCCI engine fueled by biodiesel/methanol. Energy Convers Manage 89:798–807. https://doi.org/10.1016/j.enconman.2014.10.054

Zhu L, Cheung CS, Zhang WG, Huang Z (2011) Combustion, performance and emission characteristics of a DI diesel engine fueled with ethanol–biodiesel blends. Fuel 90:1743–1750. https://doi.org/10.1016/j.fuel.2011.01.024

Part III
Biogas for Sustainable Mobility

Chapter 9
Ammonia for Decarbonized Maritime Transportation

Burak Zincir

Nomenclature

AFC	Alkaline fuel cell
CCS	Carbon capture and storage
CI	Compression ignition
DME	Dimethyl ether
FC	Fuel cell
GHG	Greenhouse gas
H-B	Haber-Bosch
HCCI	Homogeneous charge compression ignition
HFO	Heavy fuel oil
ICE	Internal combustion engine
IMEP	Indicated mean effective pressure
IMO	International Maritime Organization
LBG	Liquefied biogas
LHV	Lower heating value
LNG	Liquefied natural gas
LPG	Liquefied petroleum gas
LSHFO	Low-sulfur heavy fuel oil
MBM	Market-based measures
MCFC	Molten carbonate fuel cell
MDO	Marine diesel oil
MGO	Marine gas oil
OSHA	US Occupation Safety and Health Administration
PAFC	Phosphoric acid fuel cell
PEMFC	Proton exchange membrane fuel cell

B. Zincir (✉)
Maritime Faculty, Istanbul Technical University, İstanbul, Turkey
e-mail: bzincir@itu.edu.tr

G. Di Blasio et al. (eds.), *Clean Fuels for Mobility*, Energy, Environment, and Sustainability, https://doi.org/10.1007/978-981-16-8747-1_9

PPC	Partially premixed combustion
RCCI	Reactivity controlled compression ignition
SCR	Selective catalytic reactor
SFC	Specific fuel consumption
SOFC	Solid oxide fuel cell
SOI	Start of injection

9.1 Introduction

Maritime transportation is the most essential and efficient transportation mode that carries cargo across the countries. According to the report of the United Nations Conference on Trade and Development (UNCTAD), the number of commercial ships which are 100 gross tons and above was 98,140 in January 2020 (United Nations Conference on Trade and Development (UNCTAD) 2020). Approximately 99% of the propulsion unit of the worldwide fleet is internal combustion engines (Rattazzi et al. 2021). The majority of fuel is consumed by ship engines is fossil fuels and the annual consumption amount is 300 million tons (International Maritime Organization (IMO) 2015). Moreover, 72% of total fuel consumption was poor quality fuel, heavy fuel oil (HFO), 26% from marine diesel oil (MDO), and 2% from liquefied natural gas (LNG) (Gray et al. 2021). The high amount of fossil fuel consumption results in increased ship-based emissions. The recent report of the International Maritime Organization (IMO) indicates that the total greenhouse gas (GHG) emissions, which are carbon dioxide (CO_2), methane (CH_4), and nitrous oxide (N_2O), raise from 977 million tons to 1076 million tons, CO_2 emissions from 962 million tons to 1056 million tons, nitrogen oxide (NO_X) emissions from 19.7 million tons to 20.2 million tons, sulfur oxide (SO_X) emissions from 10.8 million tons to 11.4 million tons, and total particulate matter (PM) emissions from 3187 thousand tons to 3316 thousand tons from 2012 to 2018 (International Maritime Organization (IMO) 2020).

Despite the efforts of the IMO about more strict emission regulations (for details see (Zincir and Deniz 2021)) and the improvements on the technologies of maritime transportation, the emissions keep increasing due to the increment of the ship numbers. Especially, the increment of the GHG and CO_2 emissions draws the attention. For instance, if the CO_2 emissions of maritime transportation are compared with the countries, maritime transportation would be the sixth-largest country (Balcombe et al. 2019). These emissions contribute to speeding up global warming and prevent reaching the goals of the United Nations' Paris Agreement for climate change. IMO announced its Initial GHG Strategy in 2018 to reduce maritime transportation-based GHG and CO_2 emissions and it started decarbonization action. The Strategy has two aims; the first one is to reduce CO_2 emissions per transport work at least 40% by 2030 and 70% by 2050, compared to 2008, the second one is to diminish GHG emissions to 50% by 2050, compared to 2008 (International Maritime Organization (IMO) 2018). This is the first time that maritime transportation is in line with the goals of the Paris Agreement (American Bureau of Shipping (ABS) 2020a). The short-term (2018–2023), mid-term (2023–2030), and long-term (2030+) candidate measures were determined to reach the aims of the Initial GHG Strategy. The candidate measures are shown in Table 9.1. There is no restriction on the implementation of the candidate measures. One or a combination of more than one candidate measure can be used to achieve decarbonization in maritime transportation. The way to the decarbonization is left to ship owners/operators.

IMO Initial GHG Strategy contains operational measures, policy measures, and technology-based measures in general. These measures can be but are not limited to more efficient ship designs, condition monitoring, efficient shipboard operations, usage of alternative fuels or fuel cells (Dere and Deniz 2020). The use of alternative fuels, instead of conventional fossil fuels, is the technology-based measure that is a common point at the short-, mid-, and long-term measures. The Strategy supports projects on alternative low-carbon and zero-carbon fuels in the short-term, it encourages implementation of alternative low-carbon and zero-carbon fuels at mid-term, and lastly, it pursues the development and implementation of zero-carbon or fossil-free fuels at long-term. There are various alternative fuels that can be used in maritime transportation, for instance, LNG, liquefied petroleum gas (LPG), methanol, ethanol, hydrogen, dimethyl ether, biodiesel, and liquefied biogas (LBG) (Brinks and Hektor 2020). Recently, LNG is the most used alternative low-carbon fuel in maritime transportation. The Fourth IMO GHG Study indicates that LNG is the third most used fuel after HFO and MDO (International Maritime Organization (IMO) 2020). Besides, methanol is the fourth most used fuel in maritime transportation. According to the database of DNV GL, there are 198 LNG-fuelled ships in operation and 303 ships on order, 10 methanol-fuelled ships in operation and 15 ships on order, and 4 LPG-fuelled ships in operation and 72 ships on order (DNV GL 2021).

In addition to the above-mentioned alternative fuels, ammonia is also considered alternative marine fuel. IMO states that ammonia is one of the most promising alternative fuels in long term (2050) (Rattazzi et al. 2021). Also, the maritime industry focuses on ammonia and one of the biggest global shipping companies, Maersk, added ammonia to their list as one of the most promising ways for zero-carbon maritime transportation (Korberg et al. 2021). Since ammonia is one of the most promising alternative fuel options for maritime transportation, this chapter concentrates on ammonia for decarbonized maritime transportation.

This chapter highlights the short history of ammonia, properties of ammonia, ammonia production methods, ammonia-fueled studies, and ammonia projects and maritime industry developments. And then discussion is made about the barriers and facilitators for ammonia and the future of ammonia in maritime transportation by comparing it with other promising alternative marine fuels. Lastly, the chapter is summarized and further studies are suggested.

Table 9.1 IMO initial GHG study candidate measures (International Maritime Organization (IMO) 2018)

Short-term measures (2018–2023)	Mid-term measures (2023–2030)	Long-term measures (2030+)
Improvement of the existing energy efficiency framework	Implementation programme for the effective uptake of alternative low-carbon and zero-carbon fuels	Pursue the development and provision of zero-carbon or fossil-free fuels
Development of technical and operational energy efficiency measures for both new and existing ships	Operational energy efficiency measures for both new and existing ships	Encourage and facilitate the general adoption of other possible new/innovative emission reduction mechanism(s)
Establishment of an existing fleet improvement programme	New/innovative emission reduction mechanism(s), market-based measures (MBMs)	
Speed optimization/speed reduction	Further continuation and enhancement of technical cooperation and capacity-building	
Measures for volatile organic compounds	Development of a feedback mechanism to enable lessons learned on implementation of measures	
Development and update of national action plans		
Continuing and enhancing technical cooperation and capacity-building		
Measures for port developments and activities		
Initiation of research and development activities on marine propulsion, alternative low-carbon and zero-carbon fuels, and innovative technologies		
Incentives for first movers of new technologies		
Development of a sufficient lifecycle GHG/carbon intensity guidelines for fuels		
Active promotion of the work of the IMO		
Undertake additional GHG emission studies		

9.2 Ammonia: History, Production, Properties, and Applications

This section consists of the history of ammonia as fuel, production of ammonia, properties of ammonia, ammonia-fuelled studies, and ammonia projects and industry developments.

9.2.1 History of Ammonia as a Fuel

The first proposal of ammonia as fuel is given by Sir Goldsworthy Gurney in 1822 (Porter 1998). An ammonia gas-fuelled small locomotive engine was developed and applied by him. Later in 1872, a US patent was tested on streetcars in New Orleans (Brinks and Hektor 2020). The patented system had a fireless engine that utilized the energy of low-pressure ammonia and the ammonia exhaust was captivated in a solvent. In the 1890s, the inventor of the diesel engine was worked on ammonia engines until his serious injury at an explosion during their experiments before his game-changer invention.

In 1905, Ammonia Casale Ltd. has developed the first small-scale motor and the patents were taken in 1935 and 1936 (Valera-Medina et al. 2018). The same company got another patent of ammonia and hydrogen-fuelled internal combustion engine in 1938 and it is applied on about 100 vehicles in the 1940s (Dimitriou and Javaid 2020). The 1940s during World War II was an important period for ammonia-fuelled internal combustion engines. The oil refineries and depots of axis forces were destroyed by the allied forces and this resulted in low oil stock worldwide (Werrell 1986). After then, ammonia was an alternative fuel for the internal combustion engines at that period. The buses were used liquid ammonia in Belgium in 1943 to continue public transportation operative during World War II (MAN Energy Solutions 2019). It was the first time that liquid ammonia with compressed coal gas was used as an internal combustion engine fuel. It is applied on more than 100 buses between 1941 and 1943 (Kroch 1945). Despite the periods of slowdown in researches, the interest in ammonia as fuel was started to increase in the 1930s and continued its ascent until the 1960s (Valera-Medina et al. 2018). In addition to the use of ammonia as a fuel for internal combustion engines, there were studies on fuel cells. In 1980, it was the first time that ammonia was used as a fuel for solid oxide fuel cells (SOFC) in history (Wan et al. 2021). The ammonia fuel studies were continued in the 2000s. A car fuelled by the mixture of ammonia and gasoline was successfully built and it crossed the USA in 2007 (Brinks and Hektor 2020). The maritime industry focuses on ammonia as an alternative marine fuel and there are various studies and ongoing projects. These are explained in the further section is named “2.5 Ammonia projects and maritime industry developments”.

9.2.2 Production of Ammonia

Worldwide annual ammonia production was 150 million tons in 2019 (Al-Aboosi et al. 2021). Global ammonia producers are China (31%), Russia (10%), the US (8.9%), and India (7.9%). Approximately 80% of the production is for fertilizer and the remaining areas are some chemicals, plastics, refrigerants, synthetic fibers and resins, and explosives (Brinks and Hektor 2020). Ammonia production was defined under three general terms: brown ammonia, blue ammonia, and green ammonia (Cardoso et al. 2021). Figure 9.1 shows the production process of brown, blue, and green ammonia. Brown ammonia is the term for ammonia which is produced from fossil fuels and emits intensive carbon emissions. Blue ammonia is the term for ammonia which is still generated from fossil fuels but a carbon capture and storage (CCS) system is used to capture carbon during the ammonia production process. Green ammonia is the term for ammonia which is produced by using renewable energy (solar, wind, hydro, etc.) and a totally carbon-free process.

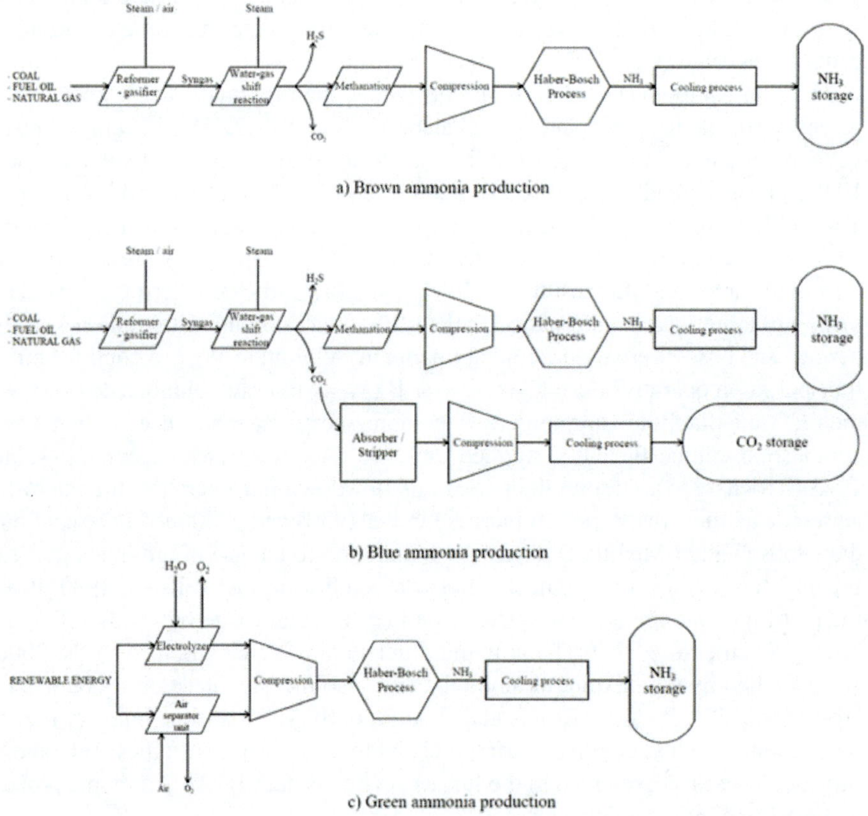

Fig. 9.1 Ammonia production methods (figure reproduced and adapted) (BA Zincir 2020; Ayvalı et al. 2021a)

Currently, the majority of the ammonia production is done by the Haber–Bosch (H-B) process which is the invention of Fritz Haber and Carl Bosch almost 100 years ago (Aziz et al. 2020). The first step of the H-B process is the steam reforming method. It is the method that a hydrocarbon fuel is reacted with steam and air at 400–500 °C and 100–300 bar with the assist of an iron-based catalyst. Various hydrocarbon fuels such as coal, oil, natural gas, etc. are the feedstock for the steam reforming method. The main feedstock at the ammonia production process is natural gas (72%), coal (22%), heavy fuel oil (1%), naphtha (1%), coke oven gas, and refinery gas (1%) (Bicer et al. 2017). The product of steam reforming is the mixture of hydrogen, nitrogen, carbon monoxide (CO), and water. After the steam reforming, at the second step, a water gas shift reaction is performed to produce CO_2 and hydrogen from the mixture. The main components of ammonia are hydrogen and nitrogen. To produce ammonia, the product mixture of two steps of the process passes from the reactor approximately at 300 bar and 400–500 °C, and ammonia is captured and condensed at the reactor (Brinks and Hektor 2020). CO and CO_2 are emitted into the atmosphere after the H-B process. This is the production process of brown ammonia. If the same fossil fuels are used at the H-B process but the carbon emissions captured and stored by the CCS system, this is the process of blue ammonia.

The needed hydrogen for the production of ammonia by the H-B process can be provided by the electrolysis of water by using renewable energy for electricity. Only 1% of hydrogen is produced by electrolysis (Brinks and Hektor 2020). Nitrogen for the H-B process is gathered from the air by an air separation unit. The technologies for air separation are absorption, adsorption, membrane, and cryogenic (Smith and Klosek 2001). Ammonia which is produced in that way can be considered as electro fuel since the process uses renewable energy and electrolysis of water for the hydrogen production and nitrogen from the air. This is the production process of green ammonia. If the ammonia production plant uses fossil fuels and renewable energy for electricity, this type of production is called hybrid green ammonia (Alfa et al. 2020).

Ammonia production is an energy-intensive process and it accounts for 2% of energy consumption and 1% of CO_2 emissions worldwide (McKinlay et al. 2020). When the evaluation is made about emissions of the ammonia production from fossil fuels, they are 1.6 ton CO_2/ton of ammonia for natural gas, 2.4 ton CO_2/ton of ammonia for oil, and 4.0 ton CO_2/ton of ammonia for coal. Despite ammonia has a carbon-free structure and does not emit carbon emissions during its combustion, its production from fossil fuels emits CO_2 emissions. The CO_2 emissions of ammonia produced from natural gas are almost similar to the CO_2 emissions of combustion of marine gas oil (MGO) and higher than the CO_2 emissions of LNG and LPG-fuelled ships (Brinks and Chryssakis 2017). Thus, the ammonia produced from fossil fuels does not have a carbon mitigation advantage, instead, it increases costs. However, the green ammonia production does not emit any carbon emission and from production until the end of the use of ammonia is the zero-carbon process. Green ammonia is the promising alternative fuel to achieve IMO's 2050 GHG reduction targets (American Bureau of Shipping (ABS) 2020b).

9.2.3 Properties of Ammonia

This subsection gives broad information about structural properties of ammonia, combustion properties of ammonia, and risks of ammonia. The properties of conventional marine fuels (MDO and low sulfur heavy fuel oil (LSHFO)), ammonia, and other promising alternative marine fuels are given in Table 9.2.

Table 9.2 Properties of fuels in maritime transportation (Zincir and Deniz 2021; Erdemir and Dincer 2020; Xing et al. 2021a)

Properties	MDO	LSHFO	Ammonia	Hydrogen (gaseous/liquid)	LNG	LPG	Methanol
Chemical formula	C_{10}–C_{15}	C_8–C_{25}	NH_3	H_2	CH_4	C_3H_8	CH_3OH
Auto-ignition temperature (°C)	210	230	651	571–585	450–540	450–470	464
Flashpoint (°C)	73.8	73.8	−33	N/A	−184.4	−87.7	12
Boiling point (°C)	>180	>180	−33	−253	−162	−42	65
Latent heat of vaporization (kJ/kg)	47.86	~47.86	1369	0/N/A	104.8	44.4	1103
Lower heating value (MJ/kg)	43.5	~43.5	18.5	120.1	38.1	45.7	19.9
Flammability limits in air (Vol%)	0.6–7.5	0.6–7.5	16–25	4.0–75	5.0–15.0	2.1–9.5	6.7–36
Flame speed (cm/s)	87	~87	14	270	38	40	50
Cetane/Octane number	>35/–	>20/–	– /120	– />130	– /130	– /94–112	<5/109
Fuel density (kg/m^3)	796–841	975–1010	602.8	17.5/71.1	430–470	1898	792
Energy density (MJ/m^3)	36,403	~36,403	11,333	2101/8539	22,020	25,300	15,600
Storage method	Liquid	Liquid	Comp. liquid	Comp. gas/cryo. liquid	Cryo. liquid	Comp. liquid	Liquid
Storage temperature (°C)	25	25	25	25/−253	−162	25	25
Storage pressure (bar)	1	1	10	248/1	1	8.5	1

9.2.3.1 General Properties

Ammonia is a colorless gas with a sharp odor that can be detected at 5 ppm in the air (Ayvalı et al. 2021b). It consists of one nitrogen atom and three hydrogen atoms. Since it is carbon-free and sulfur-free, the products of ammonia combustion are free of carbonaceous emissions (CO_2, CO, unburned hydrocarbons (HC)), SO_X, and PM emissions (Zincir 2020). The combustion of ammonia forms water, nitrogen, NO_X, and a small amount of nitrous oxide (N_2O) slip in some cases. Ammonia is a hydrogen energy carrier with a hydrogen content of 17.6% wt and the volumetric hydrogen density of liquid ammonia is higher than liquid hydrogen, gasoline, methanol, and ethanol (Dimitriou and Javaid 2020).

9.2.3.2 Storage

Ammonia can be stored as a liquid either at −33 °C at 1 bar or at room temperature at 10 bar (Ayvalı et al. 2021b). Despite it is easy to store ammonia on a ship at room temperature at 10 bar, according to the study of Kim et al., ammonia-fuelled ship needs between 1.4 and 1.6 times more weight and 3.5–5.2 times more volume than the HFO-fuelled ship (Kim et al. 2020a). Another report stated that ammonia needs about 2.4 times larger tank volume than HFO (American Bureau of Shipping (ABS) 2020b). Moreover, ammonia requires more volume on a ship than alternative fuels (LNG and methanol) on the list of IMO (DNV GL 2019). Because of similar properties of LPG and ammonia, the LPG-fuelled ships can store ammonia in their fuel tanks at the same conditions and use ammonia as a fuel (Kobayashi et al. 2019). When ammonia is compared with another promising carbon-free alternative fuel, hydrogen, its storage on a ship is far easier since hydrogen is stored as a compressed gas in high pressurized tanks (300 bar) or as a cryogenic liquid at a temperature of −253 °C (Deniz and Zincir 2016).

9.2.3.3 Combustion Properties

Ammonia has a low lower heating value (LHV) of 18.5 MJ/kg than conventional marine fuels. On the other hand, it has a low stoichiometric air/fuel ratio of 6.06 (Lesmana et al. 2019), which means more fuel can be burned with the same air amount than conventional marine fuels.

Ammonia is a low reactivity fuel. It has a high auto-ignition temperature of 651 °C and the flammability limits in air are narrow with 16–25% by volume in air. The burning velocity of ammonia is 14 cm/s which is lower than LNG (38 cm/s), LPG (40 cm/s), methanol (50 cm/s), and hydrogen (270 cm/s) (Zincir 2019). Additionally, ammonia has a low adiabatic flame temperature (1800 °C) while it is 1950 °C for methane, 2000 °C for propane, and 2110 °C for hydrogen (Aziz et al. 2020). These combustion properties of ammonia lead to unstable combustion at low engine loads and high engine speeds (Valera-Medina et al. 2018) because the in-cylinder temperature does not high enough to overwhelm high auto-ignition resistance at low loads and the burning duration does not long enough to combust ammonia completely by the reason of high auto-ignition and low flame velocity. Furthermore, the narrow flammability limit of ammonia can result in misfires under very lean or rich in-cylinder mixtures (Heywood 1988) and limits the operation range of internal combustion engines. This makes marine diesel engines suitable for ammonia combustion (with the assist of pilot diesel injection) due to their large displacement volume, constant high loads, and low engine speeds (Klüssmann et al. 2020).

Contrarily, the low adiabatic flame temperature and high latent heat of vaporization of ammonia provide lower maximum combustion temperature and heat loss, and the possibility of slowing down of the NO_X formation which is also dependent on ammonia's nitrogen bonds (American Bureau of Shipping (ABS) 2020a). Ammonia-fuelled diesel engines can have higher thermal efficiency than conventional-fuelled diesel engines (Niki et al. 2019). Another important thing for achieving complete combustion and high engine efficiency is fuel viscosity. It affects fuel flow turbulence and fuel spray formation. Ammonia has a low viscosity that can be benign for proper high-pressure fuel atomization (Lesmana et al. 2019).

The combustion of ammonia produces significant NO_X emissions depending on high in-cylinder temperature and pressure and fuel nitrogen content. Thus, NO_X emissions can be eliminated by using a selective catalytic reactor (SCR) as an after-treatment system on internal combustion engines. Moreover, ammonia combustion emits N_2O which is an extremely important emission with a global warming potential about 300 times more than CO_2 emissions (Zincir 2020). Different combustion techniques have to be applied to eliminate N_2O formation as a combustion product. Proven combustion techniques at various low reactivity fuels such as double or triple injection strategies (Ianniello et al. 2021; Shamun et al. 2020) or an advanced combustion technique, reactivity controlled compression ignition (RCCI), (Belgiorno et al. 2019) can be used to overcome this issue. Ammonia slip is another important emission type that is conditional to the combustion conditions. Unburned ammonia is toxic and if the amount is high, it threatens human life. It is caught by the SCR or it can react and mitigate NO_X emissions in the exhaust (Brinks and Hektor 2020).

9.2.3.4 Risks of Ammonia

Using ammonia as a fuel has some risks associated with the fuel properties. Ammonia is toxic, corrosive and its spill is hazardous for aquatic life. Thus, these properties have to be well-known by the ship crew and additional precautions have to be done onboard.

Ammonia is a toxic substance that has a toxicity level more than three times that of diesel, gasoline, and methanol (Aziz et al. 2020; McKinlay et al. 2020). Ammonia—human contact results in three main types of injuries. These are dehydration, caustic burning, and freezing (Ayvalı et al. 2021b). Dehydration occurs because ammonia extracts water from body tissue. Caustic burning is formed due to the corrosive property of ammonia. And freezing happens since ammonia removes heat away from the human body during evaporation (the high latent heat of vaporization). Furthermore, the low concentrations of ammonia can affect eyes, lungs, and skin; and if human exposes to the high concentrations, it is fatal for life (American Bureau of Shipping (ABS) 2020b). The exposure limits that are determined by the US Occupation Safety and Health Administration (OSHA) are 25 ppm for an 8-h exposure and 35 ppm for a 15 min exposure, and these limits are 20 ppm and 30 ppm for EU, respectively. The OSHA determines the maximum exposure limit for irreversible health effects as 300 ppm (Brinks and Hektor 2020).

Another important property of ammonia is its corrosiveness. Ammonia is incompatible with some materials and the moisture in the environment increases the chemical reactivity of ammonia. The product of ammonia combustion is water and this affects engine parts if incompatible materials are used. These materials are copper, brass, zinc, aluminum, and magnesium alloys (Klüssmann et al. 2020). Ammonia changes the color of the incompatible surfaces to a greenish/blue color. Also, it leads to cracking (stress corrosion) in the storage tanks and fuel supply lines from steel, nickel steel, or carbon-manganese materials (American Bureau of Shipping (ABS) 2019). The oxygen contamination in the liquid ammonia increases the risk of stress corrosion. Thus, the air in the tanks and fuel supply lines has to be purged before ammonia bunkering (Brinks and Hektor 2020). To prevent corrosion risk of ammonia, tanks, pipelines, and specific parts of the main and auxiliary engines should be made from steel, iron, and non-ferrous alloys on a ship (American Bureau of Shipping (ABS) 2020b; Yapicioglu and Dincer 2019).

The toxicity of ammonia is well-known and the spill of ammonia into the sea can cause serious risk for aquatic life. Ammonia dissolves in the water and forms ammonium hydroxide (NH_4OH) and ammonium ions (NH_4^+) (Ayvalı et al. 2021b; Vries 2019). Undissolved ammonia evaporates into the atmosphere with a sharp odor. Ammonia is biodegradable in the long term, but direct exposure to results in damages and fatal for organisms. The aquatic life restores itself by the nitrogen cycle in a long term.

In addition to the mentioned risks, ammonia has a low risk of fire when it is compared with other potential alternative marine fuels. Ammonia has a narrow flammability range between 16 and 25% by volume in air, 2–3 times more ignition energy than conventional hydrocarbon fuels, and low burning speed (four times lower than methane) (American Bureau of Shipping (ABS) 2020b). These properties lower the risk of fire on a ship. On the other hand, ammonia has chemical reactivity with halogens and oxidizers that may result in explosions. Ammonia has to be stored away from these substances and ignition sources, and well-ventilation has to be provided.

9.2.4 Ammonia-Fuelled Studies

The interest of researchers in ammonia as fuel was shown in the section "2.1 History of ammonia as a fuel". The studies started in the 1800s and continue until today. The study areas are compression ignition (CI) engines and fuel cells. Nowadays, the research focus is given to internal combustion engines and gas turbines (Morlanes et al. 2021). This section focuses on studies on CI engines in the literature in close history and the progress at direct ammonia fuel cells.

9.2.4.1 Compression Ignition Engine Studies

The first well-known study on ammonia-fuelled CI engines in close history was conducted by Reiter and Kong in 2010 (Reiter and Kong 2010). They did their experiments on a 4-cylinder diesel engine by applying the dual-fuel combustion concept. Diesel and ammonia were used as fuel. The diesel fuel was injected into the cylinder by the injectors and ammonia vapor was given from the intake port by the fumigation method. The purpose of the study was to observe the maximum energy replacement of diesel to ammonia. They achieved 95% energy replacement. Secondary findings of the study were higher specific fuel consumption (SFC), lower CO_2 emissions, and higher unburned hydrocarbon (HC) emissions with the increase of ammonia fraction in the total fuel energy. Another observation was if the ammonia fraction is lower than 60% the NO_X emissions were lower than the sole diesel condition. But the NO_X emissions increased after the ammonia fraction was higher than 60%.

In 2011, Reiter and Kong extended their ammonia dual-fuel CI engine studies (Reiter and Kong 2011). They used the same CI engine that they used in their previous study. Ammonia fuel fraction was between 40 and 60% in total fuel energy. They found that CO and HC emissions were at a higher value than the sole diesel combustion. Nonetheless, a significant reduction in soot emissions was observed. Lower NO_X emissions were produced when the ammonia energy fraction was lower than 40%. The important finding of the study was the ammonia combustion resulted in unburned ammonia between 1000 and 3000 ppm in the exhaust.

Gill et al. did experimental studies by using a single-cylinder ammonia dual-fuel CI engine in 2012 (Gill et al. 2012). The study aimed to find a better combustion concept for ammonia. Gaseous ammonia, dissociated ammonia, and hydrogen were tried to be fumigated from the intake manifold and combusted with the diesel fuel. The fumigation of fuels led to a significant reduction in CO_2 emissions. The engine stability and brake thermal efficiency were higher at high load during the ammonia fumigation. The fuel consumption was increased since ammonia has a low LHV. The emissions of HC, NO_X, and N_2O were raised with ammonia fumigation.

Gross and Kong conducted a dual-fuel CI engine study with ammonia—dimethyl ether (DME) mixtures in 2013 (Gross and Kong 2013). This was the first study in the literature that used liquid ammonia—DME mixtures in a CI engine. The experiments were done on a single-cylinder diesel engine and the fuel system was modified for working with liquid ammonia—DME mixture. The fuels are used at the experiments were 100% DME, 20 wt.% ammonia—80 wt.% DME, and 40 wt.% ammonia—60 wt.% DME mixtures. The presence of ammonia in the mixture resulted in longer ignition delays and load limitations. The high auto-ignition and low flame speed were the reason for this. Higher ammonia fraction in the mixture led to higher CO, HC, and NO_X emissions, on contrary, soot emissions were lower. They also found a few hundred ppm unburned ammonia under most conditions. It was indicated that a higher injection pressure can cause better combustion and extends the operating load range of the engine.

In 2014, another ammonia—DME mixture study was done by Ryu et al. (2014). They aimed to examine combustion and emissions characteristics of 100% DME, 60% DME—40% ammonia, and 40% DME—60% ammonia mixtures by weight. The findings of the study were the engine performance decreased by the increase of ammonia fraction in the fuel mixture. Moreover, the engine speed and engine power were limited. When the engine was operated with a 40% DME—60% ammonia mixture, the combustion concept was homogeneous charge compression ignition (HCCI), rather than conventional diesel combustion. The reason was since the reactivity and burning speed of ammonia is low, the injection timing was highly advanced. Higher NO_X, CO, and HC emissions with ammonia—DME mixtures than 100% DME were observed. They also measured unburned ammonia in the exhaust and it was increased by the increase of ammonia fraction in the fuel mixture. On contrary, soot emissions were extremely low during using ammonia—DME mixtures.

In 2017, Tay et al. conducted an experimental study with kerosene—diesel injection and ammonia fumigation CI engine and they detailed their study by numerical investigation (Tay et al. 2017). The fumigated ammonia was burned by the pilot 100% diesel, 100% kerosene, and 50% kerosene—50% diesel mixture by mass. The results showed that ammonia decreased CO and CO_2 emissions. NO_X emissions were reduced when the ammonia fumigation was low (below 60% energy fraction), but it rose with a higher ammonia fumigation rate. Another important finding was when they advanced start of injection (SOI) timing, the residual fuel adjacent to the cylinder wall and in the crevice was burned which resulted in more complete ammonia combustion.

An HCCI combustion concept study was done by using various ammonia—hydrogen mixtures by Pochet et al. (2017). A single-cylinder, 16:1 compression ratio HCCI engine was used in the experiments. The intake pressure range was between 1 and 1.5 bar, and the intake temperature range was between 428 and 473 K. They achieved stable combustion with 70 vol.% ammonia fraction in the fuel mixture at the intake pressure at 1.5 bar and the intake temperature at 473 K. They observed NO_X emissions of 6 ppm for hydrogen-only combustion and between 1000 and 3500 ppm for ammonia—hydrogen mixtures. Other findings of the study were the in-cylinder temperature should be above 1300 K to achieve combustion efficiencies close to pure hydrogen combustion, and below 1400 K there was a significant amount of N_2O formation. To get higher indicated mean effective pressure (IMEP), efficiency, and low requirement of intake temperature, a higher compression ratio and intake pressure was needed. Ammonia—hydrogen mixture resulted in longer combustion duration that increases brake efficiency with lower mechanical and heat losses.

In 2018, Niki et al. conducted studies on an ammonia-fumigated diesel engine and aimed to develop a new combustion strategy to reduce unburned ammonia and N_2O emissions (Niki et al. 2018). The fumigated ammonia amount varied from 0 l/min (sole diesel) to 35 l/min which was equal between 0 and 30% heat fractions. The results showed that unburned ammonia, CO, and N_2O emissions were increased while CO_2 emissions were decreased by the increase of ammonia fraction. Other findings of the study were ammonia fumigation reduced maximum in-cylinder pressure and a higher in-cylinder temperature was needed to reduce N_2O formation. Moreover, a higher in-cylinder temperature decreased unburned ammonia in the exhaust.

Niki et al. continued their ammonia fuel studies in 2019 (Niki et al. 2019). They stated that ammonia can be a more suitable fuel than hydrogen for marine diesel engines. They did studies on a single-cylinder diesel engine to observe the results of ammonia combustion with pilot diesel injection. In their study, they tried pilot and post-diesel fuel injection to control the unburned ammonia and N_2O production. The findings were the advanced diesel fuel injection slightly decreased unburned ammonia and N_2O emissions, the pilot diesel fuel injection decreased unburned ammonia but increased N_2O emissions, the post-diesel fuel injection reduced unburned ammonia, and slightly increased N_2O emissions.

In 2020, an ammonia-hydrogen HCCI engine study was done by Pochet et al. (2020). They used a single-cylinder, 22:1 compression ratio diesel engine in their studies. The fuel which was used at the experiments was sole hydrogen to various mixture fractions up to 94% ammonia—6% hydrogen mixture. The intake temperature was varied from 50 to 240 °C. To combust 94% ammonia fraction fuel mixture, the intake temperature was 240 °C. The in-cylinder temperature had to be above 1400 K to avoid N_2O formation and 1800 K was required for complete burning of ammonia.

A study on an ammonia–diesel dual-fuel diesel engine was conducted by Niki in 2021 (Niki 2021). Combustion strategies were proposed to reduce unburned ammonia and N_2O in the exhaust. A single-cylinder, naturally aspirated diesel engine was used in the experiments, and partially premixed combustion (PPC) and RCCI combustion strategies (for more information on the combustion strategies (Zincir and Deniz 2021)) were applied. A gas injector for ammonia was used to deliver ammonia to the intake and diesel fuel was injected into the cylinder. These combustion strategies aimed to mix diesel fuel into ammonia—air mixture to promote combustion and reduce the unburned ammonia and formation of N_2O emissions by advancing the pilot diesel fuel injection timing. The combustion timing was arranged by changing the amounts of diesel fuel and ammonia. The results showed that the proposed combustion strategies were successful to reduce unburned ammonia and N_2O emissions, but CO, HC, and NO_X emissions were increased. It was also stated that a split pilot injection could decrease CO and HC emissions.

A recent dual–fuel experimental study, which was combined by the modeling study, was done with aqueous ammonia and diesel fuels (Frost et al. 2021). The single-cylinder Ricardo Hydra diesel engine was used in the study. The diesel fuel was injected into the cylinder while the aqueous ammonia solution was delivered from the intake manifold. The findings of the study are a longer ignition delay period with the addition of the aqueous ammonia solution, a higher CO, PM, and unburned ammonia emissions. On the other hand, NO_X emission firstly increased until the ammonia load share to 20% but then it was lower than the diesel-only combustion (Table 9.3).

9.2.4.2 Fuel Cell Studies

Fuel cells are the units that produce electricity from fuels by chemical reactions inside the stacks. There are five major types of fuel cells; alkaline fuel cell (AFC), phosphoric acid fuel cell (PAFC), molten carbonate fuel cell (MCFC), proton exchange membrane fuel cell (PEMFC), and solid oxide fuel cell (SOFC) (Inal and Deniz 2020). Table 9.4 shows the operating conditions of the fuel cell types.

AFCs and SOFCs can use ammonia directly in their stacks to produce electricity. AFCs are affected by CO_2 in the air since durability is low (Wan et al. 2021). SOFCs have fuel flexibility and high power capacity (Inal and Deniz 2021). Therefore, SOFCs are preferred in ammonia studies. SOFCs have the characteristics of high efficiency, low emissions, fuel options, compactness, and modularity (Alemu and Ilbas 2020). The first direct ammonia SOFC was operated in 1980 (Vayenas and Farr 1980). Recently, one good example of direct ammonia SOFC application was done by Kishimoto et al. (2020). They investigated a 30 celled ammonia-SOFC which has a 1 kW power output with an efficiency of 57%. The application was achieved 1000 h of operation.

Direct ammonia SOFCs are under development and applications in the literature are scarce. There are numerical studies that model operating conditions of direct ammonia SOFCs (Baniasadi and Dincer 2011; Tan et al. 2018; Al-Hamed and Dincer 2021; Mukelabai et al. 2021). Moreover, review studies explain the advantages and challenges of direct ammonia SOFCs (Wan et al. 2021; Cheliotis et al. 2021; Xing et al. 2021b).

Direct ammonia SOFCs are not commercial recently and the durability of fuel cell stacks should be improved. However, they will be one of the promising solutions for zero-carbon shipping in the future, and fuel cells are on the agenda of IMO.

9.2.5 Ammonia Projects and Industrial Developments

Ammonia has taken the attention of companies and investors in various industries. The countries started initiatives, consortiums and groups were formed, projects were started, and some applications were done until now. Japan is one of the active countries worldwide. An action plan was established to use ammonia in electricity generation. Ammonia-fuelled gas turbine program was started to produce 1% of electricity consumption by ammonia (Erdemir and Dincer 2020). Other studies of Japan are the formation of the ammonia supply chain by 2030, direct ammonia fuel cells,

Table 9.3 Ammonia-fuelled CI engine studies in the literature

Authors and year	Combustion strategy	Findings
Reiter and Kong (2010)	Dual—fuel (ammonia—diesel)	– Achieved 95% ammonia energy – Higher SFC – Higher HC emissions – Lower CO_2 and NO_X emissions (if ammonia energy < 60%)
Reiter and Kong (2011)	Dual—fuel (ammonia—diesel)	– Higher CO and HC emissions – Significantly lower soot emissions – Lower NO_X emissions (if ammonia energy < 60%) – Unburned ammonia between 1000 and 3000 ppm
Gill et al. (2012)	Dual—fuel (ammonia—diesel)	– Significant CO_2 reduction – Higher engine stability and brake thermal efficiency at high load – Higher SFC – Higher HC, NO_X, and N_2O emissions
Gross and Kong (2013)	Dual—fuel (ammonia—DME)	– The first liquid ammonia—DME mixture study – Higher CO, HC, and NO_X emissions with a higher ammonia fraction – Lower soot emissions – Few hundred ppm unburned ammonia
Ryu et al. (2014)	HCCI (ammonia—DME)	– Higher CO, HC, and NO_X emissions – Lower soot emissions – Increased unburned ammonia by the increase of ammonia fraction
Tay et al. (2017)	Dual—fuel (kerosene—diesel mixture—ammonia)	– Lower CO and CO_2 emissions – Lower NO_X emissions (if ammonia energy < 60%) – Advanced SOI resulted in more complete ammonia combustion
Pochet et al. (2017)	HCCI (ammonia—hydrogen)	– Higher NO_X emissions with a higher ammonia fraction – Higher brake efficiency – Minimum in-cylinder temperature = 1300 K for similar combustion efficiency with sole hydrogen combustion

(continued)

Table 9.3 (continued)

Authors and year	Combustion strategy	Findings
Niki et al. (2018)	Dual—fuel (ammonia—diesel)	– Higher unburned ammonia, CO, and N_2O emissions – Lower CO_2 emissions – Higher in-cylinder temperature requirement for lower unburned ammonia and N_2O emissions
Niki et al. (2019)	Dual—fuel (ammonia—diesel)	– Slightly decreased unburned ammonia and N_2O emissions with the advanced diesel fuel injection – Higher unburned ammonia and N_2O emissions with the post diesel fuel injection
Pochet et al. (2020)	HCCI (ammonia—hydrogen)	– Stable combustion with 94% ammonia—6% hydrogen fuel blend – Lower N_2O at above 1400 K in-cylinder temperature – Complete ammonia combustion at above 1800 K
Niki (2021)	PPC, RCCI (ammonia—diesel)	– Lower unburned ammonia and N_2O emissions with advanced pilot injection – Combustion timing determined by the pilot diesel and ammonia fuel amounts – Higher CO, HC, and NO_X emissions
Frost et al. (2021)	Dual—fuel (aqueous ammonia—diesel)	– Longer ignition delay period – Higher CO, PM, and unburned ammonia emissions – Lower NO_X emission after the ammonia load share more than 20%

and ammonia-coal-fired power plants (Aziz et al. 2020). European Union, USA, UK, and Australia are other countries that have started funding research programs for ammonia to hydrogen or ammonia to electricity applications. Companies, for instance, Siemens, ThyssenKrupp, Yara, JGC, and Haldor Topsoe, are the consortium members in various CO_2-free ammonia production plants (Morlanes et al. 2021).

The maritime industry has also shown its interest in ammonia. One of the largest maritime transportation companies, Maersk, added ammonia to the alternative fuel list for zero-emission shipping (Korberg et al. 2021). The marine engine manufacturers, MAN Energy Solutions and Wartsila have been working on ammonia-fuelled two-stroke and four-stroke engines. They achieved good combustibility of ammonia and it is expected that the first ammonia-fuelled engine will in operation by 2024

Table 9.4 Fuel cell types and operating conditions (Wan et al. 2021; Inal and Deniz 2020)

Fuel cell type	Fuels	Operating temperatures (°C)	Efficiency (%)
Alkaline fuel cell (AFC)	Ammonia (direct), hydrogen	50–230	50–60
Phosphoric acid fuel cell (PAFC)	Ammonia (hydrogen carrier), diesel, hydrogen, natural gas	150–220	40–50
Molten carbonate fuel cell (MCFC)	Ammonia (hydrogen carrier), diesel, hydrogen, natural gas	600–700	30–70
Proton exchange membrane fuel cell (PEMFC)	Ammonia (hydrogen carrier), hydrogen	50–130	40–60
Solid oxide fuel cell (SOFC)	Ammonia (direct), diesel, hydrogen, natural gas	500–1000	40–70

(Ayvalı et al. 2021a). The ammonia-fuelled SOFC project, ShipFC, has been ongoing. The aim is to convert one of the gensets of the ship named, Viking Energy, to an ammonia-fuelled SOFC. The first tests of 100 kW ammonia-fuelled SOFC will be on land, and the installation of scaled SOFC to the ship in late 2023 (Brinks and Hektor 2020).

Table 9.5 shows the maritime industry developments and news on ammonia as a fuel in 2021. Only the year 2021 is shown at the table because there have been a large number of developments and news since 2019. The interest in ammonia has speeded up and it is expected that ammonia will one of the promising solutions for zero-carbon maritime transportation. It can be seen from the table that every stakeholder of the maritime industry involves in the consortiums or groups to work on building

Table 9.5 Maritime industry developments and news on ammonia as a fuel in 2021 (Ammonia Energy Association (AEA) 2021; Spectrum 2021; ITOCHU Corporation 2021; OCI NV 2021)

Article date	Developments and news	Source
21 January 2021	– An ammonia-fuelled tanker will be ready by 2024	Ammonia Energy Association
18 February 2021	– The world's first ammonia ready vessel is under construction	Ammonia Energy Association
23 February 2021	– Ammonia-fuelled marine diesel engine tests are planned to begin by Wartsila – The first ammonia-fuelled oil tanker will be ready by 2024 – Viking Energy will be the first ammonia-fuelled fuel cell vessel by 2024	IEEE Spectrum

(continued)

Table 9.5 (continued)

Article date	Developments and news	Source
5 March 2021	– A partnership for the commercialization of ammonia-fuelled ships	OCI N.V.
10 March 2021	– A group is formed and will work on the ammonia potential for Singapore, supply, bunkering, and safety of ammonia – Supply chain studies for green ammonia at the Port of Singapore	Ammonia Energy Association
18 March 2021	– A consortium will develop ammonia infrastructure for ships as fuel. The consortium will build and operate ammonia-fuelled ships – An approval for the design of a 40,000 m^3 ammonia-fuelled gas carrier was given by Lloyd's Register	Ammonia Energy Association
24 March 2021	– Green ammonia production plant was developed by the addition of solid-oxide electrolyzer cells. The first part of the facility will start to produce 300 tons/day in 2024	Ammonia Energy Association
31 March 2021	– A new green hydrogen and ammonia production plant will be established in Oman. The plant will use solar energy to produce 2200 tons of green ammonia in a day	Ammonia Energy Association
19 May 2021	– A project group will work on improving PEM electrolyzers to achieve commercial green ammonia production	Ammonia Energy Association
20 May 2021	– The Class Society, Bureau Veritas (BV), formed ammonia-prepared notation for the new build ships. The BV also prepared rules for ammonia as a marine fuel – The first ammonia-fuelled car carrier ship will operate in 2023. MAN ES designed the main engine of the ship for ammonia fuel, and DNV GL will give ammonia-ready notation to the ship – Collaboration will be done to focus on the supply of blue and green ammonia between Australia and Japan	Ammonia Energy Association
26 May 2021	– Blue ammonia production plant will be done in Ruwais by 2025 to produce a million-ton ammonia per year – Green ammonia production plant will be powered by 800 MW solar panels at Abu Dhabi Port to produce 200,000 tons of ammonia per year	Ammonia Energy Association

(continued)

Table 9.5 (continued)

Article date	Developments and news	Source
1 June 2021	– A collaboration of MIT, the University of Lisbon, and the Polytechnic Institute of Portalegre will focus on a 100% ammonia-fuelled internal combustion engine	Ammonia Energy Association
8 June 2021	– M/S NoGAPS (Nordic Green Ammonia-Powered Ship) will operate in the deep sea by 2025	Ammonia Energy Association
11 June 2021	– A large consortium with 23 companies is formed to focus on ammonia fuel safety assessment, safety assessment of ammonia bunkering, ammonia fuel properties, and CO_2 production during ammonia production	Itochu

ammonia-fuelled ships, developing ammonia-fuelled engines, establishing a supply chain, and forming rules and standards for ammonia-fuelled ships.

9.3 Discussion

Ammonia is one of the prominent alternative fuels for the decarbonization of maritime transportation. However, there are some barriers and facilitators that will determine the place of ammonia in future maritime transportation.

9.3.1 Barriers and Facilitators

Global ammonia production is about 150 million tons annually and 80% of the production is for fertilizer production. Although, ammonia is one of the most produced substances worldwide it is not produced as fuel recently. Thus, industrial-scale production of ammonia to meet the fuel consumption of ships is hard and expensive (American Bureau of Shipping (ABS) 2019). Maritime transportation requires 2.5 times larger production of ammonia to achieve sustainable decarbonization (Ayvalı et al. 2021b). In addition to this, ammonia production has to be in a green way since well-to-wheel emissions of brown ammonia are almost equal to well-to-wheel emissions of MGO and do not provide any GHG emission reduction advantage. This is one of the barriers to ammonia as a fuel for maritime transportation because there is no commercial green ammonia production plant today. Nonetheless, it is expected that there will be green ammonia demonstration plants by 2030 (Ayvalı et al. 2021a). Despite there is an established supply chain and infrastructure for ammonia, these are for the existing fertilizer industry and chemical industry. There is no supply chain and infrastructure for ammonia as fuel (American Bureau of Shipping (ABS) 2019, 2020b), and this is another barrier for ammonia at maritime transportation recently. However, the existing ammonia supply chain and infrastructure facilitate the establishment of new supply chains for ammonia as fuel and new infrastructure for ammonia fuel production and bunkering. Another barrier is a lower LHV of ammonia. Larger storage tanks for ammonia than conventional marine fuels are required for the same voyage. This leads to either a shorter voyage distance or a reduction in the cargo-carrying capacity of a ship. Ship designers/builders need to think of new ship designs to overcome this situation and ship owners/operators need to consider their commercial strategy during the decarbonization of maritime transportation with using ammonia as fuel.

One of the facilitators for ammonia usage is it can be used on diesel engines without major modifications or changes (Dimitriou and Javaid 2020). Ammonia fuel supply systems and storage tanks are the main modification areas. Another positive thing is marine diesel engines fuelled by pilot diesel fuel and ammonia (dual-fuel) is suitable for ammonia combustion since the combustion stability of ammonia is good at a constant low speed with high loads and large displacement volume engines (Klüssmann et al. 2020). On the other hand, there is a negative side to the pilot diesel-ammonia dual-fuel engines. The previous experimental studies used a higher pilot fuel fraction to improve the combustion quality (American Bureau of Shipping (ABS) 2020b; Hansson et al. 2020). On contrary, there are studies that achieved good combustion stability and efficiency with 94–95% ammonia fuel fraction (Reiter and Kong 2010; Pochet et al. 2020), which facilitates the ammonia-fuelled marine diesel engine studies. Ammonia-fuelled engine studies have resulted in significantly lower CO_2 and soot emissions, but higher NO_X emissions, ammonia slip, and N_2O emissions in the exhaust are other negative things and possible barriers for ammonia-fuelled engine studies. The marine engines have to comply with the NO_X emission limit. Additionally, the toxicity of ammonia slip and the global warming potential of N_2O have to be considered. The marine engine manufacturers should focus on reducing these emissions by studying ammonia catch systems from the exhaust or using ammonia slip at the SCR as a reducing agent to reduce both ammonia slip and NO_X emissions (Brinks and Hektor 2020). N_2O emissions can be decreased by maintaining optimum in-cylinder temperature for a lower N_2O formation rate.

Ammonia has been experienced in the maritime industry for many years. It has been carried as cargo, used as a refrigerant, and as a NO_X reducing agent in SCRs (Kim et al. 2020a). The previous experience in ammonia operations facilitates the use of ammonia as fuel. Moreover, the safety procedures were formed according to the experiences and the risks and hazards of ammonia are controllable (Wan et al. 2021). Thus, the safety issues related to ammonia does not a barrier to ammonia use in maritime transportation.

Lastly, there are no specific international maritime rules or regulations for ammonia as fuel. Despite, this can be seen as a barrier, small amendments to the

International Code of the Construction and Equipment of Ships Carrying Liquefied Gases in Bulk (IGC Code) and the International Code of Safety for Ships using Gases or other Low-flashpoint Fuels (IGF Code) can solve this incompetence (American Bureau of Shipping (ABS) 2020b). The IGC Code regulates the minimum design standards and procedures for carrying liquefied gas cargoes while the IGF Code regulates the minimum design standards and procedures for using gases or low-flashpoint alternative fuels on ships. Furthermore, the Class Societies, recognized organizations of Flag States to maintain acceptable condition standards for ships, has been working on rules and guidelines for ammonia as a marine fuel.

Ammonia has advantages and disadvantages that are similar to other alternative marine fuels. Table 9.6 shows these advantages and disadvantages. Ammonia usage results in a significant reduction of carbonaceous emissions, SO_X, and PM emissions. It can be produced from renewable energy for a carbon–neutral process. It is easy to store and transport and has low fire and explosion risk. Ammonia can be used at either internal combustion engines (ICE) or fuel cells (FC) that provides wider design options. There are some disadvantages due to the properties of ammonia. Toxicity, corrosiveness, and environmental impact at a spill are the parts that have to be paid attention to. Also, the engine manufacturer has to focus on the complete and efficient combustion of ammonia, ammonia slip, and higher NO_X emissions. It is expected that the engine will be ready 2–3 years later and the maritime community needs to wait a little bit more for the technology will become mature. The fuel cost should

Table 9.6 Advantages and disadvantages of ammonia as a marine fuel (American Bureau of Shipping (ABS) 2020a, b; Kim et al. 2020b)

Advantages	Disadvantages
Carbon-free and sulfur-free structure (no carbonaceous emissions, SO_X and PM emissions)	Toxicity
Option for production from renewable energy	Corrosiveness to the some materials
Mature commercial product	Low energy content
Relatively easy storage and transport	Poor combustion characteristics for internal combustion engines
Low fire and explosion risk	Possibility of requirement for high portion of pilot fuel
Application flexibility to the combustion engines and fuel cells	Possibility of unburned ammonia slip
Direct ammonia fuel cells and ammonia-fuelled dual fuel engines under development	Possibility of higher NO_X emissions and SCR requirement
Can be separated to H_2 and N_2 easily	Need of 2–3 years for the engine development
	Environmental impact of ammonia spill
	Fuel cost
	Missing regulations

Table 9.7 Comparison of the most promising alternative marine fuels and conventional marine fuels (Gray et al. 2021; Korberg et al. 2021; Xing et al. 2021a; Hansson et al. 2020)

	HFO	MDO	Ammonia	Hydrogen	LNG	Methanol
Application	ICE	ICE	ICE/FC	ICE/FC	ICE/FC	ICE/FC
Safety	+++	+++	++	+	+	+++
Maturity	+++	+++	+	+	+++	++
Availability	+++	+++	+	+	++	+
Voyage distance	+++	+++	++	+	++	++
Capital cost	+++	+++	++	+	++	++
Fuel cost	+++	++	+	+	++	++
Storage cost	+++	+++	++	+	++	++
Well-to-wheel emissions	+	+	$+^{a}/++^{b}/+++^{c}$	$+^{a}/++^{b}/+++^{c}$	+	$+^{a}/++^{b}$

One plus to three plusses refers to the worst from the best. [a]Brown production way, [b]blue production way, [c]green production way

be competitive with the conventional fuels since the LHV of ammonia is lower than these fuels and a larger amount of fuel will be consumed.

9.3.2 Comparison of Promising Alternative Marine Fuels

According to the IMO Fourth GHG Study, HFO and MDO, conventional marine fuels, are the most used fuels; LNG and methanol, alternative marine fuels, are other fuels that were mentioned in the Study (International Maritime Organization (IMO) 2020). In addition to LNG and methanol, ammonia and hydrogen have been considered as promising alternative marine fuels by maritime stakeholders. For this reason, Table 9.7 includes these fuels. The purpose of the table is to observe the strong and weak sides of ammonia compared to conventional marine fuels and other promising alternative marine fuels.

Ammonia and other alternative marine fuels can be applied to ICE and FC. Ammonia-fuelled marine diesel engines, direct ammonia fuel cells, or ammonia as a hydrogen carrier at hydrogen-fuelled fuel cells are the options. Ammonia has a lower safety level than conventional marine fuels, but the level is higher than gaseous fuels, hydrogen, and LNG (Deniz and Zincir 2016; Inal et al. 2021). The storage of ammonia does not need high pressure or cryogenic temperatures, thus it does not create important safety issues. Ammonia got two plusses at the safety because toxicity and corrosiveness are the properties that have to be taken into consideration on ships. The maturity level of ammonia as fuel is low. There is no ammonia-fuelled diesel engine or fuel cell recently, and it will take 2–3 years for the development. LNG is the most mature alternative marine fuel and methanol has been chasing. Ammonia is one of the most produced substances worldwide, but it is mostly used by the fertilizer industry. There is no production as fuel, and the supply chains and bunkering infrastructures are not available. The supply chains have to spread worldwide and bunkering infrastructures have to be established. Voyage distance depends on the LHV of fuel. Ammonia has a lower LHV than conventional marine fuels, but similar to LNG and methanol. Therefore, ammonia is competitive with LNG and methanol. Capital cost is related to the properties of the fuel. HFO and MDO are well-known mature marine fuels that are easy to store and do not need special precautions. On the other hand, ammonia requires 10 bar pressure to store at room temperature as liquid, compatible materials have to be used to avoid corrosion, and double-walled supply lines and additional ventilation systems are required for safety. Besides, dual-fuel diesel engines require additional equipment on the engine. Ammonia fuel requires almost similar fuel storage and supply systems with LNG and methanol. Again, ammonia is competitive with LNG and methanol. Ammonia fuel cost is expensive than the other fuels, except hydrogen. Ammonia is mostly produced for the fertilizer industry and increasing demand by using it as fuel increases the fuel price. In addition to this, low LHV of ammonia leads to higher fuel consumption which raises the fuel cost. As mentioned before, storing ammonia in a liquid state requires 10 bar pressure at room temperature. Hence, the storage cost is lower than the conventional marine fuels, but it is the same as LNG and methanol. The well-to-wheel emissions of conventional marine fuels and LNG are the worst because they are fossil fuels. Ammonia can be produced as brown, blue, or green ammonia. If ammonia is produced in a brown way, well-to-wheel emissions are similar to MDO and get one plus. If it is produced in a green way, green ammonia gets three plusses.

Ammonia is compared with conventional fuels and the most promising alternative marine fuels. As a result, it can be seen that ammonia is advantageous to hydrogen and is competitive with LNG and methanol. If the maturity improves, availability increases, fuel cost decreases, and ammonia is produced in a green way, using ammonia as fuel is one of the most important routes to achieve full decarbonization of maritime transportation.

9.4 Summary

This chapter focused on ammonia as marine fuel for decarbonization of maritime transportation. It highlighted the short history, properties, production method of ammonia, ammonia-fueled studies, and ammonia project and maritime industry developments. Later the barriers and facilitators for ammonia were discussed, and the comparison with other promising alternative fuels was made to show the place of ammonia at maritime transportation. The main outcomes of this chapter are:

- Ammonia production was defined under three general terms: brown ammonia, blue ammonia, and green ammonia. For zero-carbon shipping, green ammonia has to be used on ships since well-to-wheel emissions of brown ammonia are almost equal to well-to-wheel emissions of MGO and do not provide any GHG emission reduction advantage. Blue ammonia is an effective option during the transition to green ammonia production from renewable energy sources.

- Ammonia cannot be burned completely as a sole fuel due to its poor combustion properties and required pilot high-reactivity fuel to promote the combustion. On the other hand, the previous studies on CI engines showed that it is possible to combust up to 95% ammonia energy fraction in the fuel mixture.
- A significant reduction in CO_2 and soot emissions has been achieved, but higher NO_X and N_2O emissions and ammonia slip have been important issues to be improved at the further engine studies.
- The direct-ammonia fuel cell studies have been continuing and the direct-ammonia fuel cells will be one of the promising solutions for zero-carbon shipping in the future.
- Maritime industry stakeholders combined their research and development power and the interest is high to ammonia. The projects and developments are under process to improve the supply chain and infrastructure for ammonia delivery and bunkering. The engine manufacturers have been conducting tests on their ammonia-fueled marine diesel engines.

This review study shows that despite there are some barriers to ammonia as a marine fuel, the existing experience of the maritime industry, supply chains and infrastructures of the fertilizer industry, low modification requirement on engines, and achievement of significantly lower CO_2 and soot emissions and almost zero SO_X emissions (from the pilot fuel only) will facilitate the use of ammonia. When ammonia is compared with the other promising alternative marine fuels, it has similar advantages and disadvantages. Ammonia is advantageous to hydrogen and is competitive with LNG and methanol. Since ammonia is on the list of IMO and maritime industry stakeholders as an alternative marine fuel, it can be one of the options for full decarbonized maritime transportation by 2050.

References

Al-Aboosi FY, El-Halwagi MM, Moore M, Nielsen RB (2021) Renewable ammonia as an alternative fuel for the shipping industry. Curr Opin Chem Eng 31:100670. https://doi.org/10.1016/j.coche.2021.100670

Alemu MA, Ilbas M (2020) Direct ammonia powered solid oxide fuel cells: challenges, opportunities, and future outlooks. J Eng Appl Sci Technol 2(3):1–11

Alfa L, Hafnia, Haldor T, Vestas, Siemans G (2020) Ammonfuel—an industrial view of ammonia as a marine fuel. Industrial report, Aug 2020

Al-Hamed KHM, Dincer I (2021) A novel ammonia solid oxide fuel cell-based powering system with on-board hydrogen production for clean locomotives. Energy 220:119771. https://doi.org/10.1016/j.energy.2021.119771

American Bureau of Shipping (ABS) (2019) Setting the course to low carbon shipping. 2030 Outlook/2050 Vision

American Bureau of Shipping (ABS) (2020a) Setting the course to low carbon shipping. Pathways to sustainable shipping

American Bureau of Shipping (ABS) (2020b) Ammonia as marine fuel. Sustainability whitepaper, Oct 2020

Ammonia Energy Association (AEA) (2021) Articles. https://www.ammoniaenergy.org/articles/

Ayvalı T, Tsang SCE, Van Vrijaldenhoven T (2021a) The position of ammonia in decarbonising maritime industry: an overview and perspective: part I, technological advantages and the momentum towards ammonia-propelled shipping. Johnson Matthey Technol Rev 65(2):275–290. https://doi.org/10.1595/205651321X16043240667033

Ayvalı T, Tsang SCE, Van Vrijaldenhoven T (2021b) The position of ammonia in decarbonising maritime industry: an overview and perspective: part II, costs, safety and environmental performance and the future prospects for ammonia in shipping. Johnson Matthey Technol Rev 65(2):291–300. https://doi.org/10.1595/205651321X16127941688787

Aziz M, Wijayanta AT, Nandiyanto ABD (2020) Ammonia as effective hydrogen storage: a review on production, storage and utilization. Energies 13:3062. https://doi.org/10.3390/en13123062

Balcombe P, Brierley J, Lewis C, Skatvedt L, Speirs J, Hawkes A, Staffell I (2019) How to decarbonise international shipping: options for fuels, technologies and policies. Energy Convers Manag 182:72–88. https://doi.org/10.1016/j.enconman.2018.12.080

Baniasadi E, Dincer I (2011) Energy and exergy analyses of a combined ammonia-fed solid oxide fuel cell system for vehicular applications. J Hydrogen Energy 36:11128–11136. https://doi.org/10.1016/j.ijhydene.2011.04.234

Belgiorno G, Di Blasio G, Beatrice C (2019) Advances of the natural gas/diesel RCCI concept application for light-duty engines: comprehensive analysis of the influence of the design and calibration parameters on performance and emissions. In: Srinivasan K, Agarwal AK, Krishnan S, Mulone V (eds) Natural gas engines. Energy, environment, and sustainability. https://doi.org/10.1007/978-981-13-3307-1_9

Bicer Y, Dincer I, Vezina G, Raso F (2017) Impact assessment and environmental evaluation of various ammonia production process. Environ Manag 59:842–855. https://doi.org/10.1007/s00267-017-0831-6

Brinks HW, Chryssakis C (2017) LPG as a marine fuel. DNV GL strategic research and position paper

Brinks HW, Hektor EA (2020) Ammonia as a marine fuel. DNV GL, Group Technology & Research, White Paper 2020

Cardoso JS, Silva V, Rocha RC, Hall MJ, Costa M, Eusebio D (2021) Ammonia as an energy vector: current and future prospects for low-carbon fuel applications in internal combustion engines. J Cleaner Prod 296:126562. https://doi.org/10.1016/j.clepro.2021.126562

Cheliotis M, Boulougouris E, Trivyza NL, Theotokatos G, Livanos G, Mantalos G, Stubos A, Stamatakis E, Venetsanos A (2021) Review on the safe use of ammonia fuel cells in the maritime industry. Energies 14:3023. https://doi.org/10.3390/en14113023

de Vries N (2019) Safe and effective application of ammonia as a marine fuel. Delft University of Technology, Delft

Deniz C, Zincir B (2016) Environmental and economical assessment of alternative marine fuels. J Cleaner Prod 296:438–449. https://doi.org/10.1016/j.clepro.2015.11.089

Dere C, Deniz C (2020) Effect analysis on energy efficiency enhancement of controlled cylinder liner temperatures in marine diesel engines with model based approach. Energy Convers Manag 220:113015. https://doi.org/10.1016/j.enconman.2020.113015

Dimitriou P, Javaid R (2020) A review of ammonia as a compression ignition engine fuel. Int J Hydrogen Energy 45:7098–7118. https://doi.org/10.1016/j.ijhydene.2019.12.209

DNV GL (2019) Comparison of alternative fuels. DNV GL AS Maritime, Hovik, Norway. https://sea-lng.org/wp-content/uploads/2019/09/19-09-16_Alternative-Marine-Fuels-Study_final_report.pdf

DNV GL (2021) Alternative fuels insight. https://afi.dnvgl.com/Statistics?repld=0

Erdemir D, Dincer I (2020) A perspective on the use of ammonia as a clean fuel: challenges and solutions. Int J Energy Res 2020:1–8. https://doi.org/10.1002/er.6232

Frost J, Tall A, Sheriff AM, Schönborn A, Hellier P (2021) An experimental and modeling study of dual fuel aqueous ammonia and diesel combustion in a single cylinder compression ignition engine. Int J Hydrogen Energy (in Press). https://doi.org/10.1016/ijhydene.2021.08.089

Gill SS, Chatha GS, Tsolakis A, Golunski SE, York APE (2012) Assessing the effects of partially decarbonising a diesel engine by co-fuelling with dissociated ammonia. Int J Hydrogen Energy 37:6074–6083. https://doi.org/10.1016/j.ijhydene.2011.12.137

Gray N, McDonagh S, O'Shea R, Smyth B, Murphy JD (2021) Decarbonising ships, planes and trucks: an analysis of suitable low-carbon fuels for the maritime, aviation and haulage sectors. Adv Appl Energy 1:100008. https://doi.org/10.1016/j.adapen.2021.100008

Gross CW, Kong S (2013) Performance characteristics of a compression-ignition engine using direct-injection ammonia-DME mixtures. Fuel 103:1069–1079. https://doi.org/10.1016/j.fuel.2012.08.026

Hansson J, Fridell E, Brynolf S (2020) On the potential of ammonia as fuel for shipping, a synthesis of knowledge. Lighthouse reports. Swedish Maritime Competence Centre

Heywood JB (1988) Internal combustion engine fundamentals. McGraw-Hill, New York

Ianniello R, Belgiorno G, Di Luca G, Beatrice C, Di Blasio G (2021) Ethanol in dual-fuel and blend fueling modes for advanced combustion in compression ignition engines. In: Shukla PC, Belgiorno G, Di Blasio G, Agarwal AK (eds) Alcohol as an alternative fuel for internal combustion engines. Energy, Environment, and Sustainability. Springer, Singapore. https://doi.org/10.1007/978-981-16-0931-2_2

IEEE Spectrum (2021) Why the shipping industry is betting big on ammonia. https://spectrum.ieee.org/transportation/marine/why-the-shipping-industry-is-betting-big-on-ammonia

Inal OB, Deniz C (2020) Assessment of fuel cell types for ships: based on multi-criteria decision analysis. J Cleaner Prod 265:121734. https://doi.org/10.1016/j.clepro.2020.121734

Inal OB, Deniz C (2021) Emission analysis of LNG fuelled molten carbonate fuel cell system for a chemical tanker ship: a case study. Mar Sci Tech Bull 10(2):118–133. https://doi.org/10.33714/masteb.827195

Inal OB, Zincir B, Deniz C (2021) Hydrogen and ammonia for the decarbonization of shipping. In: 5th international hydrogen technologies congress (IHTEC-2021), 26–28 May 2021 (Online)

International Maritime Organization (IMO) (2015) Third greenhouse gas study 2014

International Maritime Organization (IMO) (2018) Adoption of the initial IMO strategy on reduction of GHG emissions from ships and existing IMO activity related to reducing GHG emissions in the shipping sector

International Maritime Organization (IMO) (2020) Fourth greenhouse gas study 2020

ITOCHU Corporation (2021) ITOCHU and 22 industry players kick off studying ammonia as an alternative marine fuel. https://www.itochu.co.jp/en/news/press/2021/210611.html

Kim K, Roh G, Kim W, Chun K (2020a) A preliminary study on an alternative ship propulsion system fueled by ammonia: environmental and economic assessments. J Mar Sci Eng 8:183. https://doi.org/10.3390/jmse8030183

Kim H, Koo KY, Joung T (2020b) A study on the necessity of integrated evaluation of alternative marine fuels. J Int Mar Saf Environ Aff Shipping 4(2):26–31. https://doi.org/10.1080/25725084.2020.1779426

Kishimoto M, Muroyama H, Suzuki S, Saito M, Koide T, Takahashi Y et al (2020) Development of 1 kW-class ammonia-fuelled solid oxide fuel cell stack. Fuel Cells 20:80–88. https://doi.org/10.1002/fuce.201900131

Klüssmann JN, Ekknud LR, Ivarsson A, Schramm J (2020) Ammonia application in CI engines. Special report, a report from the advanced motor fuels technology collaboration programme

Kobayashi H, Hayakawa A, Somarathne KDKA, Okafor EC (2019) Science and technology of ammonia combustion. Proc Combust Inst 37:109–133. https://doi.org/10.1016/j.proci.2018.09.029

Korberg AD, Brynolf S, Grahn M, Skov IR (2021) Techno-economic assessment of advanced fuels and propulsion systems in future fossil-free ships. Renew Sustain Energy Rev 142:110861. https://doi.org/10.1016/j.rser.2021.110861

Kroch E (1945) Ammonia—a fuel for motor buses. J Inst Pet 31:213–223

Lesmana H, Zhang Z, Li X, Zhu M, Xu W, Zhang D (2019) NH_3 as a transport fuel in internal combustion engines: a technical review. J Energy Resour Technol 141:070703–070711. https://doi.org/10.1115/1.4042915

MAN Energy Solutions (2019) Engineering the future two-stroke green-ammonia engine. Technical paper, Nov 2019

McKinlay CJ, Turnock SR, Hudson DA (2020) A comparison of hydrogen and ammonia for future long distance shipping fuels. LNG/LPG and alternative fuels, 29–30 Jan 2020. The Royal Institution of Naval Architects, London

Morlanes N, Katikaneni SP, Paglieri SN, Harale A, Solami B, Sarathy SM, Gascon J (2021) A technological roadmap to the ammonia energy economy: current state and missing technologies. Chem Eng J 408:127310. https://doi.org/10.1016/j.cej.2020.127310

Mukelabai MD, Gillard JM, Patchigolla K (2021) A novel integration of a green power-to-ammonia to power system: reversible solid oxide fuel cell for hydrogen and power production coupled with an ammonia synthesis unit. J Hydrogen Energy 46:18546–18556. https://doi.org/10.1016/j.ijhydene.2021.02.218

Niki Y (2021) Reductions in unburned ammonia and nitrous oxide emissions from an ammonia-assisted diesel engine with early timing diesel pilot injection. J Eng Gas Turbines Power 143:091014–091021. https://doi.org/10.1115/1.4051002

Niki Y, Nitta Y, Sekiguchi H, Hirata K (2018) Emission and combustion characteristics of diesel engine fumigated with ammonia. In: Proceedings of the ASME 2018 internal combustion engine division fall technical conference ICEF2018, 4–7 Nov 2018, San Diego, CA, USA

Niki Y, Nitta Y, Sekiguchi H, Hirata K (2019) Diesel fuel multiple injection effects on emission characteristics of diesel engine mixed ammonia gas into intake air. J Eng Gas Turbines Power 141:061020–061021. https://doi.org/10.1115/1.4042507

OCI NV (2021) News. https://www.oci.nl/news/2021-oci-nv-1/

Porter DH (1998) The life and times of sir Goldsworthy Gurney, gentleman scientist and inventor: 1793–1875. Lehigh University Press

Pochet M, Truedsson I, Foucher F, Jeanmart H, Contino F (2017) Ammonia-hydrogen blends in homogeneous-charge compression-ignition engine. SAE technical papers 2017–24–0087, pp 1–10. https://doi.org/10.4271/2017-24-0087

Pochet M, Jeanmart H, Contino F (2020) A 22:1 compression ratio ammonia-hydrogen HCCI engine: combustion, load, and emission performances. Front Mech Eng 6:43. https://doi.org/10.3389/fmech.2020.00043

Rattazzi D, Rivarolo M, Massardo AF (2021) An innovative tool for the evaluation and comparison of different fuels and technologies onboard ships. In: E3S web conferences 238 08001, 16 Feb 2021. https://doi.org/10.1051/e3sconf/202123808001

Reiter AJ, Kong S (2010) Diesel engine operation using ammonia as a carbon-free fuel. In: Proceedings of the ASME 2010 internal combustion engine fall technical conference ICEF2010, USA

Reiter AJ, Kong S (2011) Combustion and emissions characteristics of compression-ignition engine using dual ammonia-diesel fuel. Fuel 90:87–97. https://doi.org/10.1016/j.fuel.2010.07.055

Ryu K, Zacharakis-Jutz GE, Kong S (2014) Performance characteristics of compression-ignition engine using high concentration of ammonia mixed with dimethyl ether. Appl Energy 113:488–499. https://doi.org/10.1016/j.apenergy.2013.07.065

Shamun S, Belgiorno G, Di Blasio G (2020) Engine parameters assessment for alcohols fuels application in compression ignition engines. In: Singh PS, Sharma YC, Mustafi NN, Agarwal AK (eds) Alternative fuels and their utilization strategies in internal combustion engines. Energy, Environment, and Sustainability. Springer, Singapore. https://doi.org/10.1007/978-981-15-0418-1_8

Smith AR, Klosek J (2001) A review of air separation technologies and their integration with energy conversion processes. Fuel Process Technol 70(2):115–134. https://doi.org/10.1016/S0378-3820(01)00131-X

Tan WC, Iwai H, Kishimoto M, Brus G, Szmyd JS (2018) Numerical analysis on effect of aspect ratio of planar solid oxide fuel cell fueled with decomposed ammonia. J Power Sources 384:367–378. https://doi.org/10.1016/j.jpowsour.2018.03.011

Tay KL, Yang W, Li J, Zhou D, Yu W, Zhao F, Chou SK, Mohan B (2017) Numerical investigation on the combustion and emissions of a kerosene-diesel fueled compression ignition engine assisted by ammonia fumigation. Appl Energy 204:1476–1488. https://doi.org/10.1016/j.apenergy.2017.03.100

United Nations Conference on Trade and Development (UNCTAD) (2020) Review of maritime transport 2020

Valera-Medina A, Xiao H, Owen-Jones M, David WIF, Bowen PJ (2018) Ammonia for power. Prog Energy Combust Sci 69:63–102. https://doi.org/10.1016/j.pecs.2018.07.001

Vayenas CG, Farr RD (1980) Co-generation of electric energy and nitric-oxide. Science 208:593–594. https://doi.org/10.1126/science.208.4444.593

Wan Z, Tao Y, Shao J, Zhang Y, You H (2021) Ammonia as an effective hydrogen carrier and a clean fuel for solid oxide fuel cells. Energy Convers Manag 228:113729. https://doi.org/10.1016/j.enconman.2020.113729

Werrell KP (1986) The strategic bombing of Germany in world war II: costs and accomplishments. J Am Hist 73(3):702–713. https://doi.org/10.2307/1902984

Xing H, Stuart C, Spence S, Chen H (2021a) Alternative fuel options for low carbon maritime transportation: pathways to 2050. J Cleaner Prod 297:126651. https://doi.org/10.1016/j.clepro.2021.126651

Xing H, Stuart C, Spence S, Chen H (2021b) Fuel cell power systems for maritime applications: progress and perspectives. Sustainability 13:1213. https://doi.org/10.3390/su13031213

Yapicioglu A, Dincer I (2019) A review on clean ammonia as a potential fuel for power generators. Renew Sustain Energy Rev 103:96–108. https://doi.org/10.1016/j.rser.2018.12.023

Zincir B (2019) An alternative fuel assessment model for ships and experiments on the effect of methanol on diesel engines. Dissertation thesis. Department of Maritime Transportation Engineering, Istanbul Technical University

Zincir BA (2020) Comparison of the carbon capture systems for onboard application and voyage performance investigation by a case study. M.Sc. thesis. Department of Maritime Transportation Engineering, Maritime Faculty, Istanbul Technical University

Zincir B (2020) A short review of ammonia as an alternative marine fuel for decarbonised maritime transportation. In: Proceedings of international conference on energy, environment and storage of energy (ICEESEN2020), 19–21 Nov 2020, Kayseri-Turkey

Zincir B, Deniz C (2021) Methanol as a fuel for marine diesel engines. In: Shukla PC, Belgiorno G, Di Blasio G, Agarwal AK (eds) Alcohol as an alternative fuel for internal combustion engines. Energy, environment, and sustainability. Springer. https://doi.org/10.1007/978-981-16-0931-2_4

Chapter 10
Biogas as a Sustainable and Renewable Energy Source

Wojciech Czekała

10.1 Introduction

Sustainable development is one of the most crucial humanity challenges in the twenty-first century (Eisenmenger et al. 2020). Continuous economic growth and increased consumerism cause changes in the environment (Austgulen 2016) that may become irreversible. Access to energy is an indispensable development element, necessary in both the municipal and industrial sectors (Gunnarsdottir et al. 2021; Mika et al. 2021). The main sources of energy are fossil fuels, nuclear energy and renewable energy sources. Fossil fuels are the most popular energy production source globally (Hassan et al. 2021). This group includes, i.a. hard coal, lignite, crude oil, and natural gas. Even though the mentioned energy sources are popular worldwide, a systematic reduction in their consumption should be expected due to the negative impact on the environment when using them (Kibria et al. 2019; Pellegrini et al. 2021).

Nuclear energy is considered a clean source of energy (Sadekin et al. 2019). In some countries, for example, in France, this type of energy is dominant in the country's overall balance (Velasquez et al. 2020). However, due to the disasters that took place in the past, e.g., Chernobyl or Fukushima, the discussed source of energy still raises some concerns for part of the society (Muellner et al. 2021).

Renewable energy sources including mainly: wind (Sun et al. 2020), sun (Silvi 2008), water (Chidambaram et al. 2021), and biomass (Destek et al. 2021), are becoming more and more popular. Solid biofuels play a special role, mainly due to their ease of obtaining and further processing (Larsson and Samuelsson 2017). This group includes, i.a., pellets, briquettes, or raw—unprocessed biomass. Nevertheless, liquid and gaseous fuels are also popular and often used in many countries (Chanthawong and Dhakal 2016; Panuccio et al. 2016). This is supported not only by legal

W. Czekała (✉)
Poznań University of Life Sciences, Wojska Polskiego 28, 60-637 Poznań, Poland
e-mail: wojciech.czekala@up.poznan.pl

G. Di Blasio et al. (eds.), *Clean Fuels for Mobility*, Energy, Environment, and Sustainability, https://doi.org/10.1007/978-981-16-8747-1_10

regulations aimed at e.g., increasing the share of renewable energy sources in the balance of primary energy production, energy consumption or the share of liquid biofuels in transport. The production and processing of liquid and gaseous biofuels have many other advantages, such as many possibilities of their use. Mention is made of direct combustion, cogeneration or use in transport.

Biogas, which is a mixture of gases with methane as the main component, can be distinguished as renewable fuels (Abbas et al. 2020; Lu and Gao 2021). Biogas is produced in agricultural biogas plants (Igliński et al. 2020), biogas plants in a sewage treatment plant (Mejdoub and Ksibi 2015), and installations located in landfills (Jeswani et al. 2013). Its production is related to the anaerobic digestion process (Kowalczyk-Juśko et al. 2020). It is worth emphasizing that biogas is often produced from waste. This should definitely be considered one of the most important advantages (Mugodo et al. 2017). Therefore, some specialists classify biogas plants not as energy production installations but primarily as installations to treat biodegradable waste (Marks et al. 2020). The substrates used in its production include, among others natural fertilizers such as manure or slurry. The agri-food industry is also an important source of waste. Additionally, biogas can be produced from all kinds of wasted food (Malik et al. 2020). In this case, not only energy is recovered in the form of biomethane or electricity and heat, but also waste that may have a negative impact on the environment is also managed.

Due to the variety of substrates processed from biogas plants, these plants are becoming more and more popular. Scientific research in the biogas field includes the selection of substrates in terms of their availability and their impact on the biogas plants economic balance. Another important aspect is to determine the impact of various substrates on the anaerobic digestion process, especially high and stable biogas production. Due to the systematic development of the market, new substrates that can be processed in biogas plants are taken into account (Kushkevych et al. 2018). In addition, the technology used may have a significant impact on the efficiency of the process. Therefore, in addition to the selection of substrates, research is carried out on the pre-treatment of substrates and technology used in biogas plants (Kupryaniuk et al. 2020). By converting waste for biogas production, it is possible to get a double benefit—for waste management and the sale of biogas or produced energy. The aim of the study is to discuss the agricultural biogas production process as a sustainable and renewable energy source and the possibilities of biogas use in transport.

10.2 Agricultural Biogas Plants and the Anaerobic Digestion Process

Agricultural biogas plants are installations where energy is produced in an environmentally friendly manner (Pizarro-Loaiza et al. 2021). The production of biogas occurs thanks to the activity of anaerobic microorganisms as a result of the anaerobic digestion process. In order for biogas to be produced, technical and technological requirements must be met that allow for the production of, among others, methane. Lack of oxygen during anaerobic digestion differs from composting, which is also a process of biological waste treatment but takes place under aerobic conditions (Żukowska et al. 2019). From the energy point of view, the most significant difference is that in the anaerobic digestion process, energy is released in the form of a chemical (methane). Methane is easy to store and transport, while in the process of composting organic waste, energy is released in the form of heat—difficult to recover and store (Lalander et al. 2018; Sołowiej et al. 2021).

10.2.1 Anaerobic Digestion

Anaerobic digestion is a complicated process under the influence of microorganisms (Christy et al. 2014). The essence of the process consists of the decomposition of the substrates used in the fermentation process into simple chemical compounds (Kozłowski et al. 2018). The final products are methane, carbon dioxide, and other trace gases that are part of the biogas.

This process occurs in four steps:

- Hydrolysis—degradation of organic compounds such as sugars, proteins, and fats into simpler compounds. Proteins hydrolyze to amino acids, polysaccharides to di- and monosaccharides, and fats to alcohols and fatty acids. This process occurs thanks to enzymes produced by anaerobic hydrolytic bacteria.
- Acidogenesis—conversion of hydrolysis products mainly to short-chain organic acids (C1–C6) (formic, acetic, propionic, butyric, valeric and hexanoic), alcohols, carbon dioxide, and hydrogen.
- Octanogenesis—conversion of acids produced in acidogenesis into acetic acid, carbon dioxide, and hydrogen by bacteria.
- Methanogenesis—the stage in which methane is produced from acetates and as a result of the CO_2 reduction process.

Anaerobic digestion, which is a biological process involving microorganisms, takes place under strictly defined conditions. Factors that respond to the reaction environment determine the amount of biogas produced. In the case of inadequate parameters, inhibition conditions may even occur. As a result, the production of biogas can be limited or even completely stopped.

Oxygen. Biogas production as a biological waste treatment process is carried out in completely anaerobic conditions. Methanogenic bacteria are obligate anaerobes. This factor distinguishes anaerobic digestion from the composting process (Wolna-Maruwka and Dach 2009). Proper operation of an agricultural biogas plant is possible only with completely impermeable fermentation chambers.

pH. The pH value is an important parameter influencing the anaerobic digestion process. The appropriate value of the discussed parameter is responsible for creating the proper conditions for microorganisms for growth. Methanogenic bacteria prefer the pH from 6.8 to 7.5, so it is recommended that the substrates fermented mixture should have such a value. The substrates used for the production of biogas have an influence on the pH value.

C:N. One of the key parameters determining the course of the anaerobic digestion process is the C:N ratio (Shahbaz et al. 2020). The correct value of this parameter should range from 15:1 to 25:1. The correct proportion between these two elements will provide the bacteria with nutrients, and at the same time, will not increase ammonia emission and increase the pH, acting as an inhibitor of the process.

Temperature. The temperature has a significant impact on the rate of decomposition of substrates under anaerobic conditions. This parameter influences the biogas efficiency also. In agricultural biogas plants, the temperature should be constant. This is due to the necessity to provide bacteria with appropriate conditions for growth. Depending on the temperature in the fermentation chamber, psychrophilic, mesophilic, and thermophilic fermentation are mentioned. Usually, biogas highest production takes place under thermophilic conditions, although deviations from this rule may be observed. On the other hand, anaerobic digestion methods in low and medium temperature conditions are more often used.

Hydraulic retention time. The hydraulic retention time is a parameter that indicates how long the substrate remains in the fermentation chamber until it is degraded. This value depends, i.a., on the substrates used for the biogas production, the temperature of the process, or mixing.

Process inhibitors. Process inhibitors are all kinds of toxic substances to bacteria that decompose compounds rich in an organic matter (Nsair et al. 2020). The effect of the inhibitors may be to limit or completely stop the biogas production process. This group includes, inter alia, ammonia, hydrogen sulfide, heavy metals, or antibiotics.

10.2.2 Components of a Biogas Plant

The construction and components of each agricultural biogas plant can be significantly different. This is due to, inter alia, economic reasons and the need to adapt individual elements to the process, which is primarily determined by the types of substrates used for energy production. The amount of processed raw materials also influences it and whether they are waste. For example, slaughterhouse waste requires special treatment before it is placed in digesters. The method of the biogas treatment plant and the way of its use are also important.

The biogas installation includes numerous devices that allow the production of biogas from substrates, including problematic waste. The elements of the installation that allow to purify biogas, pump it into the network, or generate electricity and heat in cogeneration are also extremely important (Fig. 10.1).

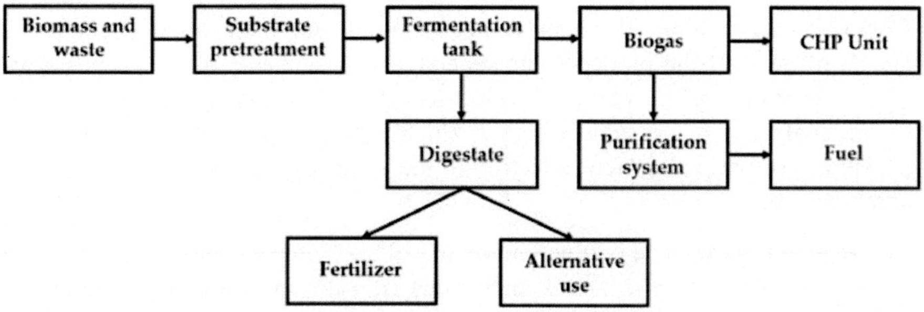

Fig. 10.1 Biogas: production and selected directions of the application

Devices for the storage and pretreatment of substrates. The number of substrates used for biogas production is huge. The easiest way is to categorize them into solid and liquid forms. Due to the state of aggregation, it is precisely necessary to have places to store all used waste at the installation. In the case of solid substrates, all kinds of silos are most often used. On the other hand, liquid waste is stored in closed tanks, underground, and above the surface. The ideal solution is when the waste is placed in the fermentation chamber without storage before. This reduces the space and costs for storage and will reduce the environmental burden that may result from uncontrolled emissions during storage. Such situations are possible, but these situations require appropriate logistics, a wide supply base, and appropriate contracts with companies supplying biogas plants with the substrate.

Fermentation tanks. Fermentation tanks, also known as bioreactors or digesters, are the main components of an agricultural biogas plant. In these chambers, the anaerobic digestion process takes place and biogas is produced. The number of chambers and their volume varies from plant to plant. It depends, i.a., on the amount of substrates, their type, and anaerobic digestion method. Among the materials used for the production of chambers, concrete and steel should be mentioned. It is important that the tank have to be tight to prevent oxygen ingress and insulation, which will allow reducing heat loss.

Biogas tank. As the name suggests, biogas tanks are intended to store of the gas mixture produced in the process. In practice, biogas is most often stored in an impermeable tank located above the fermentation chamber. Gas is continuously extracted and used from this container. For logistical and safety reasons, biogas should be used systematically.

Biogas purification system. The obtained biogas may require purification. It all depends on its quality and the direction of further management. Example pollution is hydrogen sulphide, from which biogas is treated in special beds. Purified biogas can be directed to a cogeneration engine, where it will be burnt, or used as a transport fuel.

Cogeneration system. The cogeneration system, mainly includes an engine, which allows to burn off the produced biogas containing methane. It will be possible to obtain electricity and heat both. The electrical efficiency is usually around 44%, and the thermal efficiency is slightly higher. The advantages of cogeneration include the possibility of optimal system selection to the needs of the installation, small size, and low losses.

The digestate tank. The production of agricultural biogas is directly related to the production of digestate, which is the residue after the anaerobic digestion process. This material is collected and stored in tanks, then it is taken for further use, mainly as fertilizer. Due to legal regulations, it may be necessary to store digestate even for a several months. For this reason, the volume of the tank should be appropriately adapted to the installation. In this tank, biogas is still produced. That is why collecting and use the biogas generated at this stage it is also recommended.

10.3 Substrates for Biogas Production

The essence of an agricultural biogas plant is to supply the installation with substrates from which biogas is produced every day (Wang et al. 2021). All the substrates and waste from which biogas is produced arise directly or indirectly as a result of the photosynthesis process. The process is based on the reaction in which organic compounds are formed from carbon dioxide and water with the sun's participation.

The number of substrates, waste and by-products that can be successfully used in the anaerobic digestion process is large (Nwokolo et al. 2020). They include most of the substrates and waste of plant and animal origin. Due to the fact that the input to biogas plants is remarkably differentiated, the very course of the decomposition process will be different. Particular substrates differ in terms of their decomposition time, or biogas and methane efficiency. However, it is necessary that they biodegrade in the absence of oxygen, which is characteristic of the processes taking place in biogas plants. Each of the substrates will differ in the biogas efficiency and the methane concentration in the produced biogas. This fact is conditioned primarily by the content of organic matter and the forms of chemical compounds that are part of this matter. For example, substrates such as fruits and vegetables are high in sugar. Consequently, the decomposition time will be relatively short. The concentration of methane will be relatively low and will be around 50% (Czekała et al. 2016, 2018). Methane in the mixture will allow obtaining substrates rich in fats and proteins. First of all, slaughter waste should be mentioned. However, it should be emphasized that the decomposition time of this waste in the fermentation tank will be relatively long.

Another factor influencing the efficiency of the substrates used in the bioconversion process in biogas plants is the water content. As the water content in the substrate increases, the amount of available dry matter, including organic matter, decreases. Therefore, the methane efficiency will be limited. With this in mind, it is recommended to use substrates with a high percentage of dry matter and organic matter. However, it should be remembered that water is also needed in the anaerobic digestion process (Basumatary et al. 2021). The first reason is that it provides a condition for bacteria that degradation substrates and produce methane. On the other hand, it affects the technical conditions of the installation—pumping capacity of substrates and digestate between individual elements of the installation. In practice, 15% is the upper limit of the dry matter content in the substrates used for biogas production.

10.3.1 Agricultural Sector

Agriculture is a very important sector from which substrates for the production of agricultural biogas can be obtained (Czekała et al. 2017; Tagne et al. 2021). It is related to plant and animal production. Among the plants used for biogas production in Europe and many other countries, maize is often used. In practice, maize silage is most often used for the production of methane. This is mainly due to the high biogas efficiency of the discussed substrate and its well-known cultivation, harvesting and ensiling technology. In addition to the mentioned plant, other plants can be used to produce biogas, e.g., energy plants (Waliszewska et al. 2018), straw, or grain that does not meet the requirements.

Intensive animal production around the world has become a global problem. It concerns not only food and feed but also environmental issues. All kinds of animal faeces are a popular and important substrate used in agricultural biogas plants. Manure (Abdelsalam et al. 2021), slurry (Marchetti and Vasmara 2020) and chicken manure (Jarwar et al. 2021) should be mentioned. These substrates are characterized by a different degree of hydration or organic matter content. However, all of the above-mentioned natural fertilizers can be successfully used in methane fermentation. In addition to the available organic matter, the discussed substrates are a living environment for microorganisms used in the anaerobic digestion process. Slurry is an inoculum and a diluent for substrates in an anaerobic process (Wawrzyniak et al. 2021). Additionally, it should be emphasized that the slurry after digestion is characterized by better fertilizing properties and can be successfully used in the fields. This is extremely important for environmental reasons, especially for reducing greenhouse gas emissions and the carbon footprint.

10.3.2 Agri-Food Sector

An agri-food sector is a place where a variety of products, by-products and waste are produced. Each waste and by-product should be managed following the regulation. The most common solution for the management of the discussed waste is its storage or fertilization. The anaerobic digestion process can successfully process the vast majority of them. Landfilling is a popular solution, but the amount of waste treated with this method decreases in developed countries. Since these activities are often not carried out properly, harmful gases and leachates may be emitted into the environment. Among the gases emitted into the environment are greenhouse gases, inter alia, methane or carbon dioxide. In addition, emissions of compounds such as ammonia, hydrogen sulfide and odors can cause odors in the area where the waste is deposited. Leachate is a threat to surface and ground waters. Therefore, agricultural biogas plants seem to be a proper place where waste, can be managed with the additional benefit of producing biogas. Due to a wide possibility of a negative impact of agri-food waste on the environment, it is necessary to develop and apply technologies allowing not only to reduce the negative impact, but also to use the potential of the waste (Rejeb et al. 2021).

Places related to the agri-food sector where waste are generated include, among others: fruit and vegetable processing plants, slaughterhouses, dairies (Kozłowski et al. 2019), distilleries, breweries, and bakeries. Because food production takes place in each country, in each of the regions, it is also possible that biogas plants will operate close to such plants (Pochwatka et al. 2020). The amount and variety of generated waste positively influence the possibility of selecting the appropriate mixtures for the fermentation process. On the other hand, the biogas plant, will be an ideal place where the generated waste can be managed. This will bring both environmental, economic, and social benefits (Czekała et al. 2021).

10.3.3 Innovative Substrates

The amount and variety of substrates that can be used for biogas production is very wide. The substrates from agriculture and the agri-food industry are definitely dominant. However, it should be emphasized that these are not the only sectors from which waste can be obtained, or biogas input from special crops. Due to the economic and environmental benefits, apart from waste, other substrates for the production of biofuels are sought. The innovative solutions include the use of *Hermetia illucens* insect larvae as a substrate for biogas production. The conducted research at the Poznań University of Life Sciences concluded that the insect larvae and fractions related to its processing can be successfully used to produce biogas (Czekała 2017; Czekała et al. 2020).

10.4 Products of the Anaerobic Digestion Process

10.4.1 Biogas

The main product of the anaerobic digestion process is biogas. It is a mixture of gases with a predominant share of methane—a source of energy. The percentage share of this compound in the mixture most often varies within the range of 50–65% and it depends on many factors. First of all, these factors include, the type of substrates and the technological conditions of the process. As mentioned earlier, efficient substrates will be slaughterhouse waste with high fat content. On the other hand, the substrates that allow obtaining a relatively low methane efficiency will be especially those with a high degree of hydration, such as liquid manure or technological sewage from some sectors of the agri-food industry. The second gas, in terms of concentration, after methane, will be carbon dioxide, which is a product of the decomposition of organic matter in both thermal and biological processes. Apart from methane and carbon dioxide, in smaller amounts, biogas includes, among others, hydrogen, nitrogen, ammonia and hydrogen sulfide.

10.4.2 Digestate

The digestate is the second product of the anaerobic digestion process taking place in agricultural biogas plants. The digestate, also known as the digested pulp is the residue of the process found in the fermentation chambers (Baştabak and Koçar 2020). The digestate consists mainly of organic compounds that are not decomposed in the process, mineral compounds, the biomass of living organisms and water.

Each day, an appropriate portion of the substrates must be supplied to the process, and each day the appropriate amount of digestate is pumped out of the fermentation tanks. The type and amount of substrates used and the technology used in the anaerobic digestion process have an impact on the digestate composition and properties. The pretreatment of the substrates before their transfer to the tanks also has a significant impact on their decomposition process, and thus on the properties of the digestate. Additionally, an important factor will be the hydraulic retention time of the substrates.

Due to properties such as high content of macronutrients, micronutrients and the content of elements in ionic forms, digestate is by far the most commonly used as a fertilizer. This fertilizer can be used directly as well as after its processing. The methods of digestate use include composting the solid fraction of the digestate. The solid fraction of the digestate can also be formed (briquetting, pelletization), which will allow obtaining organic fertilizer in the form of briquettes or pellets. An alternative to fertilizing the use of digestate is energy use. Own research showed that the digestate, especially its solid fraction, may have high biogas and methane efficiency, which in turn will allow it to be used for energy production (Czekała 2019).

10.5 Application of Biogas

Biogas, a product of the anaerobic digestion process, has many possibilities of use (Abatzoglou and Boivin 2009). They are mainly conditioned by methane concentration in the gas mixture in question and the amount and type of other gases (Kapoor et al. 2019). The direction of biogas application is also often determined by legal regulations and financial instruments for energy efficiency and renewable energy.

Table 10.1 Energy content of various fuels (Börjesson and Mattiasson 2008)

Fuel	Energy content (MJ per dm^3)
Petrol	31.3
Diesel	35.6
Ethanol	21.2
Biodiesel (RME)	33.1
Methane (per m^3)	35.3

The simplest possible solution is the direct use of biogas as cooking fuel. This method is popular i.a. in China and Vietnam. The most common practice in biogas management is combustion in a cogeneration engine. This is due to the high efficiency of the discussed solution, which is about 90%. Biogas burning will generate electricity and heat. An alternative to the presented solution is pumping biogas into the gas network. This solution is gaining importance in many countries around the world. However, in order for biogas to be used in such a way, it must first be cleaned of carbon dioxide, ammonia and hydrogen sulphide.

Agricultural biogas meeting the standards can also be successfully used as fuel for transport purposes (Hosseinipour and Mehrpooya 2019). As in the case of feeding biogas into the network, at first, biogas has to be cleaned. After meeting the requirements, including cleaning to the quality of natural gas, it is possible to use the fuel for vehicles equipped with gas installations (Larsson et al. 2016). However, there is a requirement for a specific fuel distribution infrastructure.

The processing of waste into biogas and the application of biogas for transport fuels is a solution that has many benefits. Methane is characterized by high energy content (Table 10.1). The use of purified agricultural biogas as a sustainable fuel used in transport is clearly in line with the requirements related to sustainable development. Vehicles using biomethane are responsible for lower emissions of harmful gases into the atmosphere. As a result, it is possible to improve air quality, including reducing smog in cities. Purified agricultural biogas should become a significant alternative to petrol and diesel oil for environmental and economic reasons.

10.6 Conclusion

Renewable energy development, including the biogas sector, is largely conditioned by climate and energy policy goals. Regardless of this, the growing awareness of the society in environmental protection and the limited resources of fossil fuels resulted in an increase in energy production from these sources.

Under controlled conditions, biogas may be produced in agricultural biogas plants, biogas plants located in wastewater treatment plants and landfills. Biogas plants as installations converting waste into energy should be included in the growing sector of renewable energy sources. The share of energy obtained from biogas, compared to solid and liquid biofuels, is becoming more and more important. This is mainly due to the possibility of using waste as an energy source. There is also a wide range of possibilities for managing the produced biogas, including, e.g., cogeneration, injection into the grid, or use as fuel for transport.

Irrational use of raw materials and improperly implemented waste management may increase the consumption of raw materials, energy and water. Using waste as a source of biogas as a fuel, contributes to reducing the amount of waste going to landfills and reducing fossil fuels used in transport. Providing the necessary amount of substrates for biogas production is a great challenge. For typical biogas plants, the daily input is at least several dozen megagrams per day. Hence, it is so important to make an appropriate inventory before starting the building of a biogas plant.

Legal regulations and environmental protection requirements often condition the development of renewable energy sources worldwide. This also applies to the increase in the amount of energy produced from biogas. Biogas production is a response to the challenges of sustainable development—environmental, economic and social. It should be emphasized that the production of all biofuels, including agricultural biogas, is stable. Having access to appropriate substrates and carrying out the process stably, it is possible to achieve efficiency close to the maximum throughout the entire period of operation of the biogas plant. It is also worth mentioning that the production of biofuels is based on diversification, not centralization. That is why the benefits of using this type of solutions are visible primarily locally.

All activities for environmental protection, waste management and the production of biofuels and fertilizers should be undertaken in the global, national, provincial and local areas. Considering the environmental and economic benefits, the use of biogas as a sustainable fuel seems to be the direction of increasing importance. The use of waste for biogas is certainly also in line with the circular economy trend.

References

Abatzoglou N, Boivin S (2009) A review of biogas purification processes. Biofuels Bioprod Biorefin 3:42–71

Abbas I, Liu J, Noor RS, Faheem M, Farhan M, Ameen M, Shaikh SA (2020) Development and performance evaluation of small size household portable biogas plant for domestic use. Biomass Convers Biorefinery

Abdelsalam EM, Samer M, Amer MA, Baher MA (2021) Amer biogas production using dry fermentation technology through co-digestion of manure and agricultural wastes. Environ Dev Sustain 23:8746–8757

Austgulen MH (2016) Environmentally sustainable textile consumption—what characterizes the political textile consumers? J Consum Policy 39:441–466

Baştabak B, Koçar G (2020) A review of the biogas digestate in agricultural framework. J Mater Cycles Waste Manag 22:1318–1327

Basumatary S, Das S, Kalita P, Goswami P (2021) Effect of feedstock/water ratio on anaerobic digestion of cattle dung and vegetable waste under mesophilic and thermophilic conditions. Bioresour Technol Rep 14:100675

Börjesson P, Mattiasson B (2008) Biogas as a resource-efficient vehicle fuel. Trends Biotechnol 26(1):7–13

Chanthawong A, Dhakal S (2016) Liquid biofuels development in Southeast Asian countries: an analysis of market, policies and challenges. Waste Biomass Valorization 7:157–173
Chidambaram PK, Thamilarasan K, Kumar JB, Mary LA (2021) A review on turbines in power production using wind and hydro Energy. In: Materials today: proceedings (in Press)
Christy PM, Gopinath LR, Divya D (2014) A review on anaerobic decomposition and enhancement of biogas production through enzymes and microorganisms. Renew Sustain Energy Rev 34:167–173
Czekała W (2017) Concept of in-oil project based on bioconversion of by-products from food processing industry. J Ecol Eng 18(5):180–185
Czekała W (2019) Biogas production from raw digestate and its fraction. J Ecol Eng 20(6):97–102
Czekała W, Gawrych K, Smurzyńska A, Mazurkiewicz J, Pawlisiak A, Chełkowski D, Brzoski M (2017) The possibility of functioning micro-scale biogas plant in selected farm. J Water Land Dev 35(1):19–25
Czekała W, Janczak D, Cieślik M, Mazurkiewicz J, Pulka J (2020) Food waste management using the hermetia illucens insect. J Ecol Eng 21(4):214–216
Czekała W, Tarkowski F, Pochwatka P (2021) Social aspects of energy production from renewable sources. Problemy Ekorozwoju 16(1):61–66
Czekała W, Smurzyńska A, Cieślik M, Boniecki P, Kozłowski K (2016) Biogas efficiency of selected fresh fruit covered by the Russian embargo. Int Multi Sci GeoConference Surveying Geol Min Ecol Manag SGEM 3(BOOK 4):227–233
Czekała W, Cieślik M, Janczak D, Czekała A, Wojcieszak D (2018) Fruit and vegetable waste from markets as a substrate for agricultural biogas plant. Int Multi Sci GeoConference Surveying Geol Min Ecol Manag SGEM 18(4.3):473–478
Destek MA, Sarkodie SA, Asamoah EF (2021) Does biomass energy drive environmental sustainability? An SDG perspective for top five biomass consuming countries. Biomass Bioenergy 149:106076
Eisenmenger N, Pichler M, Krenmayr N, Noll D, Plank B, Schalmann E, Wandl MT, Gingrich S (2020) The sustainable development goals prioritize economic growth over sustainable resource use: a critical reflection on the SDGs from a socio-ecological perspective. Sustain Sci 15:1101–1110
Gunnarsdottir I, Davidsdottir B, Worrell E, Sigurgeirsdottir S (2021) Sustainable energy development: history of the concept and emerging themes. Renew Sustain Energy Rev 141:110770
Hassan A, Ilyas SZ, Jalil A, Ullah Z (2021) Monetization of the environmental damage caused by fossil fuels. Environ Sci Pollut Res 28:21204–21211
Hosseinipour SA, Mehrpooya M (2019) Comparison of the biogas upgrading methods as a transportation fuel. Renew Energy 130:641–655
Igliński B, Piechota G, Iwański P, Skarzatek M, Pilarski G (2020) 15 years of the Polish agricultural biogas plants: their history, current status, biogas potential and perspectives. Clean Technol Environ Policy 22:281–307
Jarwar AI, Laghari AQ, Maitlo G, Qureshi K, Bhutto AW, Shah AK, Jatoi AS, Ahmed S (2021) Biological assisted treatment of buffalo dung and poultry manure for biogas generation using laboratory-scale bioreactor. Biomass Convers Biorefinery
Jeswani HK, Smith RW, Azapagic A (2013) Energy from waste: carbon footprint of incineration and landfill biogas in the UK. Int J Life Cycle Assess 18:218–229
Kapoor R, Ghosh P, Kumar M, Vijay VK (2019) Evaluation of biogas upgrading technologies and future perspectives: a review. Environ Sci Pollut Res 26:11631–11661
Kibria A, Akhundjanov SB, Oladi R (2019) Fossil fuel share in the energy mix and economic growth. Int Rev Econ Financ 59:253–264
Kowalczyk-Juśko A, Pochwatka P, Zaborowicz M, Czekała W, Mazurkiewicz J, Mazur A, Janczak D, Marczuk A, Dach J (2020) Energy value estimation of silages for substrate in biogas plants using an artificial neural network. Energy 202:117729
Kozłowski K, Dach J, Lewicki A, Cieślik M, Czekała W, Janczak D, Brzoski M (2018) Laboratory simulation of an agricultural biogas plant start-up. Chem Eng Technol 41:711–716

Kozłowski K, Pietrzykowski M, Czekała W, Dach J, Kowalczyk-Juśko A, Jóźwiakowski K, Brzoski M (2019) Energetic and economic analysis of biogas plant with using the dairy industry waste. Energy 183:1023–1031

Kupryaniuk K, Oniszczuk T, Combrzyński M, Czekała W, Matwijczuk A (2020) The influence of corn straw extrusion pretreatment parameters on methane fermentation performance. Materials 13(3003):1–16

Kushkevych I, Vítězová M, Vítěz T, Kováč J, Kaucká P, Jesionek W, Bartoš M, Barton L (2018) A new combination of substrates: biogas production and diversity of the methanogenic microorganisms. Open Life Sci 13(1):119–128

Lalander C, Nordberg Å, Vinnerås B (2018) A comparison in product-value potential in four treatment strategies for food waste and faeces—assessing composting, fly larvae composting and anaerobic digestion. GCB Bioenergy 10:84–91

Larsson SH, Samuelsson R (2017) Prediction of ISO 17831-1: 2015 mechanical biofuel pellet durability from single pellet characterization. Fuel Process Technol 163:8–15

Larsson M, Gronkvist S, Alvfors P (2016) Upgraded biogas for transport in Sweden—effects of policy instruments on production, infra-structure deployment and vehicle sales. J Cleaner Prod 112:3774–3784

Lu J, Gao X (2021) Biogas: potential, challenges, and perspectives in a changing China. Biomass Bioenergy 150:106127

Malik W, Mohan C, Annachhatre AP (2020) Community based biogas plant utilizing food waste and cow dung. Mater Today Proc 28(3):1910–1915

Marchetti R, Vasmara C (2020) Co-digestion of deproteinized dairy waste with pig slurry: effect of recipe and initial pH on biogas and volatile fatty acid production. BioEnergy Res 13:643–658

Marks S, Dach J, Morales FJF, Mazurkiewicz J, Pochwatka P, Gierz Ł (2020) New trends in substrates and biogas systems in Poland. J Ecol Eng 21:19–25

Mejdoub H, Ksibi H (2015) Regulation of biogas production through waste water anaerobic digestion process: modeling and parameters optimization. Waste Biomass Valorization 6:29–35

Mika K, Minna M, Noora V, Luukkanen J, Kaivo-oja J, Alexeeva A, Chea E, Va D, Kallio M, Hogarth N (2021) Situation analysis of energy use and consumption in Cambodia: household access to energy. Environ Dev Sustain

Muellner N, Arnold N, Gufler K, Kromp W, Renneberg W, Liebert W (2021) Nuclear energy—the solution to climate change? Energy Policy 155:112363

Mugodo K, Magama PP, Dhavu K (2017) Biogas production potential from agricultural and agro-processing waste in South Africa. Waste Biomass Valorization 8:2383–2392

Nsair A, Cinar SO, Alassali A, Qdais HA, Kuchta K (2020) Operational parameters of biogas plants: a review and evaluation study. Energies 13:3761

Nwokolo N, Mukumba P, Obileke K, Enebe M (2020) Waste to energy: a focus on the impact of substrate type in biogas production. Processes 8:1224

Panuccio MR, Attinà E, Basile C, Mallamaci C, Muscolo A (2016) Use of recalcitrant agriculture wastes to produce biogas and feasible biofertilizer. Waste Biomass Valorization 7:267–280

Pellegrini L, Arsel M, Orta-Martínez M, Mena CF, Muñoa G (2021) Institutional mechanisms to keep unburnable fossil fuel reserves in the soil. Energy Policy 149:112029

Pizarro-Loaiza CA, Antón A, Torrella M, Torres-Lozada P, Palatsi J, Bonmatí A (2021) Environmental, social and health benefits of alternative renewable energy sources. Case study for household biogas digesters in rural areas. J Cleaner Prod 297:126722

Pochwatka P, Kowalczyk-Juśko A, Sołowiej P, Wawrzyniak A, Dach J (2020) Biogas plant exploitation in a middle-sized dairy farm in Poland: energetic and economic aspects. Energies 13:6058

Rejeb A, Rejeb K, Zailani S (2021) Big data for sustainable agri-food supply chains: a review and future research perspectives. J Data Inf Manag

Sadekin S, Zaman S, Mahfuz M, Sarkar R (2019) Nuclear power as foundation of a clean energy future: a review. In: Proceedings of the energy procedia, vol 160. Elsevier Ltd., pp 513–518

Shahbaz M, Ammar M, Korai RM, Ahmad N, Ali A, Khalid MS, Zou D, Li X (2020) Impact of C/N ratios and organic loading rates of paper, cardboard and tissue wastes in batch and CSTR anaerobic digestion with food waste on their biogas production and digester stability. SN Appl Sci 2:1436

Silvi C (2008) History and future of renewable solar energy. Development 51:409–414

Sołowiej P, Pochwatka P, Wawrzyniak A, Łapiński K, Lewicki A, Dach J (2021) The effect of heat removal during thermophilic phase on energetic aspects of biowaste composting process. Energies 14:1183

Sun H, Khan AR, Bashir A, Alemzero DA, Abbas Q, Abudu H (2020) Energy insecurity, pollution mitigation, and renewable energy integration: prospective of wind energy in Ghana. Environ Sci Pollut Res 27:38259–38275

Tagne RFT, Dong X, Anagho SG, Kaiser S, Ulgiati S (2021) Technologies, challenges and perspectives of biogas production within an agricultural context. The case of China and Africa. Environ Dev Sustain

Velasquez CE, Estanislau FBGL, Costa AL, Pereira C (2020) Assessment of the French nuclear energy system—a case study. Energy Strategy Rev 30:100513

Waliszewska H, Zborowska M, Waliszewska B, Borysiak S, Antczak A, Czekała W (2018) Transformation of Miscanthus and Sorghum cellulose during methane fermentation. Cellulose 25(2):1207–1216

Wang X, Li Z, Cheng S, Ji H, Shi J, Yang H (2021) Multiple substrates anaerobic co-digestion: a farm-scale biogas project and the GHG emission reduction assessment. Waste Biomass Valorization 12:2049–2057

Wawrzyniak A, Lewicki A, Pochwatka P, Sołowiej P, Czekała W (2021) Database system for estimating the biogas potential of cattle and swine feces in Poland. J Ecol Eng 22(3):111–120

Wolna-Maruwka A, Dach J (2009) Effect of type and proportion of different structure-creating additions on the inactivation rate of pathogenic bacteria in sewage sludge composting in a cybernetic bioreactor. Arch Environ Prot 35:87–100

Żukowska GZ, Mazurkiewicz J, Myszura M, Czekała W (2019) Heat energy and gas emissions during composting of sewage sludge. Energies 12(24):4782

Chapter 11
Natural Gas as a Clean Fuel for Mobility

Ayat Gharehghani and Amir Hossein Fakhari

11.1 Introduction

There are three major fossil fuel resources: coal, oil, and natural gas (NG). The use of NG dates back to the middle of the twentieth century when the share of NG in the energy market was gradually raised. These three fossil fuel resources make up more than 85% of the world's primary energy. Due to population growth and the development of industries and transportation facilities, the consumption of fossil fuel is constantly increasing, so that the rise of energy consumption over the past decade is 38%. In addition, global energy consumption in 2011 was increased by 2.5%, as compared to 2010 (Srinivasan et al. 2004). Figure 11.1 depicts the world's primary energy demand from fuels between 1980 and 2035. It is clear that before 2030 the consumption of natural gas gets higher than coal, and by 2035 the consumption of this fuel becomes as high as one-quarter of global energy demand. Additionally, by 2030, it is predicted that consumption of natural gas grows by 2% annually, which is higher than the annual growth of total energy demand (1.2%) (International Energy Agency 2011).

Based on the outlook of the International Energy Agency (IEA) in 2011, NG has the fastest growth in primary energy resources. The consumption of NG is expected to double between 2020 and 2040, with the highest growth in demand in developing countries (U.S. Energy Information Administration 2013). NG is widely available in many parts of the world at reasonable prices. This fuel is cleaner than oil or coal and it is not as challenging as nuclear energy (U.S. Energy Information Administration 2013). The combustion of this fuel produces relatively lower levels of carbon dioxide (CO_2), and since 23% of the total CO_2 production in the European Union belongs to the transportation section (Belgiorno et al. 2019), the use of this fuel for applications in internal combustion engines (ICEs) can be effective (Kakaee and Paykani 2013).

A. Gharehghani (✉) · A. H. Fakhari
School of Mechanical Engineering, Iran University of Science and Technology, Tehran, Iran
e-mail: ayat_gharehghani@iust.ac.ir

G. Di Blasio et al. (eds.), *Clean Fuels for Mobility*, Energy, Environment, and Sustainability, https://doi.org/10.1007/978-981-16-8747-1_11

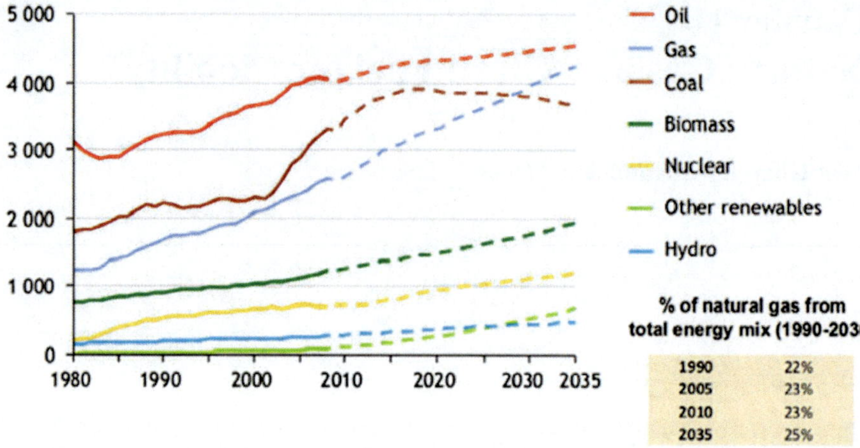

Fig. 11.1 World primary energy demand by fuel (International Energy Agency 2011)

In a study (Napolitano et al. 2014) which was carried out on a diesel engine in the dual-fuel mode with NG fuel and transient driving condition, it was observed that compared to the conventional fossil fuels such as diesel and gasoline, it is possible to obtain a sustainable reduction in CO_2 production with an average substitution ratio of 47% for the NG fuel and the new European driving cycle. Also, using this strategy, the CO_2 production is reduced by 5%, as compared to turbo-charged gasoline-fueled engines. NG which is a mixture of gaseous hydrocarbons is mainly composed of methane (Carvalho et al. 2003; Kato et al. 1999), but it usually contains different amounts of higher alkanes and a small percentage of CO_2, nitrogen, hydrogen sulfide, and helium. The composition of this fuel depends on different factors, such as climate and location. This fuel has been employed as an energy source for various applications such as electricity generation, cooking, and heating. It also has application as a vehicle fuel and raw material for the production of plastic and other commercial organic chemicals (Gharehghani 2019). The NG which is produced from fossil fuels is considered non-renewable. Various compositions of NG have been presented in detail in Table 11.1 (Shasby 2004) but based on the statement of Srinivasan et al. (2004) more than 98% of the NG composition is methane. This fuel can be compressed for storage or usage as Compressed Natural Gas (CNG). By using a high compression pressure of around 200 bar or 2900 psi, CNG can store the same mass of NG with a much higher volume. NG is safer than gasoline in many aspects (Semin 2008). The ignition temperature of this fuel is higher than gasoline and diesel fuel. Additionally, NG is lighter than air, and in case of leakage, it is easily dispersed. On the other hand, gasoline and diesel fuels accumulate on the ground and increase the risk of fire. CNG is non-toxic and in case of leakage, it does not contaminate groundwater. CNG-fueled engines have significant advantages over conventional gasoline or diesel engines (Kakaee and Paykani 2013). This non-renewable fuel has environmental advantages over gasoline and diesel fuel. For instance, this fuel is cleaner regarding

Table 11.1 Natural gas composition (Shasby 2004)

Composition	Formula	Volume fraction (Economides and David 2009)	Volume fraction (International Energy Agency 2011)	Volume fraction (U.S. Energy Information Administration 2013)	Volume fraction (Belgiorno et al. 2019)
Methane	CH_4	94.00	92.07	94.39	91.82
Ethane	C_2H_6	3.30	4.66	3.29	2.91
Propane	C_3H_8	1.00	1.13	0.57	–
Iso-Butane	i-C_4H_{10}	0.15	0.21	0.11	–
N-Butane	n-C_4H_{10}	0.20	0.29	0.15	–
Iso-Pentane	i-C_5H_{12}	0.02	0.10	0.05	–
N-Pentane	n-C_5H_{12}	0.02	0.08	0.06	–
Nitrogen	N_2	1.00	1.02	0.96	4.46
Carbon dioxide	CO_2	0.30	0.26	0.28	0.81
Hexane	C_6 + C_6H_{14}	0.01	0.17	0.13	–
Oxygen	O_2	–	0.01	<0.01	–
Carbon monoxide	CO	–	<0.01	<0.01	–
Total	**–**	**100**	**100**	**100**	**100**

the greenhouse gases. CNG is considered a cleaner alternative for other fossil fuels (Kakaee and Paykani 2013; Parikh 1999).

NG can be sorted as CNG or liquefied NG (LNG) at low temperatures around 111 K. This fuel cannot be used as the fuel for transportation at ambient pressure and temperature, thus it should be compressed or liquefied to increase its bulk energy density (Demirbas 2002). There are three major types of NG-fueled engines: dedicated NG-fueled engines or vehicles which only use CNG or LNG fuels, the gasoline/NG bi-fuel engines that are based on spark-ignition (SI), and the diesel/NG dual-fuel engines that work based on compression-ignition (CI). The self-ignition of NG happens at a high temperature and this fuel cannot be easily ignited by compression. Initially, gasoline-based engines were converted to gasoline/NG bi-fuel engines, by adding a CNG cylinder. The bi-fuel engines make use of two separate fuel supply systems and can operate with either one of these fuels. The cetane number of NG is higher than gasoline, therefore it has a better anti-knock performance (Kakoee et al. 2018). However, when NG fuel is used in bi-fuel engines the thermal efficiency cannot be improved by increasing the comparison ratio (CR) while meeting, the requirement for preventing knock in the gasoline-fueled mode (Chen et al. 2019). In other words, the advantage of NG cannot be employed in bi-fuel engines. Based on this, the engines that operate solely using the NG fuel have higher CR (11–13) compared to the CR of the gasoline-fueled engines (9.5–11), as a result, the thermal efficiency is enhanced in these engines (Frailey et al. 2000). Theoretically, the CR of NG-fueled engines can be increased if they only use NG fuel. NG fuel can also be used in diesel engines in dual-fuel mode (Mtui and Philip 1996). Similar to bi-fuel engines, dual-fuel engines have two fuel supply systems, but these systems should work simultaneously since NG is combusted by diesel fuel that is ignited by compression. Therefore, the CR of dual-fuel engines is usually over 16 (Frailey et al. 2000). Some challenges should be coped with when NG is used as a vehicle fuel: The first and most important issue is the storage of NG fuel (Reynolds et al. 2005). Nowadays, NG is stored in CNG cylinders on vehicles. These cylinders store NG at very high pressure, but in terms of bulk energy density, they are not much compact. As a result, most converted NG-fueled engines have a limited driving range. The acceleration of the dedicated NG-fueled engines is lower than that of gasoline-fueled engines. Moreover, the engines that work in bi-fuel mode should be adjusted between the optimal configurations of both modes, therefore they lose more performance in NG or gasoline-fueled modes (Cho and He 2007). With motivations regarding emissions, strategic advantages, and cost-effectiveness, researchers (Brombacher 1999; Aslam et al. 2006), investigated the use of NG as an alternative fuel. Several alternative fuels have been identified that have significant potential in producing a lower level of pollutants compared to gasoline and diesel fuels. For various reasons, among these fuels, NG is known as one of the main candidates for transportation (Shasby 2004; Srinivasan et al. 2006; Gharehghani et al. 2015a). Shasby (2004) identified tree reasons for this: availability of this fuel, being eco-friendly due to its fast absorption, and being usable in both conventional gasoline and diesel-fueled engines (Poulton 1994). Operating costs are another reason why theoretically, NG-fueled vehicles have advantages over other engines. There are four reasons for the attractiveness of CNG:

1. This is the only fuel that is cheaper than diesel and gasoline.
2. Inherently it produces low levels of emission.
3. It produces a lower amount of greenhouse gases.
4. Using this fuel lifts the pressure from petroleum sources.

CNG problems arise from low vehicle performance, fuel storage, infrastructure cost, and ensuring sufficient supply. According to Lave et al. (2000), CNG-fueled vehicles show a considerable reduction in gas emissions and PM emissions. However, there are problems such as low bulk energy density for storage, knock phenomenon at high loads, and high methane and CO emission levels at low loads. These challenges can be coped with using fuel management, proper design, and exhaust treatment methods. The characteristics of CNG have been tabulated in Table 11.2.

Most existing CNG-fueled vehicles have gasoline-fuel engines that are modified by after-market retrofit conversion to acquire a bi-fuel option. These converted engines have problems such as power loss and drivability problems that are caused due to their lack of proper design (Poulton 1994). On the other hand, dual-fuel engines that are based on diesel-fueled engines, maintain the high efficiency of diesel engines, while they offer advantages such as reduction of gaseous, PM, and CO_2 emissions. Also, single-fueled engines that are optimized for using CNG appear to have a high performance and are more cost-effective. According to Poulton (1994), a single-fuel CNG-based vehicle should be equally efficient or even more efficient, and they should produce higher levels of pollutants compared to an equivalent gasoline-fueled engine. The driving range of this vehicle would be considerably lower than an equivalent gasoline-fueled vehicle unless the fuel tank becomes significantly large which has disadvantages such as weight, space, performance loss, and cost. If LNG is used instead of CNG the driving range limitation gets significantly modified. The octane number of NG is around 130, this means that the engine can work with a CR of 16:1 without knock or explosion. Most importantly, NG can decrease CO_2 emission by 20–25% compared to gasoline, since the chemical structure of NG is simple (by containing methane (CH_4) with one carbon) as compared to diesel ($C_{15}H_{23}$) and gasoline (C_8H_{18}) (Srinivasan et al. 2006). Having fewer carbon and hydrogen atoms in the CH_4 molecule makes it lighter than air, thus it can be easily blended to reduce the vehicle emissions up to 50%. The characteristics of CNG as a fuel have been presented in Table 11.3.

Table 11.2 CNG properties (Durell et al. 2000)

CNG properties	Value
Density at 15 °C and 1 atm (kg/m^3)	0.75
Flammability limits (volume% in air)	4.3–15
Auto-ignition temperature in air	0.4–1.6
Minimum ignition energy (MJ)	723
Flame velocity (ms^{-1})	0.28
Adiabatic flame temperature (k)	0.38
Quenching distance (mm)	2214
Stoichiometric fuel/air mass ratio	2.1
Stoichiometric volume fraction (%)	0.069
Lower heating value (MJ/kg)	9.48
Heat of combustion (MJ/kg)	45.8

Table 11.3 CNG fuel characteristics (Srinivasan et al. 2006)

CNG characteristics	Value
Auto-ignition (°C)	700
Octane rating	130
Boiling point (Atm. Press) (°C)	−162
Air–fuel ratio (Weight)	17.24
Chemical reaction with rubber	No
Storage pressure (Mpa)	20.6
Fuel air mixture quality	Good
Emission CO-HC-No_x	Very low
Flame speed m per s	0.63
Combust. ability with air (%)	4–14

The technology of engine conversion has a good level of stability and the conversion apparatus is easily accessible. For spark-ignition engines, bi-fuel conversion and dedicated CNG-fueled engines are the available choices. For the bi-fuel conversion, if the engine can be modified to control the gas flow and revise the ignition timing for the NG-mode, the original engine management systems can be employed. For this purpose, the engine should be equipped with a CNG control system. When the engine is set to work on gas, the fuel injectors should be disabled, however, to provide the cooling in this mode, the fuel should be running in the injector line, and then it should directly go through the return fuel line. According to Poulton (1994), gasoline-fueled SI engines in various sizes can be easily converted to bi-fuel engines by installing a carburetor/mixer, regulator, shut-off valves, control system, and NG storage tank. When the gasoline-fueled mode is used, a bi-fuel arrangement is created, but this arrangement prevents the optimization of engines for the high-octane gaseous fuel. The dedicated NG-fuel engines work only with NG fuel and they are optimized for this fuel. The design of these engines can be obtained from gasoline-fueled engines or they can be designed specifically for this purpose. For the diesel engines that are designed or converted to work with NG fuel, there are two major choices: the first choice is dual-fuel engines. These CI engines work with a blend of NG and diesel fuels. NG has a low cetane number, therefore it is not suitable for CI, but if diesel is injected into the mixture of NG/air, natural ignition can be initiated. While operating in this mode, 50–75% of diesel fuel consumption can be substituted by NG. This engine can also return to using only diesel fuel (100% diesel fuel consumption). The second type is dedicated NG-fueled engines that are optimized for using NG fuel. The design of these engines can be obtained from gasoline-fueled engines or they can be designed for this purpose. These engines are manufactured by converting CI engines to SI engines and until the original equipment (OE) of the manufacturers are not easily accessible, this conversion (which involves substituting diesel fuel equipment with a gas carburetor and adding an ignition system and spark plug) will continue. Figure 11.2 presents different combustion systems of NG. It can be observed that each combustion system has specific strategies regarding the preparation of the mixture, combustion control, and reduction of emissions. However, the common objective in all of these systems is to increase the power output, while reducing fuel consumption and emissions (Li and Zhang 2019).

This chapter aims to review the influence of using NG fuel in vehicle engines and the capability of this fuel for obtaining sustainable mobility. Thus, in the subsequent sections, various strategies for the use of NG fuel in CI, SI, and dual-fuel engines were separately investigated in terms of performance and emission levels.

11.2 Natural Gas in SI Engines

The emission of NG-fueled SI engines can be reduced by lean-burn and stoichiometric engines with a three-way catalyst (Gharehghani et al. 2015b). In lean-burn NG-fueled engines, the fuel-to-air ratio is considerably important. As the mixture is

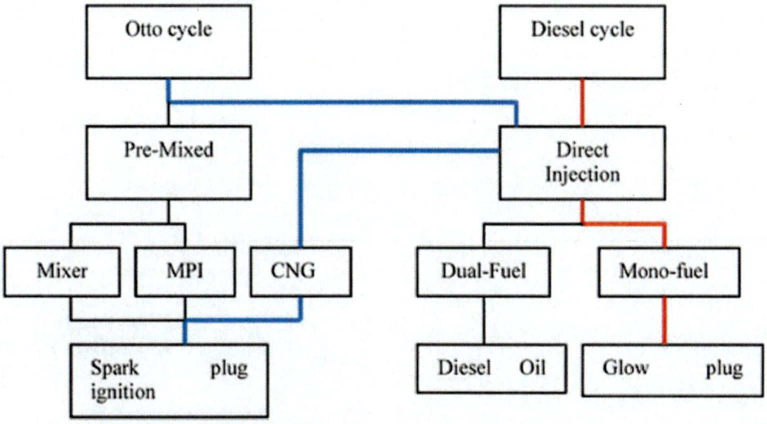

Fig. 11.2 Various combustion systems of the NG fuel (Li and Zhang 2019)

leaned out to past a point that suppresses the NO_x emission, the burning rate decreases as compared to that in the stoichiometric condition, which leads to longer combustion time and increased thermal loss to the cylinder walls and reduces the thermal efficiency (Kato et al. 2000). The low propagation speed of flame results in slower heat release and creates instability in the combustion with high cyclic changes (Das 1997). Therefore, the obtained efficiency with the lean operation cannot be realized. When the flame propagation stops, it is possible that misfire cycles and incomplete combustion form in the end zone, which can lead to high HC emissions (Chen and John 2001). On the other hand, slight errors in fuel-to-air ratio near the lean limits can mislead the engine, leading to a significant rise in emission levels, engine roughness, and poor throttle response (Franklin et al. 1994). Several factors affect the lean misfire limits, including cylinder air motion, existing combustion energy, the composition of NG, the mixture temperature during the combustion, residual fraction, and water from moisture in the fuel–air mixture. In addition, the influence of mixture preparation is important on the homogeneity of the in-cylinder charge. Figure 11.3 illustrates the operating range in different conditions. The limitation of EGR (λ) in the NG-fueled engines is based on managing the increase of fuel consumption under lean-burn conditions and controlling NO_x, knock, and rise of fuel consumption in the rich condition of fuel (Beck et al. 1997). With more excess air, the combustion temperature decreases, which leads to higher CO and HC emissions in the engine, due to slower propagation of flame.

NG is usually injected through the inlet manifold, therefore it presents lower volumetric efficiency and power. On the other hand, with the higher octane number of NG fuel, higher CRs are usable. By employing turbochargers, superchargers, and internal coolers, a 50% rise in power output and thermal efficiency can be obtained. The proper adjustment of NG-fueled SI engines results in lower NO_x, CO_2, and non-methane HC emissions, but it increases methane emissions compared to when gasoline fuel is used (Gharehghani 2013).

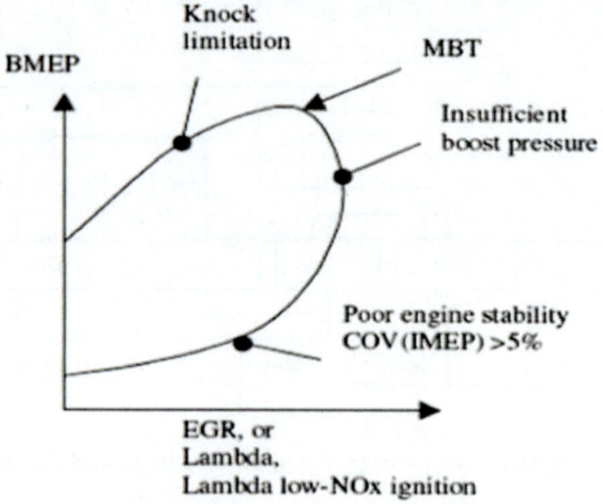

Fig. 11.3 Operating range of load and fuel concentration for engine mapping (Einewall and Bengt 2005)

11.2.1 Performance

Previous studies (Cho and He 2007; Zeng et al. 2006) reported the NG-fueled engines can produce 5% higher thermal efficiency compared to the gasoline-fueled engine. This can be because of various factors. The first cause is the high octane number of this fuel, which facilitates the use of high CRs. Previous works (Gharehghani et al. 2015b; Huang and Crookes 1998) indicate that by using NG, the CR of the engine can be raised from 8:1 to 13:1, while higher CRs lead to the knock phenomenon. The increase of CR in SI engines can considerably increase the thermal efficiency. For example, making the CR double can lead to 13% better thermal efficiency. The inlet pressure in the NG-fueled SI engines is higher (almost 10% higher) compared to gasoline fuel engines, thus lower pumping losses are created in these engines (Thurnheer et al. 2009). These lower losses are due to the wider throttle setting which decreases the restrictions of the inlet manifold and has a positive influence on the thermal efficiency. The power loss in the naturally aspirated gasoline-fueled engine that operates with NG fuel can be compensated by raising the CR (Huang and Crookes 1998) or by using a supercharger or turbocharger (Bach et al. 2004; Kesgin 2005). As the result of the high octane number for NG fuel, NG-fueled engines can be operated with high CRs without knock phenomenon. Forced injection in naturally-aspirated leads to higher power output. In a study (Bach et al. 2004) that was conducted on gasoline/NG bi-fuel 1.0 L SI engine that worked with and without a supercharger, it was observed that NG fuel produces 10–15% less torque compared to gasoline fuel. On the other hand, raising the CR while employing a supercharger (in the boosted pressure of 0.6 bar) increases the torque up to 50% compared to the naturally-aspirated gasoline-fueled mode. The NG-fueled engine can operate with a leaner fuel–air mixture compared to gasoline-fueled engines. The cycle-to-cycle changes of the NG-fueled engines have the same order as gasoline-fueled engines. The Coefficient of Variation (COV) in the Indicated Mean Effected Pressure (IMEP)

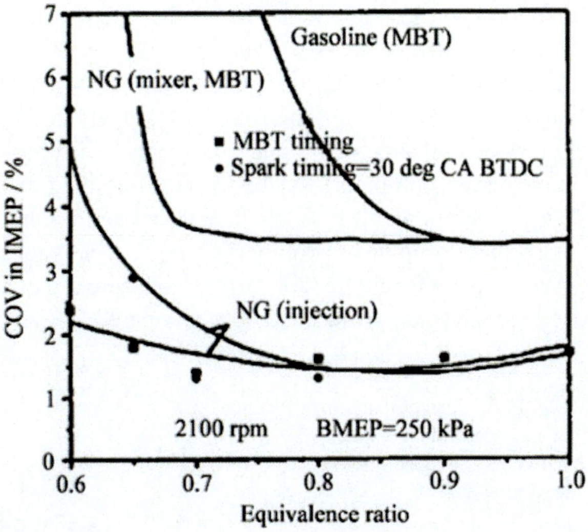

Fig. 11.4 COV in IMEP with different equivalence ratios (Cho and He 2007)

of this similarity can be observed in Fig. 11.4. Figure 11.4 also shows that the lean operation limit for the NG-fueled mode has increased around 28% compared to the gasoline-fueled mode. However, when NG is injected through the inlet manifold, significantly lower cycle-to-cycle variation of IMEP is created. This is due to more accurate control and distribution of fuel-to-air ratio, which assist keeping a relatively constant fuel–air mixture from cycle to cycle.

11.2.2 Emissions

NG-fueled engines generally produce less CO and non-methane HC emissions than gasoline-fueled engines (Dondero and Goldemberg 2005). Lower equivalence ratios in the NG-fueled engines lead to the production of lower levels of CO emissions (between 50 and 90%) compared to gasoline-fueled engines. However, the production of CO and HC emissions rises due to the reduction of combustion quality (Mello et al. 2006). Also, these engines reduce unburnt HC emissions by up to 55%. The higher level of NO_x emission in NG-fueled engines compared to the conventional gasoline-fueled SI engines is due to the higher pressure and temperature in the combustion chamber, which are the result of the advanced fuel injection timing and higher CRs. This raises the production of NO_x emission during the combustion, as the compression stroke progresses, which leads to the rise of temperature and pressure in these engines as compared to the conventional gasoline-fueled engines. Higher oxygen concentration in the lean fuel mixtures intensifies the production and emission of NO_x. This trend can be traced back to the higher CR of NG-fueled engines compared to gasoline-fueled engines. When the CR of both these engine types are the same, probably the lean fuel–air mixture in the NG-fueled and gasoline-fueled engines leads to the production of lower NO_x emission levels as the result of lower combustion temperature compared to the stoichiometric performance of these engines (Varde et al. 1995). The vast majority of NG-fueled engines are pre-charged SI engines (Kar et al. 2021; Chiu et al. 2004). While SI engines have significant advantages over CI engines in terms of PM and NO_x emissions, they have several drawbacks in the performance (see Figs. 11.5 and 11.6). In addition, SI engines have a high pumping loss due to throttling the intake air at partial load conditions. These factors result in a 15–30% decrease in volumetric efficiency in comparison with diesel engines (Brombacher 1999; Narayanaswamy and Christopher 2004). According to Ouellette (2000), high-pressure direct injection of NG in diesel engines reduces NO_x emission up to 40% compared to conventional diesel engines.

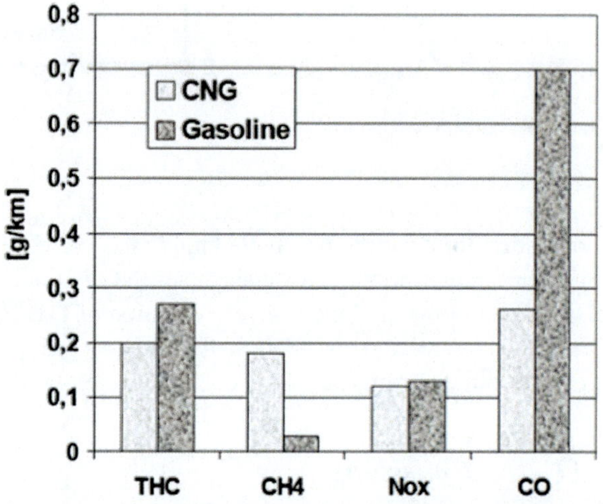

Fig. 11.5 The comparison of emissions in an engine with CNG and gasoline fuels (Kar et al. 2021)

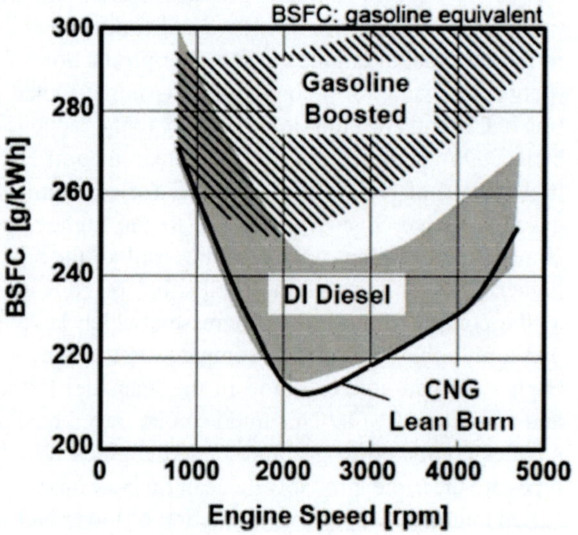

Fig. 11.6 The brake specific fuel consumption (BSFC) in various engine speeds and with various fuels (Kar et al. 2021)

11.3 NG Fuel in the CI Engines

NG fuel can be used as an alternative fuel in both SI and CI engines. CI engines have higher CRs compared to the SI engines, thus they have higher thermal efficiency. In most applications, NG is injected through the inlet manifold, therefore, it has lower volumetric efficiency. In some applications, NG is directly injected into the cylinder. The ignition quality is connected to the Octane and Cetane numbers (Korakianitis et al. 2011; Gharehghani and Yusaf 2013). For a particular fuel, a negative but non-proportional connection exists between the Cetane and Octane numbers. The Octane number of Methane, which is the most important compound of NG, is higher than gasoline and while its Cetane number is higher than diesel. The Cetane number of this compound is so low that it is generally neglected. The combustion quality of NG is considerably different than gasoline and diesel fuels. Employing NG along with injecting pilot fuel in CI engines leads to lower NO_x emission at low loads and almost the same level of this emission at high loads. Although, at low loads because of incomplete combustion in low temperatures, the CO and HC increase. On the other hand, at higher loads, these emissions are lower than the emissions of conventional CI engines. Due to the higher hydrogen-to-carbon ratio in NG, at all of the loads, the CO_2 emission level is lower (Gharehghani and Jazayeri 2012).

11.3.1 Performance

NG fuel does not self-ignite at conventional CRs of CI engines. To use this fuel, a certain mode known as the dual-fuel mode is required. In this mode, the source of ignition is the self-ignition of pilot fuel. A low amount of this high-Cetane fuel is directly injected into the chamber. Then this fuel is mixed with a premixed NG-air charge. After the initial ignition delay, the pilot fuel ignites and starts to burn. Based on the mixing level of NG and air, a second delay occurs, and after that NG ignites and most of the heat release is produced in the combustion. Dual-fuel engines have lower NO_x and soot emissions while they maintain an acceptable thermal efficiency in the operating range (Karim 2010; Crookes and Olsen 2007). The reduction of flame propagation speed in SI engines while using NG fuel is a challenge, but it is not proved that it is a major problem in CI engines. This is because, in the conventional SI engines, there is just one ignition source (spark plug). When the spark plug sparks in these engines, the ignition is started and it propagates through the charge. On the other hand, in conventional CI engines, the fuel enters the cylinder, then it is mixed with the charge. When the requirements (for example, the temperature, pressure, and mixing level) are met, these scattered droplets of fuel self-ignite. This creates several ignition points in the combustion chamber. Therefore, several flame fronts are formed that make the combustion faster (Namasivayam et al. 2009). In the dual-fuel engines (where the pilot fuel directly enters the chamber) the multiplicity of ignition points counteracts the low flame propagation speed of NG and leads to more rapid combustion. The CI engines do not have throttle valves along the intake manifold, therefore they present higher volumetric efficiencies compared to NG-fueled SI engines. The load control in the CI engines is influenced by varying the quantity of fuel that is injected into the chamber. This reduces the pumping loss that is common in SI engines. At low speeds (1000 rpm), the output of dual-fuel CI engines is compatible with the operation of conventional CI engines, but with this particular method of injection, the engine produces significantly less power at high speeds as compared to that of conventional CI engines. This is because, at higher speeds, a higher number of consumption cycles occur every second, therefore, the injection method of NG has a considerable influence on the engine performance in different conditions (Papagiannakis and Hountalas 2004). In the next section, the dual-fuel engines that use NG/diesel fuels will be examined in detail.

11.3.2 Emissions

Using the dual-fuel mode in CI engines significantly reduces the NO_x emission (Gharehghani and Jazayeri 2012; Kusaka et al. 2000; Carlucci and Laforgia 2006), and based on the engine type and conditions, at low or medium loads, almost 50% of reduction in this emission is observed compared to the conventional CI engines. Figure 11.7 shows the usual trend of NO_x emission from a dual-fuel engine with pilot fuel at two different load levels in different BMEPs. Compared to the conventional working condition of engines, employing the dual-fuel mode reduces the level of NO_x pollutant. The reduction of this emission is mainly due to low combustion temperature which is the result of the low flame propagation speed of NG (Crookes and Olsen 2007).

The soot and PM emissions are insignificant and sometimes undetectable in dual-fuel engines (Gharehghani and Yusaf 2013; Kusaka et al. 2000; Carlucci and Laforgia 2006). The combustion of NG produces an inconsiderable amount of soot because NG has a lower number of carbon atoms compared to diesel. Besides the low number of carbon atoms, methane does not have any C–C bonds. This leads to a lower tendency toward soot production (Tree and Kenth 2007). Additionally, the NG mixture is usually mixed properly before the combustion. In the dual-fuel mode and at partial loads, the unburnt HC emission is significantly higher than the conventional diesel engines at low to medium loads. Figure 11.8 shows the concentration of HC emission for the dual-fuel mode around 6000 ppm, while this value for the conventional diesel engine is 100 ppm. This may be because of the stratified nature of the pilot fuel which operates as the ignition source.

Figure 11.9 shows the changes in the CO emissions for dual-fuel and conventional CI engines in different BMEPS (Crookes and Olsen 2007). Over the entire tested speed, the CO emission is significantly higher than the conventional CI engines. Some studies approve this trend to some extent, and this even intensifies by raising the performance of the dual-fuel operating at partial loads (Gharehghani and Jazayeri 2012; Karim 2010). In the experimental data of some studies (Gharehghani and

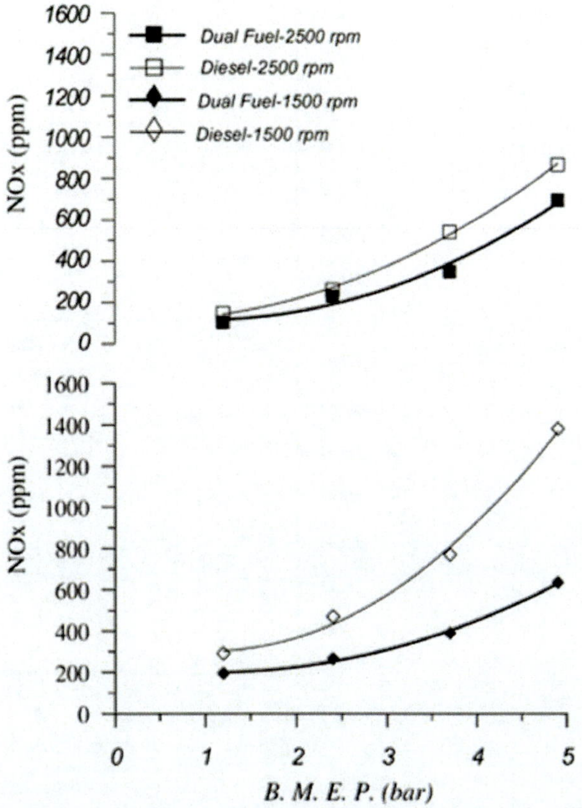

Fig. 11.7 The comparison of NO_x emission for dual-fuel and 1 CI engines based on different BMEPs (Crookes and Olsen 2007)

Jazayeri 2012; Karim 2010) at low loads, the trend of CO emission starts relatively higher than CO emission of a conventional CI engine, but by raising the equivalence ratio the CO emission level approaches the level of the conventional diesel engines. Because of the higher combustion temperature at higher loads, this indicates that the incomplete combustion of NG at low loads.

Some techniques have been developed to improve the emission of dual-fuel engines at low loads. One of them is raising the amount of injected pilot fuel which leads to the formation of a higher number of ignition points and consequently results in more complete NG combustion (Srinivasan et al. 2006; Papagiannakis and Rakopoulos 2007). The other method is injecting low levels (5% in volume) of hot EGR. The hot exhaust gas raises the temperature of the combustion chamber and therefore reduces the CO and unburnt HC emission (Pirouzpanah et al. 2007; Selim 2003; Abd-Alla et al. 2001). This is due to the increase in the equivalence ratio.

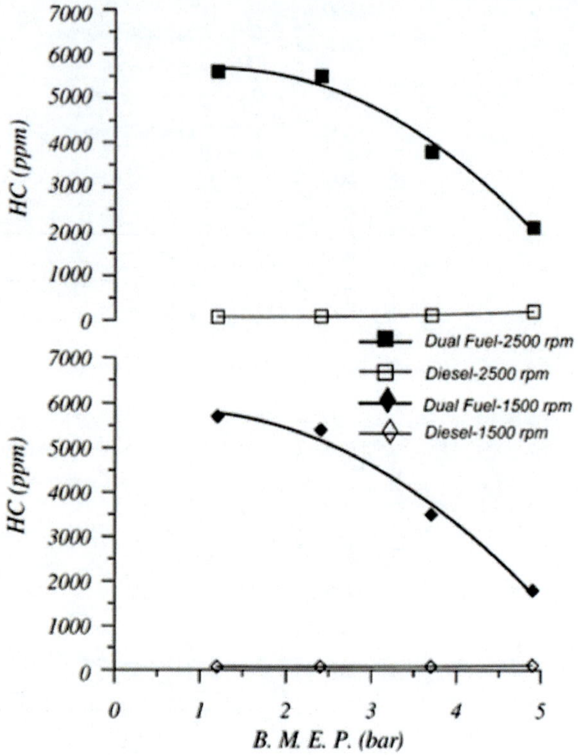

Fig. 11.8 The comparison of greenhouse gas emissions for dual-fuel and conventional CI engines based on different BMEPs (Crookes and Olsen 2007)

11.4 The NG/Diesel Dual-Fuel Engines

The combustion of the in-cylinder content is one of the processes that affect the performance and durability of the engine (Gatts et al. 2012). NG and diesel fuels have different physical and chemical characteristics. The combustion of diesel fuel in the cylinder is mainly the combustion of controlled penetration of the fuel (Zhao et al. 2017), while for the combustion of NG, the fuel is premixed with air. The addition of NG in the chamber influences the combustion characteristics because of the changes in the formation and combustion of the fuel–air mixture. Overall, the CNG/diesel dual-fuel engines reduce the peak in-cylinder pressure and the peak pressure rise (Fakhari et al. 2020). Most studies reported that at low loads, the use of NG leads to weaker performance of the engine and higher CO emissions compared to those of the conventional diesel engines (Ramadhas and Muraleedharan 2008). Generally, dual-fuel engines have higher ignition delays than diesel engines (Fakhari et al. 2019). Also, these engines present lower thermal efficiencies at low loads while they exhibit higher thermal efficiencies at medium and high loads as compared to the conventional diesel engines. The NO_x and PM emissions of these engines are lower than the diesel engines, while they have higher HC and CO emissions (Cheenkachorn et al. 2013; Zhen-Tao and Shao-mei 2004). Moreover, increased tendency towards the knock phenomenon is one of the other negative aspects of these engines (Abdelaal et al. 2013). Figure 11.10 shows a schematic of dual-fuel combustion of natural gas/diesel. The combustion characteristics of NG/diesel dual-fuel engines were examined and here are some of these studies.

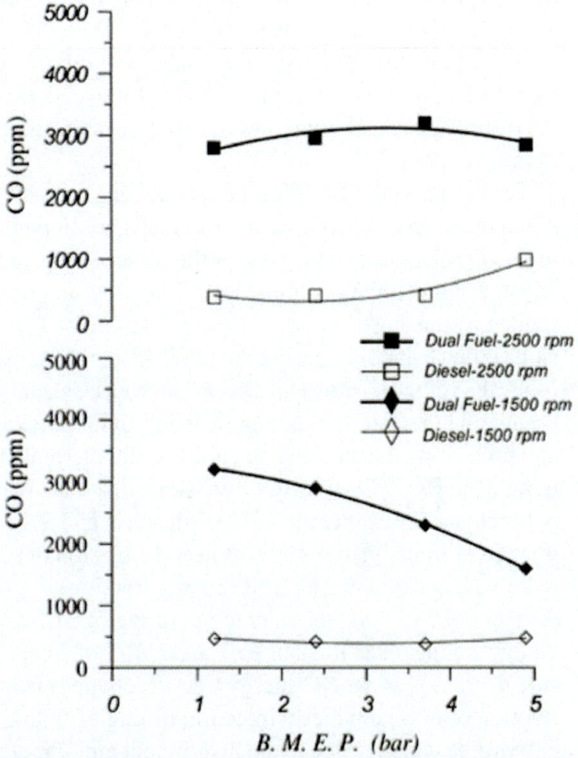

Fig. 11.9 The comparison of CO emission for dual-fuel and l CI engines based on different BMEPs (Crookes and Olsen 2007)

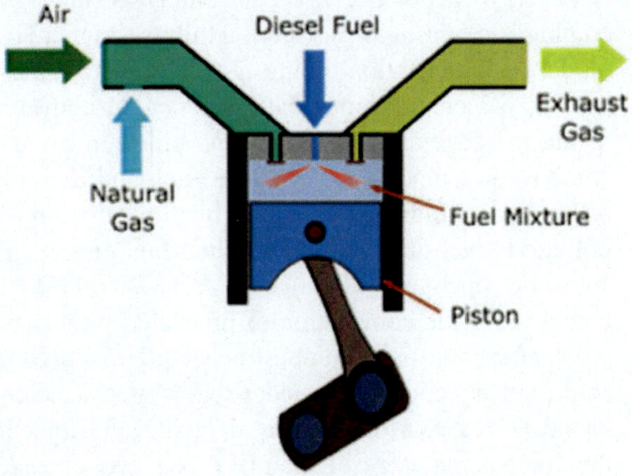

Fig. 11.10 The schematic diagram of ng/diesel dual-fuel combustion (Fakhari et al. 2019)

Li et al. (2017) studied the influence of adding the NG fuel (0–5%) in the dual-fuel mode of a heavy-duty diesel engine on the combustion characteristic such as the start of combustion (SOC), peak heat release rate (PHRR), and combustion duration. Some significant and positive effects were observed by adding an inconsiderable amount of NG to the mixture. It was observed that when the engine was working at low loads, the addition of high amounts of NG, decreased the heat release rate (HRR) and at high loads, it significantly increased the PHRR. They stated that the addition of NG retards the SOC. Also, the increased thermal efficiency was only observed at high loads and high amounts of NG. The rise of thermal efficiency was because of the decreased combustion duration and the shift of the combustion stage towards the optimal state.

In summary, the NG/diesel dual-fuel engines reduce the level of NO_x pollutants, while they increase the level of HC and CO pollutants, especially, at low and medium loads. The decrease of the NO_x pollutant is mostly due to these factors: (1) the amount of pilot fuel is quite insignificant, (2) the low-temperature combustion of NG that is ignited by the diesel fuel produces low amounts of NO_x. The rise of CO and HC emissions is due to incomplete combustion. Also, the increase in greenhouse gas pollutions is because of the narrow gap effect. The NO_x, CO, and HC emission can be reduced by preheating the entering air and by using EGR (Paykani et al. 2012). The use of EGR and preheating the entering air can raise the thermal efficiency to some extent and can reduce the HC, CO, and NO_x emissions in a dual-fuel mode (CR = 17.5) at low and partial loads (Paykani et al. 2012). Yousefi et al. (2017) investigated the effect of diesel injection timing on the combustion and emission of a dual-fuel engine at Under a load of 25%, speed of 910 RPM, NG energy fraction of 75%, diesel common rail-injection pressure of 525 bars, the inlet temperature of 313 K, and inlet pressure of 1.05 bars. At low engine loads (load of 25% and IMEP of 3 bar), the maximum in-cylinder pressure decreases as the NG energy fraction changes from 0 to 60%. With the NG energy fraction of 60% and diesel injection timing of 6 °CA BTDC, diesel ignition tends to deteriorate which is due to the insufficient amount of the pilot fuel to ignite the entire premixed fuel–air mixture. During the early stages of the NG combustion, raising the NG energy fraction decreases the total HRR. In the final combustion stage (diffusion combustion), the dual-fuel mode exhibits a higher HRR than that of the diesel-fueled mode. At the injection timing of 6 °CA BTDC, increasing the NG energy fraction leads to a decrease in thermal efficiency. Moreover, raising the NG energy fraction results in a decrease in NO_x concentration compared to diesel mode, due to the lower combustion temperature. In 2018, Yang and Zeng (2018) examined the impacts of NG injection timing and the pilot fuel injection on the combustion performance and pollution level of an NG/diesel dual-fuel engine. The results revealed that in the split-pilot operating condition in an NG/diesel dual-fuel engine, the first stage of combustion (the combustion of preheated pilot diesel) weakens, while the second stage of combustion (combustion of premixed NG) intensifies as compared with split-pilot injection. Also under this strategy, the combustion duration and engine sound decrease. Moreover, by delaying NG injection under split-pilot condition, the combustion stage of premixed NG gets considerably faster, and the combustion duration decrease by about 30%. Besides, the HC and CO emissions decrease.

The injection timing of NG and the split-pilot strategy are important parameters for affecting the combustion performance and emission characteristics in NG/diesel dual-fuel engines. The combination of retarded NG injection while keeping a constant interval between the pilot and main injection are effective techniques for optimizing the combustion performance and emission characteristics of NG/diesel dual-fuel engines. In 2019, Zheng et al. (2019) experimentally studied the influence of equivalence ratio on the combustion and emission levels of o a six-cylinder dual-fuel engine with a CR of 14.2 and NG substitution ratio of 80%. The comparison was performed between a dual-fuel and a gasoline-fueled SI engine in stoichiometric condition and with equal IMEPs. The result showed that by raising the equivalence ratio, the NG fuel consumption decreases. As a result of raising the equivalence ratio in the dual-fuel mode, the maximum heat release rate, and the exhaust gas temperature increase while the combustion duration decreases. The maximum heat release rate in the dual-fuel mode is higher compared to that of the gasoline-fueled engines, while the combustion duration in these engines is lower. Papagiannakis (2003) analyzed the effect of NG mass ratio on the combustion characteristic of an engine at three different loads and speeds. The NG mass ratio in their tests varied between 0 and 80%. The results showed that raising the equivalence ratio increases the consumption of NG regardless of injection parameter, and level of EGR. It was also observed that under the dual-fuel mode, the ignition delay and the combustion duration for all the tests are longer than those of the conventional diesel engine, while this mode leads to higher maximum in-cylinder pressure. By raising the NG mass ratio, the maximum in-cylinder pressure considerably gets lower, but the ignition delay and combustion duration rise. Abdelaal (2012) compared the combustion characteristic of the NG/diesel dual-fuel and conventional diesel modes in a single-cylinder diesel at 52 and 87% of engine load and an engine speed of 1600 rpm. They found that as compared to the conventional diesel mode, the dual-fuel mode presents 6.7 and 6.2 bar lower in-cylinder pressure, 2.8 and 5.5 °CA higher ignition delay, and 0.39 and 1.14 bar/°CA lower pressure rise rate, at 52 and 87% of engine load, respectively. Furthermore, the dual-fuel mode presents lower peak values for HRR. They also stated that the emplying EGR in the dual-fuel mode decreases the in-cylinder pressure and increases the ignition delay. The in-cylinder pressure, ignition delay, HRR, and level of pollutants are major factors that can show the combustion pattern. The findings revealed that the in-cylinder pressure is slightly lower for the dual-fuel mode throughout the compressing stage and the early combustion periods. Moreover, as can be observed in Fig. 11.11, the ignition point and the combustion phase were retarded due to the longer ignition delay.

In CI engines, the ignition delay is defined as the crank angle interval between the start of injection and the start of combustion (Gharehghani et al. 2015c). During the ignition delay and before the start of combustion, various physical and chemical processes happen. Most studies reported longer ignition delays in the dual-fuel mode, due to lower charge temperature, partial reduction of oxygen pressure, and absorption of a part of energy before the combustion, which is due to the higher specific thermal capacity of NG and chemical factors (Liu and Karim 1998). As illustrated in Fig. 11.11, the trend of combustion in the dual-fuel mode is quite different compared to conventional diesel engines. In the second combustion stage in the dual-fuel mode,

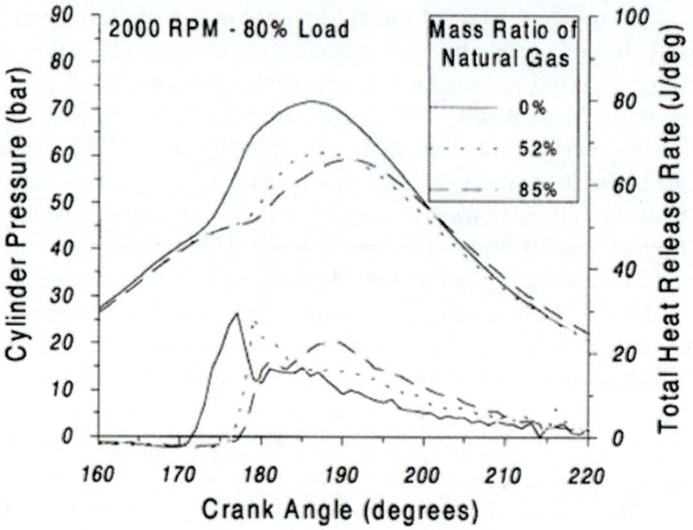

Fig. 11.11 The cylinder pressure and total HRR in the diesel and dual-fuel modes (Abdelaal 2012)

the premixed pilot fuel (diesel) starts to burn and ignite the NG. Due to the low amount of pilot fuel (diesel) and low concentration of NG/air, the heat release rate in this mode is lower than the conventional diesel mode. In the third combustion stage in the dual-fuel mode, because of the decrease in the combustion rate of NG compared to that of diesel fuel and the next combustion stage, peak heat release is generally not so high and the combustion becomes prolonged. As a result, in the fourth phase, a higher amount of fuel combusts, which may lead to higher exhaust temperature. However, as shown in Fig. 11.12, the combustion of premixed NG may get intensified due to the rise of mixture concentration, advancement of pilot fuel injection, etc. (Lounici et al. 2014).

11.4.1 Emissions

Regarding the pollution of NG/diesel dual-fuel engines, studies have been conducted on the NO_x, PM, HC, and CO emissions, the results of which can be summarized as follows.

11.4.2 NO_x

According to studies (Cheenkachorn et al. 2013), in the dual-fuel mode, NO_x emission levels rise as the amount of diesel injection and the engine load increase. And as you can see in Fig. 11.13 the level of this emission decrease as engine speed rises. The reasons for the reduction in NO_x emissions are as follows: The specific heat capacity of natural gas is significantly higher than that of air. The addition of natural gas increases the overall heat capacity of the in-cylinder mixture, which reduces the average temperature during the entire combustion process. Low combustion temperatures reduce the formation of NO_x. Moreover the longer ignition delay and

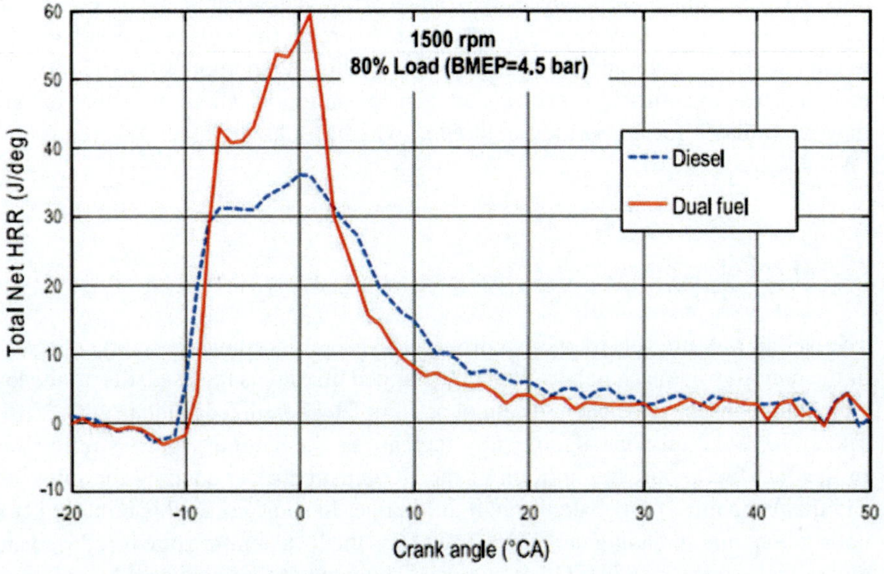

Fig. 11.12 The total HRR in the dual-fuel and diesel modes (Lounici et al. 2014)

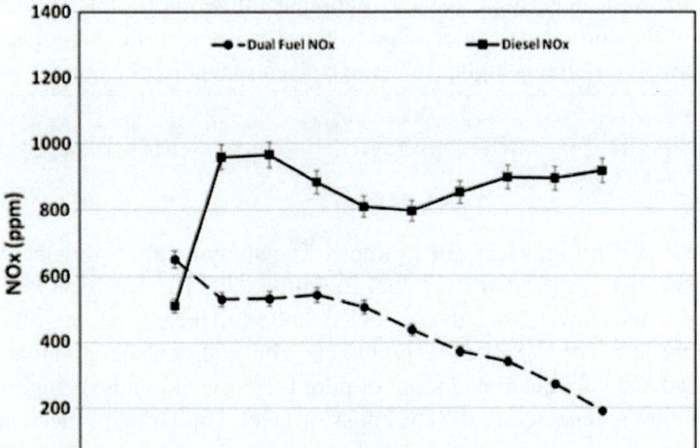

Fig. 11.13 NO_X emission in terms of engine speed changes (Cheenkachorn et al. 2013)

low combustion quality of NG in the NG/diesel dual-fuel mode reduce combustion temperature, resulting in reduced NO_x emissions. The injection of NG decreases the concentration of air and oxygen in the charge, which can lead to a negative impact on the existing oxygen for the NO_x formation and leads to lower levels of NO_x emission. As the engine speed increases, the residence time for the NO_x formation decreases, thus, NO_x emissions are reduced. It is noteworthy that factors such as the greater heat release rate in the combustion phase may increase the NO_x emission level due to the rise of the maximum combustion temperature. Better combustion quality also results in less CO emissions and less oxygen consumption. Therefore, more oxygen may be available for the formation of NO_x, which results in higher NO_X emissions (Wei and Geng 2016).

11.4.3 CO

The NG-air mixture gets trapped in the crevices, deposits, and quench layer because of the long stay in the chamber. When the trapped mixture is released, due to the low temperature during expansion, it cannot be completely oxidized (Raihan et al. 2015). Therefore, CO emissions significantly increase as the natural gas mass ratio rises. In the dual-fuel mode, the ignition of the NG-air mixture is done by the pilot fuel and the flame must propagate through the charge. In some areas, this mixture is too poor to continue spreading the flame. Therefore, the local temperature is reduced and the oxidation reactions of CO are stopped. The competition between the oxidation reactions of HC and CO may lead to higher CO emissions because the oxidation rate of HC is much higher than CO (Shoemaker et al. 2012; Polk et al. 2013). The HC may get oxidized in early processes of combustion while for the CO the oxidation happens at the end of the cycles. Due to these reasons, in the NG/diesel dual-fuel mode the CO emission is higher than the conventional diesel engines.

11.4.4 HC

As you can see in Fig. 11.14 for example, The study results (Abdelaal and Hegab 2012) show that in the NG/diesel dual-fuel mode, the HC emission level is higher compared to the conventional diesel-fueled mode and there is a trade-off connection between the levels of NO_x and HC pollutants. The studies also report that raising the engine load and the injection amount of pilot fuel as well as advancing the injection timing of pilot fuel decreases the HC emission level. Similar to the mechanism of CO emission formation, getting trapped in the crevices and flame quenching leads to the ignition of the unburned fuel in the second section of the combustion process, leading to an increased HC emission level. Also, this mixture is lean and the in-cylinder temperature is low, especially in low load conditions, thus unburned mixture may lead to higher HC production. As the engine load increases and the injection timing of pilot fuel is advanced, the combustion quality is improved and the temperature inside

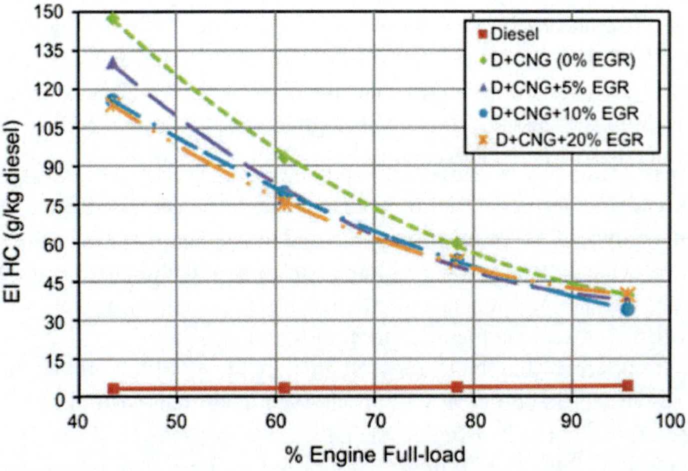

Fig. 11.14 Comparison of HC emission in dual fuel and diesel operating modes (Abdelaal and Hegab 2012)

the cylinder gets higher. Therefore, some of these effects may be weakened which results in reduced HC emissions. It is noteworthy that the D and El in Fig. 11.14 indicate diesel, and emission levels, respectively.

11.4.5 PM

In the dual-fuel mode, a great portion of diesel fuel is substituted by NG and the quantity of pilot diesel is ignorable. Thus, a lower quantity of diesel fuel is combusted in diffusion mode and most of the fuel is combusted in the premixed combustion. Therefore, soot formation decreases. The addition of NG also increases the ignition delay and provides more time for fuel and air to be mixed better. Therefore, fuel-rich areas are reduced. It is noteworthy that NG lacks C–C bonds and sulfur. Therefore, NG produces insignificant amounts of soot. Also, the combustion of a homogeneous NG-air mixture enhances the oxidation of soot throughout the combustion of diesel fuel (Polk et al. 2014). For these, it is clear that in the dual-fuel mode there is a significant reduction in the PM emission level.

11.5 Conclusions

In this chapter, the influence of using NG fuel in vehicle engines and their ability for obtaining sustainable mobility was investigated. The combustion phasing, emission characteristics, and power output were the main parameters that were considered for investigating the role of this fuel in sustainable mobility. Due to environmental growth and limitations in energy resources, natural gas is considered a promising alternative fuel. There are three options for using NG fuel in engines or vehicles: (1) dedicated NG engines or vehicles that use CNG or LNG, (2) gasoline/NG bi-fuel mode based on the gasoline-fueled SI engines, and (3) diesel/NG dual-fuel mode based on the diesel-fueled diesel engines.

- NG is a clean alternative fuel for transport vehicles, and since the combustion of this fuel reduces CO_2 production compared to conventional fuels, the substitution of this fuel can help slow down global warming and climate change. In contrast, methane (CH_4) emission is a challenge for NG-fueled engines, which can be controlled by after-treatment technologies and catalysts.
- In the studies regarding NG-fueled SI engines, it was observed that these engines can present thermal efficiency of about 5% higher than gasoline-fueled engines. Proper adjustment of NG-fueled SI engines reduces NO_x, CO_2, and non-methane HC emissions, but overall, these engines have higher CH_4 emissions compared to gasoline-fueled engines.
- During the examination of CI engines with NG fuel, it was observed that for using this fuel in CI engines, a special operating mode known as dual-fuel is required. The stable and acceptable performance of the dual-fuel engine depends on the engine design and the employed NG injection system. It is noteworthy that the use of NG along with the injection of pilot fuel in CI engines reduces NO_x at low loads. While at low loads, due to incomplete combustion which is caused by low temperatures, CO and HC emission levels increase in this mode, at high loads, the levels of these two emissions are reduced to the levels of conventional CI engines.
- Dual-fuel engines have a longer ignition delay than diesel-fueled engines. Also, at low loads, the thermal efficiency of these engines is lower than the conventional diesel engines, while at high loads these engines present higher thermal efficiencies. The NO_x and PM emission levels of dual-fuel engines are lower than the diesel-fueled engines. On the other hand, these engines have higher HC and CO emissions.
- In the fully open throttle state, by using the NG fuel, due to displacement of air by NG and lack of evaporation cooling, the torque, maximum power, and volumetric efficiency considerably decrease.

It is expected that in near future the application of NG in vehicle engines considerably rises because of the strict regulations on liquid-fuel-based engines and the development of novel combustion mechanisms and fueling systems.

References

Abd-Alla GS, Badr HA, Abd-Rabbo OA (2001) Effects of diluent admissions and intake air temperature in exhaust gas recirculation on the emissions of an indirect injection dual-fuel engine. Energy Convers Manag 42:1033–1045. https://doi.org/10.1016/S0196-8904(00)00072-8

Abdelaal MH (2012) Combustion and emission characteristics of a natural gas-fueled diesel engine with EGR. Energy Convers Manag 64:301–312. https://doi.org/10.1016/j.enconman.2012.05.021

Abdelaal MM, Hegab AH (2012) Combustion and emission characteristics of a natural gas-fueled diesel engine with EGR. Energy Convers Manag 64:301–312. https://doi.org/10.1016/j.enconman.2012.05.021

Abdelaal MMR, Hegab BA, Abdelrahman H (2013) Effect of adding oxygen to the intake air on dual-fuel engine performance, emissions, and knock tendency. Energy 61:612–620. https://doi.org/10.1016/j.energy.2013.09.022

Aslam MU, Masjuki HH, Kalam MA, Abdesselam H, Mahlia TM, Amalina MA (2006) An experimental investigation of CNG as an alternative fuel for a retrofitted gasoline vehicle. Fuel 85:717–724. https://doi.org/10.1016/j.fuel.2005.09.004

Bach C, Lammle C, Bill R, Soltic P, Dyntar D, Janner P, Boulouchos K, Onder C, Landenfeld T, Kercher L, Seel O (2004) Clean engine vehicle a natural gas driven Euro-4/SULEV with 30% reduced CO2-emissions. SAE technical paper. https://doi.org/10.4271/2004-01-0645

Beck NJ, Barkhimer RL, Johnson WP, Wong HC, Gebert K (1997) Evolution of heavy-duty natural gas engines-stoichiometric, carbureted and spark-ignited to lean burn, fuel injected and micro-pilot. SAE technical paper. https://doi.org/10.4271/972665

Belgiorno G, Di Blasio G, Beatrice C (2019) Advances of the natural gas/diesel RCCI concept application for light-duty engines: comprehensive analysis of the influence of the design and calibration parameters on performance and emissions. Natural gas engines. Springer, Singapore, pp 251–266

Brombacher EJ (1999) Flow visualisation of natural gas fuel injection. http://www.collectionscanada.ca/obj/s4/f2/dsk2/tape17/PQDD_0010/MQ33968.pdf

Carlucci AF, Laforgia A (2006) Control of the combustion behavior in a diesel engine using early injection and gas addition. Appl Therm Eng 26:2279–2286. https://doi.org/10.1016/j.applthermaleng.2006.03.016

Cheenkachorn KP, Ho Ch, Gyeung Ch (2013) Performance and emissions of a heavy-duty diesel engine fuelled with diesel and LNG (liquid natural gas). Energy 53:52–57. https://doi.org/10.1016/j.energy.2013.02.027

Chen SKB, John N (2001) Gas engine combustion principles and applications. SAE technical paper. https://doi.org/10.4271/2001-01-2489

Chen H, He J, Zhong X (2019) Engine combustion and emission fuelled with natural gas: a review. J Energy Inst 92:1123–1136. https://doi.org/10.1016/j.joei.2018.06.005

Chiu JPW, Murphy J, Kenneth E (2004) Low emissions class 8 heavy-duty on-highway natural gas and gasoline engine. SAE Trans 1843–1852. https://www.jstor.org/stable/44740894

Cho HM, He B-Q (2007) Spark ignition natural gas engines—a review. Energy Convers Manag 48:608–618. https://doi.org/10.1016/j.enconman.2006.05.023

Crookes RBM, Olsen KDH (2007) Renewable fuelling of a diesel engine in single and dual fuel operation. In: Technische Akademie Esslingen (TAE) 6th international colloquium fuels

Das AW (1997) Development of a natural gas spark ignition engine for optimum performance. Proc Instit Mech Eng (Part D). J Automob Eng 211:361–378. 10.1243%2F0954407971526506

de Carvalho RDB, Valle RM, Rodrigues VF, de Magalhaes FE (2003) Performance and emission analysis of the turbocharged spark-ignition engine converted to natural gas. SAE technical paper, 2003-01-3726

Demirbas A (2002) Fuel properties of hydrogen, liquefied petroleum gas (LPG), and compressed natural gas (CNG) for transportation. Energy Sources 24:601–610. https://doi.org/10.1080/00908312.2002.11877434

Dondero L, Goldemberg J (2005) Environmental implications of converting light gas vehicles: the Brazilian experience. Energy Policy 33:1703–1708. https://doi.org/10.1016/j.enpol.2004.02.009

Durell E, Allan J, Law D, Heath J (2000) Installation and development of a direct injection system for a bi-fuel gasoline and compressed natural gas engine. In: Proceeding ANGVA 2000 conference, Yokohama, Japan

Economides MJW, David A (2009) The state of natural gas. J Nat Gas Sci Engi 1:1–13

Einewall PT, Bengt PJ (2005) Lean burn natural gas operation vs. stoichiometric operation with EGR and a three-way catalyst. SAE technical paper. https://doi.org/10.4271/2005-01-0250

Fakhari AH, Shafaghat R, Jahanian O (2019) Numerical simulation of natural gas/diesel dual-fuel engine for investigation of performance and emission. J Therm Anal Calorim 139:2455–2464. https://doi.org/10.1007/s10973-019-08560-7

Fakhari AH, Shafaghat R, Jahanian O (2020) Numerical simulation of a naturally aspirated natural gas/diesel RCCI engine for investigating the effects of injection timing on the combustion and emissions. J Energy Resour Technol 142. https://doi.org/10.1115/1.4046470

Frailey MN, Clark P, Lyons N, Donald W (2000) An evaluation of natural gas versus diesel in medium-duty buses. SAE technical paper. https://doi.org/10.4271/2000-01-2822

Franklin MK, Leuer DB, Pipho RH (1994) A PC-based fuel and ignition control system used to map the 3-D surfaces of torque and emissions versus air-fuel ratio and ignition timing. SAE Trans 915–926. https://www.jstor.org/stable/44632843

Gatts T, Liu S, Liew C, Ralston B, Bell C, Li H (2012) An experimental investigation of incomplete combustion of gaseous fuels of a heavy-duty diesel engine supplemented with hydrogen and natural gas. Int J Hydrogen Energy 37:7848–7859. https://doi.org/10.1016/j.ijhydene.2012.01.088

Gharehghani A (2013) Experimental investigation of thermal balance of a turbocharged SI engine operating on natural gas. Appl Therm Eng 60:200–207. https://doi.org/10.1016/j.applthermaleng.2013.06.029

Gharehghani A (2019) Load limits of an HCCI engine fueled with natural gas, ethanol, and methanol. Fuel 239:1001–1014. https://doi.org/10.1016/j.fuel.2018.11.066

Gharehghani AM, Jazayeri SM (2012) Numerical and experimental investigation of combustion and knock in a dual fuel gas/diesel compression ignition engine. J Combust. https://doi.org/10.1155/2012/504590

Gharehghani AH, Yusaf R (2013) Investigation of the effect of additives to natural gas on heavy-duty si engine combustion characteristics. J Mech Eng Sci 5:677–687. https://doi.org/10.15282/jmes.5.2013.14.0065

Gharehghani AH, Mirsalim R, Yusaf M, Talal F (2015a) A comparative study on the first and second law analysis and performance characteristics of a spark-ignition engine using either natural gas or gasoline. Fuel 158:488–493. https://doi.org/10.1016/j.fuel.2015.05.067

Gharehghani AH, Mirsalim R, Yusaf M, Talal F (2015b) A computational study of operating range extension in a natural gas SI engine with the use of hydrogen. Int J Hydrogen Energy 40:5966–5975. https://doi.org/10.1016/j.ijhydene.2015.03.015

Gharehghani AH, Mirsalim R, Jazayeri M, Talal AY (2015c) An experimental study on reactivity controlled compression ignition engine fueled with biodiesel/natural gas. Energy 89:558–567. https://doi.org/10.1016/j.energy.2015.06.014

Huang JC, Crookes RJ (1998) Assessment of simulated biogas as a fuel for the spark-ignition engine. Fuel 77:1793–1801. https://doi.org/10.1016/S0016-2361(98)00114-8

International Energy Agency (2011) World Energy Outlook, OECD/IEA

Kakaee AH, Paykani A (2013) Research and development of natural-gas fueled engines in Iran. Renew Sustain Energy Rev 26:805–821. https://doi.org/10.1016/j.rser.2013.05.048

Kakoee AB, Aval Y, Motadayen S, Gharehghani A (2018) An improvement of a lean burning condition of natural gas/diesel RCCI engine with a pre-chamber by using hydrogen. Energy Convers Manag 166:489–499. https://doi.org/10.1016/j.enconman.2018.04.063

Kar T, Zhou Z, Brear M, Yang Y, Khosravi M, Lacey J (2021) A comparative study of directly injected, spark ignition engine performance and emissions with natural gas, gasoline and charge dilution. Fuel 304:121438

Karim GA (2010) Combustion in gas-fueled compression ignition engines of the dual fuel type. In: Handbook of combustion, pp 213–235 (Online)

Kato KI, Masuda M, Otsubo K, Yasuda A, Takeda K, Sato T (1999) Development of engine for natural gas vehicle. SAE Trans 108:939–947. https://www.jstor.org/stable/44743426

Kato T, Saeki K, Nishide H, Yamada T (2000) Development of CNG fueled engine with lean burn for small size commercial van. JSAE Rev 22:365–368

Kesgin U (2005) Effect of the turbocharging system on the performance of a natural gas engine. Energy Convers Manag 46:11–32. https://doi.org/10.1016/j.enconman.2004.02.006

Korakianitis T, Namasivayam AM, Crookes RJ (2011) Natural-gas fueled spark ignition (SI) and compression ignition (CI) engine performance and emissions. Progress Energy Combust Sci 37:89–112. https://doi.org/10.1016/j.pecs.2010.04.002

Kusaka J, Okamoto T, Daisho Y, Kihara R, Saito T (2000) Combustion and exhaust gas emission characteristics of a diesel engine dual-fueled with natural gas. JSAE Rev 21:489–496. https://doi.org/10.1016/S0389-4304(00)00071-0

Lave LB, Maclean H, Lankey R, Joshi S, McMichael F, Horvath A, Hendrickson C (2000) Life cycle inventories of conventional and alternative automobile fuel/propulsion systems: summary and conclusions. SAE Trans 1953–1961. https://doi.org/10.1021/es991322+

Li L, Zhang Z (2019) Investigation on steam direct injection in a natural gas engine for fuel savings. Energy 183:958–970

Li HL, Liew S, Gatts C, Wayne T, Clark S, Nuszkowski N (2017) An investigation of the combustion process of a heavy-duty dual fuel engine supplemented with natural gas or hydrogen. Int J Hydrogen Energy 42:3352–3362. https://doi.org/10.1016/j.ijhydene.2016.12.115

Liu Z, Karim GA (1998) An examination of the ignition delay period in gas-fueled diesel engines. https://doi.org/10.1115/1.2818080

Lounici MS, Loubar K, Tarabet L, Balistrou M, Niculescu DC, Tazerout M (2014) Towards improvement of natural gas-diesel dual fuel mode: an experimental investigation on performance and exhaust emissions. Energy 64:200–211. https://doi.org/10.1016/j.energy.2013.10.091

Mello P, Pelliza G, Cataluña R, da Silva R (2006) Evaluation of the maximum horsepower of vehicles converted for use with natural gas fuel. Fuel 85:2180–2186. https://doi.org/10.1016/j.fuel.2006.04.002

Mtui PLH, Philip G (1996) Ignition delay and combustion duration with natural gas fueling of diesel engines. SAE technical paper. https://doi.org/10.4271/961933

Namasivayam AC, Korakianitis RJ, Olsen T (2009) Assessment of combustion in natural gas dual-fuelled compression ignition engines with dimethyl ether and rapeseed methyl ester pilot ignition. Int J Engine Res 10:165–174. 10.1243%2F14680874JER02909

Napolitano P, Fraioli V, Beatrice C, Migliaccio M, Guido C (2014) Estimation of TTW and WTW factors for a light duty dual fuel NG-diesel EU5 passenger car (No. 2014-01-1621). SAE technical paper

Narayanaswamy KR, Christopher J (2004) Cycle simulation diesel HCCI modeling studies and control. SAE technical paper. https://doi.org/10.4271/2004-01-2997

Ouellette P (2000) High pressure direct injection (HPDI) of natural gas in diesel engines. In: Proceeding ANGVA 2000 conference, Yokohama

Papagiannakis RH (2003) Experimental investigation concerning the effect of natural gas percentage on performance and emissions of a DI dual fuel diesel engine. Appl Therm Eng 23:353–365. https://doi.org/10.1016/S1359-4311(02)00187-4

Papagiannakis RG, Hountalas DT (2004) Combustion and exhaust emission characteristics of a dual fuel compression ignition engine operated with pilot diesel fuel and natural gas. Energy Convers Manag 45:2971–2987. https://doi.org/10.1016/j.enconman.2004.01.013

Papagiannakis RH, Rakopoulos DT (2007) Theoretical study of the effects of pilot fuel quantity and its injection timing on the performance and emissions of a dual fuel diesel engine. Energy Convers Manag 48:2951–2961. https://doi.org/10.1016/j.enconman.2007.07.003

Parikh P (1999) Spark ignition producer gas engine and dedicated compressed natural gas engine-technology development and experimental performance optimization. SAE technical paper. https://doi.org/10.4271/1999-01-3515

Paykani AS, Khoshbakhti Shervani-Tabar R, Mohammadi-Kousha MT (2012) Effect of exhaust gas recirculation and intake pre-heating on performance and emission characteristics of dual-fuel engines at part loads. J Central S Univ 19:1346–1352

Pirouzpanah VS, Khoshbakhti Sohrabi R, Niaei A (2007) Comparison of thermal and radical effects of EGR gases on combustion process in dual-fuel engines at part loads. Energy Convers Manag 48:1909–1918. https://doi.org/10.1016/j.enconman.2007.01.031

Polk AC, Gibson CM, Shoemaker NT, Srinivasan KK, Krishnan SR (2013) Detailed characterization of diesel-ignited propane and methane dual-fuel combustion in a turbocharged direct-injection diesel engine. Proc Inst Mech Eng Part D J Automobile Eng 227(9):1255–1272

Polk AC, Carpenter CD, Scott Guerry E, Dwivedi U, Kumar Srinivasan K, Rajan Krishnan S, Rowland ZL (2014) Diesel-ignited propane dual fuel low temperature combustion in a heavy-duty diesel engine. J Eng Gas Turbines Power 136(9)

Poulton M (1994) Alternative fuels for road vehicles, Ashurst lodge, Ashurst, Southampton, SO40 7AA. Computational Mechanics Publications, UK

Raihan MS, Guerry ES, Dwivedi U, Srinivasan KK, Krishnan SR (2015) Experimental analysis of diesel-ignited methane dual-fuel low-temperature combustion in a single-cylinder diesel engine. J Energy Eng 141(2):C4014007

Ramadhas AJ, Muraleedharan S (2008) Dual fuel mode operation in diesel engines using renewable fuels: rubber seed oil and coir-pith producer gas. Renew Energy 33:2077–2083. https://doi.org/10.1016/j.renene.2007.11.013

Reynolds CCO, Evans RL, Andreassi L, Cordiner S, Mulone V (2005) The effect of varying the injected charge stoichiometry in a partially stratified charge natural gas engine. SAE technical paper. https://doi.org/10.4271/2005-01-0247

Selim MY (2003) Effect of exhaust gas recirculation on some combustion characteristics of the dual-fuel engine. Energy Convers Manag 44:707–721. https://doi.org/10.1016/S0196-8904(02)00083-3

Semin RAB (2008) A technical review of compressed natural gas as an alternative fuel for internal combustion engines. J Eng Appl Sci 1:302–311

Shasby BM (2004) Alternative fuels: incompletely addressing the problems of the automobile. Virginia Polytechnic Institute, and State University, The USA, pp 5–13. http://hdl.handle.net/10919/9976

Shoemaker NT, Gibson CM, Polk AC, Krishnan SR, Srinivasan KK (2012) Performance and emissions characteristics of bio-diesel (B100)-ignited methane and propane combustion in a four cylinder turbocharged compression ignition engine. J Eng Gas Turbines Power 134(8)

Srinivasan KK, Singh S, Midkiff KC, Bell SR, Gong W, Fiveland SB, Willi M (2004) The advanced injection low pilot ignited natural gas engine: a combustion analysis. J Eng Gas Turbines Power 128:213–218. https://doi.org/10.1115/1.1915428

Srinivasan KK, Krishnan SR, Singh S, Midkiff KC, Bell SR, Gong W, Fiveland SB, Willi M (2006) The advanced injection low pilot ignited natural gas engine: a combustion analysis. https://doi.org/10.1115/1.1915428

Thurnheer TS, Eggenschwiler P, Dimopoulos P (2009) SI engine fuelled with gasoline, methane, and methane/hydrogen blends: heat release and loss analysis. Int J Hydrogen Energy 34:2494–2503. https://doi.org/10.1016/j.ijhydene.2008.12.048

Tree DRS, Kenth I (2007) Soot processes in compression ignition engines. Prog Energy Combust Sci 33:272–309. https://doi.org/10.1016/j.pecs.2006.03.002

U.S. Energy Information Administration (2013) Annual energy outlook with projection to 2040

Varde KS, Patro N, Drouillard K (1995) Lean burn natural gas-fueled SI engine and exhaust emissions. SAE techical paper. https://doi.org/10.4271/952499

Wei L, Geng P (2016) A review on natural gas/diesel dual fuel combustion, emissions and performance. Fuel Process Technol 142:264–278

Yang B, Zeng K (2018) Effects of natural gas injection timing and split pilot fuel injection strategy on the combustion performance and emissions in a dual-fuel engine fueled with diesel and natural gas. Energy Convers Manag 168:162–169. https://doi.org/10.1016/j.enconman.2018.04.091

Yousefi A, Guo H, Birouk M (2017) An experimental and numerical study on diesel injection split of a natural gas/diesel dual-fuel engine at a low engine load fuel. Fuel 212:332–346. https://doi.org/10.1016/j.fuel.2017.10.053

Zeng K, Huang Z, Liu B, Jiang L, Ren Y, Wang J (2006) Combustion characteristics of a direct-injection natural gas engine under various fuel injection timings. Appl Therm Eng 26:806–813. https://doi.org/10.1016/j.applthermaleng.2005.10.011

Zhao W, Yang W, Fan L, Zhou D, Ma X (2017) Development of a skeletal mechanism for heavy-duty engines fuelled by diesel and natural gas. Appl Therm Eng 123:1060–1071. https://doi.org/10.1016/j.applthermaleng.2017.05.175

Zheng J, Wang J, Zhao Z, Wang D, Huang Z (2019) Effect of equivalence ratio on combustion and emissions of a dual-fuel natural gas engine ignited with diesel. Appl Therm Eng 146:738–751. https://doi.org/10.1016/j.applthermaleng.2018.10.045

Zhen-Tao L, Shao-mei F (2004) Study of CNG/diesel dual-fuel engine's emissions by means of RBF neural network. J Zhejiang Univ Sci A 5:960–965. https://doi.org/10.1007/BF02947606